Bioenergetics
and linear nonequilibrium
thermodynamics

Harvard books in biophysics
number 3

Bioenergetics and linear nonequilibrium thermodynamics

the steady state

S. Roy Caplan and Alvin Essig

Harvard University Press
Cambridge, Massachusetts, and London, England   1983

Copyright © 1983 by the President and Fellows of Harvard College
All rights reserved
Printed in the United States of America
10 9 8 7 6 5 4 3 2 1

This book is printed on acid-free paper, and its binding materials have been chosen for strength and durability.

Library of Congress Cataloging in Publication Data

Caplan, S. Roy.
  Bioenergetics and linear nonequilibrium thermodynamics.

  (Harvard books in biophysics; no. 3)
  Bibliography: p.
  Includes index.
    1. Bioenergetics.   2. Thermodynamics.   I. Essig, Alvin.   II. Title.   III. Title: Linear nonequilibrium thermodynamics : the steady state.   IV. Title: Nonequilibrium thermodynamics : the steady state.
  V. Series.
  QH510.C36 1983   574.19'121   82-12083
  ISBN 0-674-07352-5

Anyone writing in this area cannot but be aware of the tremendous debt owed to the late Professor Aharon Katzir-Katchalsky, whose tragic death in 1972 deprived the biophysical community of one of its most creative and inspiring figures

Preface

In approaching diverse biological phenomena from a somewhat unfamiliar point of view, the reader will inevitably have a number of questions. Is there really a need for nonequilibrium thermodynamics in studying biological processes? Can one reasonably expect a biologist to spend the time and effort necessary to learn this approach? Can the problems not be dealt with by means of the standard physical and mathematical techniques? Even if we grant the applicability of nonequilibrium thermodynamics to certain problems and its attraction for certain investigators, does it do more than force data into a certain mold? Are we then dealing only with logical constructs congenial to our own idiosyncratic attitudes and prejudices? And finally, can thermodynamics teach us anything about mechanism? In the words of one critic, "It is a legitimate approach, but as long as the preparation is treated as a black box the really interesting questions can't even be asked."

We claim that nonequilibrium thermodynamics is constructive because it provides a self-consistent, logical, and compact framework for organizing and systemizing information. The systematic correlation of diverse data underlines the strong analogy that often exists between seemingly different systems. It will often become apparent that the formulation developed in the analysis of one system is directly applicable to a quite different system. Such is the case, for example, for maintenance of an electrochemical potential gradient by an epithelial membrane and for maintenance of isometric tension by a muscle. Furthermore, within a given system the thermodynamic framework may indicate correlations among diverse phenomena that may not suggest themselves on other grounds. For example, the discrepancy between the tracer permeability coefficient and the permeability coefficient for net flow quantitatively predicts the extent of abnormality of the flux ratio. This prediction, which is perhaps not generally evident from standard approaches, follows immediately from the thermodynamic formulation.

The statement that thermodynamics teaches nothing about mechanism is a truism. However, one must be precise about the meaning of the word "mechanism." If one means the ultimate molecular details of operation of a sodium pump, for example, clearly thermodynamics alone cannot provide final answers, although even here it often sets limits which must be satisfied by appropriate kinetic models. However, if the notion of mechanism is broader, including also aspects of the energy supply to the pump and its regulation, thermodynamic considerations are invaluable.

For these reasons we feel that the nonequilibrium thermodynamic approach is an important tool in modern biophysics. To provide evidence for the validity of this point of view, this book analyzes in considerable detail the small number of systems with which we have had personal experience and which illustrate the general principles. Limitations of time and space have prevented us from analyzing much work of others that deserves close attention. Clearly, many systems can be treated in this way, and we hope that readers will enjoy applying these concepts to their own problems.

This book, which is the result of a long-standing collaboration, was written during a series of stimulating stays at various institutions. Accordingly we are grateful for the hospitality of Dr. Rui DeSousa, École de Médicine de l'Université de Genève, Drs. Barbara Banks and Charles Vernon, University College London, Drs. Franz Grün and Dieter Walz, Biozentrum der Universität Basel, Drs. Reinhard Schlögl and Friedrich Sauer, Max-Planck-Institut für Biophysik,

Frankfurt am Main, and Dr. Terrell Hill, National Institutes of Health, Bethesda, Maryland. Many of the fundamental ideas presented derived from collaborations and discussions with Dr. Ora Kedem. We thank Dr. Donald Mikulecky for careful reading and criticism of an earlier version of the manuscript.

Finally, we are deeply grateful to our respective wives, Theolea and Caroline, for their encouragement during the long-drawn-out writing of this book and for the many hours spent in proofreading intermediate and final versions of the chapters.

Contents

1 General considerations 1
2 Fundamentals of equilibrium and nonequilibrium thermodynamics 5
3 Relationships between flows and forces: the Kedem-Katchalsky equations 24
4 Effectiveness of energy conversion 54
5 The diagram method 74
6 Possible conditions for linearity and symmetry of coupled processes far from equilibrium 96
7 Energetics of active transport: theory 132
8 Energetics of active transport: experimental results 168
9 Kinetics of isotope flows: background and theory 215
10 Kinetics of isotope flows: mechanisms of isotope interaction 246
11 Kinetics of isotope flows: tests and applications of thermodynamic formulation 274
12 Muscular contraction 301
13 Energy coupling in mitochondria, chloroplasts, and halophilic bacteria 348

Afterword 389
List of symbols 393
Notes 401
References 413
Index 425

1 General considerations

Although it is obvious that most biological processes occur far from equilibrium, many people feel that the established techniques of classical equilibrium thermodynamics, in conjunction with an appropriate kinetic analysis, are sufficient to deal with these phenomena. Examples of this attitude abound in the biophysical literature. A few words as to why this point of view is incorrect would perhaps be appropriate. As Prigogine pointed out many years ago, classical thermodynamics has serious limitations as a tool for the macroscopic description of biological energetics. Its method is based on the consideration of equilibrium states and on the concept of "reversible processes" (hypothetical ideal processes that occur without disturbing equilibrium and hence without producing entropy). The question of the time taken to pass from one state to another lies outside the purview of the classical method. In contrast, nonequilibrium thermodynamics, a major extension of the thermodynamic method, is based on consideration of states removed from equilibrium and on the associated irreversible processes (real processes that do produce entropy). A prerequisite of dealing with such processes is the explicit consideration of time as a factor.

Even classical equilibrium thermodynamics deals not only with states of complete equilibrium, but also with states of partial or restricted equilibrium, such as the Donnan equilibrium, in which complete equilibration is prevented by a permeability barrier to certain components of the system. Another example of partial equilibrium is a mixture of chemical reactants at nonequilibrium concentrations, which are prevented from reacting by the absence of a catalyst. In contrast, the steady states in nonequilibrium thermodynamics are maintained solely as a consequence of the production of entropy. The calculation of the rate of entropy production is the core of the entire procedure, providing as it does a natural basis for choosing appropriate sets of flows and forces for the phenomenological characterization of the system. This characterization introduces kinetic parameters in the form of phenomenological coefficients. For a given set of coefficients, the functioning of the system and its efficiency are fully determined over a wide range of operating conditions, and its maximum efficiency is readily calculated. To obtain comparable information from a combination of classical thermodynamics and kinetic models would be a very laborious process, and practical only in the simplest cases.

The great power of the thermodynamic method lies in its ability to predict correlations among observations in the absence of detailed knowledge of the structure of the system: it establishes a framework within which patterns of behavior can be organized and tested for self-consistency. In this regard nonequilibrium thermodynamics extends and complements the classical method. Although its kinetic parameters are not derived from specific models, they are subject to certain constraints which offer great predictive value. We shall see that the most tractable processes are those which occur in the "steady state." However, under stationary conditions the state parameters of a system give no information about the processes occurring; to study these it is necessary to look at their effects on the surroundings, and indeed this is the very essence of the approach. The resulting analysis immediately determines very valuable information: the number and nature of the degrees of freedom of the system. To the experimenter this indicates how many (and what) constraints are required to define a steady state of his system, and it predicts correlations between such states. This information will often suggest meaningful new experiments. The theoretician is likewise provided with a scaffolding around which models can be constructed conforming to the proper thermodynamic restrictions.

In most of the topics to be discussed we shall assume that the forces and flows characterizing a given stationary state of a system are related to one another by a linear response matrix. Although this may not always be strictly true, we believe that in many of the systems we shall be discussing, departures from linearity will frequently be too small to be significant. This may be a consequence of regulatory mechanisms, possibly involving feedback, that at present are incompletely understood. In other cases we can study systems in the linear range near equilibrium. Even systems that at first sight appear to be highly nonlinear may in fact prove to be linear when the appropriate degrees of freedom are taken into account. A further simplifying consideration is that according to Onsager's reciprocity theorem the response matrix must be symmetrical, which reduces the number of phenomenological coefficients characterizing the system.

If we accept this approach, fully knowing its limitations, we shall see a conspicuous feature emerging from the several different studies—the great ability of nonequilibrium thermodynamics to unify diverse phenomena. Whether or not the systems are truly linear over the entire range of biological interest is therefore not crucial; providing a single coherent logical framework is much more important. Similarly, it is not necessary that the observed linearity reflect a simple linear behavior of the fundamental kinetic elements; given the possibility of elaborate regulation, linearity may well be the consequence of a complex interaction among nonlinear elements. In contrast to the unwieldy conglomerations of kinetic parameters that frequently stem from model building, the phenomenological equations have an impressive simplicity. Although they cannot describe molecular mechanisms (unless interpreted in terms of molecular parameters), they provide restrictive conditions that must be satisfied by any models considered, and invariably they clarify energetic questions.

In the following chapters we shall consider a number of generally accepted views of biological systems, some based on models, others based on classical thermodynamic considerations, which we feel to be either restrictive in scope or actually misleading. In presenting the nonequilibrium thermodynamic approach, we shall focus on considerations relevant to a wide variety of transport and other energy conversion systems. For convenience we shall be returning again and again to the systems we are most familiar with that have been characterized from the nonequilibrium thermodynamic point of view. For epithelial tissues, our emphasis is on frog skin and toad bladder; symmetric systems will be discussed largely in the context of mitochondria, chloroplasts, and muscle. Muscle will receive a good deal of attention in its own right as a mechanochemical transducer.

The question of stoichiometry will recur frequently. For example, people tend to expect that the transport of a certain number of sodium ions is associated with the consumption of a specific number of moles of oxygen. A similar expectation crops up in studies of oxidative phosphorylation. Presumably these attitudes result from thinking about these processes as analogous to stoichiometric reactions taking place in a test tube. However, stoichiometry need not necessarily be the case, as nonequilibrium thermodynamic considerations strongly suggest.

A further question that will concern us is the problem of energetics and the effectiveness of energy utilization. Here again people have tended to think in terms of kinetic models. Thus they are led to think that a molecule in the course of transport has to overcome friction, and this leads to a concept of "internal work." A similar concept arises in studies of muscular contraction. In assessing efficiency, the sum of the so-called internal work and all other work is related to the energy expenditure. This calculation usually involves both the stoichiometric ratio and the enthalpy of reaction, an energetic parameter derived from bomb calorimetry. Again, the analyses become considerably clearer when subjected to the discipline of nonequilibrium thermodynamics.

2 Fundamentals of equilibrium and nonequilibrium thermodynamics

In this chapter we shall summarize a few essential elements of equilibrium thermodynamics, then present in rather general terms the basic nonequilibrium thermodynamic approach to the analysis of a steady-state system. The core of this analysis is the evaluation of the rate of entropy production in terms of the flows and forces operating in the system. It will be shown that although entropy production occurs as a consequence of the irreversible process or processes taking place within the system, the parameters of state of the steady-state system do not change with time, and hence the rate of entropy production must be evaluated by considering changes brought about by the system on its surroundings.

2.1 Some basic principles and functions of equilibrium thermodynamics

The first law of thermodynamics concerns the notion of conservation of energy, according to which the total energy of an isolated system does not change. Energy includes kinetic energy (by virtue of the motion of a system), potential energy (by virtue of the position of a system), and internal energy (determined by the characteristics of the components of a system). In most processes of interest to biophysicists, the kinetic and potential energies of a system remain constant; consequently, we will here be concerned only with the internal energy, U. The internal energy is an example of a function of state; that is, it is completely defined by specifying the values of a number of thermodynamic parameters, such as temperature, pressure, and concentrations, sufficient to characterize the system uniquely. In a closed system, one that allows the exchange of energy, but not matter, with its environment, the first law is a statement of the dependence of the internal energy on the heat taken up by the system from its surroundings, Q, and the work done by the system on its surroundings, W. In differential form,

$$dU = dQ - dW \qquad (2.1)$$

This law is readily verifiable for any kind of change between two given states. It should be noted that work is a mode of energy transfer that takes many forms, all of which may in principle be utilized entirely to raise a weight. The internal energy, or in Bridgman's terminology the internal energy function, is a logical construct justified by the fact that the difference between dQ and dW is uniquely defined for any given change of state, whatever their individual values may be.

The second law of thermodynamics deals with the fact that although certain changes of state within a system (for example, $A \rightarrow B$) can take place spontaneously, the reverse changes (such as $B \rightarrow A$) cannot. To quote Denbigh (1966), "The possibility or impossibility of the processes $A \rightarrow B$ and $B \rightarrow A$ depends entirely on the nature of the states A and B and is thus determined by the values of the variables, such as temperature and pressure, at the beginning and end of the process. We may therefore expect to find a function of these variables whose special characteristic is to show whether it is the process $A \rightarrow B$ which is the possible one, or the converse process $B \rightarrow A$." This function is the entropy function, S, defined by $dS \equiv (dQ/T)_{rev}$, where we refer to the hypothetical limiting case of a reversible process, that is, one in which the thermodynamic parameters of the surroundings of a system differ only infinitesimally from those of the system itself. T represents the absolute temperature of the system. In a closed system the second law is then given in differential form by the statement that for any process which is in principle possible,

$$dS \geq \frac{dQ}{T} \qquad (2.2)$$

As in the case of the first law, Eq. (2.2) is readily verifiable for any change between two given states.

The two laws are readily combined. For a reversible process $dQ = TdS$, and limiting ourselves to work of expansion, $dW = pdV$, where p represents the pressure inside (and in this case also outside) the system, and V represents the volume of the system. Thus

$$dU = TdS - pdV \qquad (2.3)$$

Although Eq. (2.3) was derived from the consideration of a reversible change, it is generally applicable because all the parameters are functions of state. For an irreversible process, however, the quantities TdS and pdV can no longer be identified as heat and work, respectively; for such processes dQ is less than TdS, and dW is less than pdV. In an irreversible expansion, for example, the internal pressure p must be larger than the external pressure against which the work is done.

It may truly be said that the quantities U, S, T, p, and V introduced so far embody in their properties the whole of classical thermodynamics. These quantities are all functions of state, unlike Q and W, which depend not only on the initial and final states of a system that absorbs heat and performs work, but also on the precise path or manner in which the system moves from its initial state to its final state. For convenience only it is customary to introduce three additional functions of state that are simply combinations of the previous quantities. These are the enthalpy, H; Helmholtz free energy, F; and Gibbs free energy, G. These additional quantities are defined by the relations

$$H = U + pV \qquad (2.4)$$

$$F = U - TS \qquad (2.5)$$

$$G = H - TS \qquad (2.6)$$

8   Bioenergetics and linear nonequilibrium thermodynamics

The fundamental equation for a closed system, Eq. (2.3), may be reformulated in terms of these new quantities as follows:

$$dH = TdS + Vdp \qquad (2.7)$$

$$dF = -SdT - pdV \qquad (2.8)$$

$$dG = -SdT + Vdp \qquad (2.9)$$

The diagram in Fig. 2.1, a form of which, according to Tisza (1966), was first described by Born, is an excellent mnemonic aid, enabling one to write down Eqs. (2.3), (2.7), (2.8), and (2.9) instantaneously.[1] (The dependent parameters in these equations appear at the sides of the square, each one flanked by the appropriate pair of independent parameters at the corners with their associated signs. The coefficient associated with a given differential appears at the opposite end of the corresponding diagonal.)

The meaning of enthalpy for a system at constant *pressure* may be seen as follows: here we consider closed systems which may perform any type of work in addition to expansion. Combining the first law, Eq. (2.1), with the definition of enthalpy, Eq. (2.4), we obtain

$$dH = dQ - dW + pdV \quad \text{(constant } p\text{)}$$
$$= dQ - (dW - pdV)$$
$$= dQ - dW' \qquad (2.10)$$

W', representing work other than work of expansion performed by the system, is commonly referred to as "useful work." It includes, for

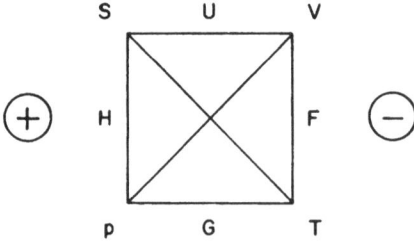

Fig. 2.1. A mnemonic aid in memorizing the thermodynamic functions of state (see text).

example, the lifting of a weight or the stretching of an elastic system.[2] In the case that no useful work is done at constant pressure,

$$dH = dQ \quad (dW' = 0) \tag{2.11}$$

It is for this reason that enthalpy is sometimes referred to as the "heat content" of a system.

The concept of useful work immediately clarifies the meaning of free energy. Thus from Eqs. (2.6) and (2.10) we have, for an incremental change in Gibbs free energy at constant temperature and pressure,

$$dG = dQ - dW' - TdS \quad \text{(constant } p \text{ and } T\text{)} \tag{2.12}$$

Consequently,

$$dQ = TdS + (dG + dW') \tag{2.13}$$

Thus, in a *reversible process* at constant p and T,

$$dW' = -dG \quad \text{(reversible process)} \tag{2.14}$$

In an *irreversible process*, $dQ < TdS$. Consequently, $dG + dW' < 0$, so that

$$dW' < -dG \quad \text{(irreversible process)} \tag{2.15}$$

From this, one can appreciate that the maximum useful work which can be obtained from a process at constant pressure and temperature is equal to the negative Gibbs free energy change, and it becomes available only in the limiting case of a hypothetical reversible process. By perfectly analogous reasoning it can be seen that the negative Helmholtz free energy change in a system corresponds to the maximum useful work which can be obtained under conditions of constant *volume* and constant temperature. Usually in biological systems the distinction between the Gibbs and Helmholtz free energies is unimportant. In any case, their utility is restricted to isothermal systems.

In nonequilibrium thermodynamics we shall be concerned with open systems, and therefore it will be important to have a thorough awareness of the classical thermodynamic treatment of such systems. For open systems, which exchange matter with their surroundings, the fundamental Eqs. (2.3), (2.7), (2.8), and (2.9) are inadequate as they stand. Since the composition of such a system may change, further variables are necessary to define its state. (This is also true of closed systems in which chemical reactions take place.) For example, expressing U as previously as a function of the independent variables S and V, we must also incorporate its dependence on the independent mole numbers n_i. Following the treatment of Prigogine and Defay (1954),

$$dU = \left(\frac{\partial U}{\partial S}\right)_{V,n_i} dS + \left(\frac{\partial U}{\partial V}\right)_{S,n_i} dV + \sum_i \left(\frac{\partial U}{\partial n_i}\right)_{S,V,n_j \neq n_i} dn_i \quad (2.16)$$

When an open system undergoes a change in which all the n_i remain constant, its state changes exactly as it would in a closed system of constant composition. Therefore, by comparison with Eq. (2.3),

$$\left(\frac{\partial U}{\partial S}\right)_{V,n_i} = T \text{ and } \left(\frac{\partial U}{\partial V}\right)_{S,n_i} = -p \quad (2.17)$$

In the terminology of Gibbs, the quantity

$$\left(\frac{\partial U}{\partial n_i}\right)_{S,V,n_j \neq n_i} = \mu_i \quad (2.18)$$

is termed the chemical potential of species i. Combining Eqs. (2.16), (2.17), and (2.18), we obtain the Gibbs equation:

$$dU = TdS - pdV + \sum_i \mu_i dn_i \quad (2.19)$$

From similar considerations the remaining three fundamental Eqs. (2.7), (2.8), and (2.9) require the addition of the term $\sum_i \mu_i dn_i$ in order that they may adequately account for compositional changes. It is seen that

$$\mu_i = \left(\frac{\partial U}{\partial n_i}\right)_{S,V,n_j} = \left(\frac{\partial H}{\partial n_i}\right)_{S,p,n_j}$$

$$= \left(\frac{\partial F}{\partial n_i}\right)_{T,V,n_j} = \left(\frac{\partial G}{\partial n_i}\right)_{T,p,n_j} \quad (2.20)$$

It should be noted that in open systems the quantity TdS can never be interpreted unambiguously as heat, even in the case of a reversible process. The point is that the very notion of a heat transfer is not well defined in open systems, since matter carries with it an associated energy.

It is useful to consider the direct integration of the Gibbs equation between two states, such that in the second state the system has been enlarged in size without changing its intensive variables: temperature, pressure, and concentrations. Under such conditions the chemical potentials remain constant. This gives

$$\Delta U = T\Delta S - p\Delta V + \sum_i \mu_i \Delta n_i \quad \text{(constant } T,p,\mu_i)$$

These changes would be brought about by the simple addition of material to the system. If there is a k-fold increase in size,

$$(k-1)U = T(k-1)S - p(k-1)V + \sum_i \mu_i(k-1)n_i$$

that is,

$$U = TS - pV + \sum_i \mu_i n_i \qquad (2.21)$$

This derivation of Eq. (2.21), which is due to Denbigh (1964), makes use of an item of physical knowledge—that intensive parameters are independent of size, while extensive parameters are proportional to size. Thus from the primary definitions Eqs. (2.4) and (2.6), Eq. (2.21) leads to the conclusion that

$$G = \sum_i n_i \mu_i \qquad (2.22)$$

The above integration also leads us to a relationship between simultaneous changes in intensive variables. If one writes down the complete differential of Eq. (2.21) and compares it with the Gibbs Eq. (2.19), one finds that

$$SdT - Vdp + \sum_i n_i d\mu_i = 0 \qquad (2.23)$$

This relation is known as the Gibbs-Duhem equation.

2.2 Some basic nonequilibrium thermodynamic concepts

To illustrate the general procedure used to evaluate the rate of entropy production, let us consider the system shown in Fig. 2.2. In the region denoted by (0), irreversible processes occur; although the processes are confined to this region, their effects are felt in the surrounding region (1). For example, region (0) might be a bacterium or a mitochondrion or a fragment of endoplasmic reticulum (or even of a synthetic membrane having enzymatic activity). In all these cases, substrates which are present in (1) enter (0), undergo reaction, and are returned to (1) in the form of products. The region (1) may be thought of as a portion of the local environment of (0) large enough to contain completely all possible effects of the processes in (0). For convenience we may consider (1) to be enclosed by rigid adiabatic walls, in other words, isolated or sealed off entirely from the world beyond, a situation which is often approximated experimentally. Alternatively one may regard (1) as the *total* surroundings of (0), in other words the rest of the universe, ignoring all occurrences irrelevant to the process of interest. In analogy with the standard treatments of

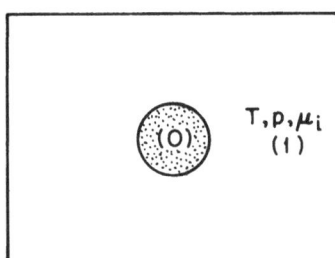

Fig. 2.2. The system: irreversible processes are confined to region (0), which is of small capacity compared to the surrounding region (1). Region (1) is enclosed by rigid adiabatic walls. T = temperature; p = pressure; μ_i = chemical potential of ith species.

classical equilibrium thermodynamics, we assume region (1) to be well mixed so that equilibrium is everywhere attained, that is, the temperature T, the pressure p, and the chemical potentials μ_i are everywhere constant. Again, this situation is often approximated experimentally.

Two possibilities arise. First, (0) may be a *closed* system, one that can exchange only heat with (1). None of the examples quoted earlier belong to this category, and it might be difficult to find a biological example which does. However, the possibility is conceptually useful as a starting point. Writing the first law for region (0), we have

$$dU^{(0)} = dQ^{(0)} - dW^{(0)} \tag{2.24}$$

where, as usual, $dU^{(0)}$ is the internal energy change of (0) as a consequence of its gaining a quantity of heat $dQ^{(0)}$ from (1) and performing a quantity of work $dW^{(0)}$ on (1), and in the process undergoing an infinitesimal change of state. Now according to the second law, region (0) obeys the general inequality

$$dS \equiv \frac{dQ_{\text{rev}}}{T} > \frac{dQ}{T} \tag{2.25}$$

where dS, the entropy gain, is determined by dQ_{rev}, the heat which would be absorbed in a reversible change. (The identity indicated exists, of course, by definition.) Since we are here considering real (irreversible) processes, TdS is always greater than dQ, the actual heat absorbed. It follows that in practice, region (0) fails to extract from its surroundings the maximum amount of heat which theoretically might be transformed into work. Instead of the inequality, it is illuminating to write for real processes

$$dS = \frac{dQ}{T} + \frac{dQ'}{T} \tag{2.26}$$

where dQ is the actual uptake of heat as before, and dQ' is a positive quantity. This is essentially the basis of nonequilibrium thermodynamics—the explicit calculation of dQ' as a function of the appropriate variables. Clearly dQ' represents additional heat that would have been absorbed from (1) if the change had been made reversibly.[3] In the actual change of state the entropy increment dS was completed by the *creation* of a quantity of entropy dQ'/T. This may represent, for example, entropy of mixing or reaction within (0).

In the usual notation of nonequilibrium thermodynamics it is customary to write Eq. (2.26) in the form

$$dS = d_e S + d_i S \qquad (2.27)$$

The terms on the right side of Eq. (2.27) correspond to those on the right side of Eq. (2.26). Thus $d_e S$ is the "exchange" contribution to the entropy change in (0), and $d_i S$ is an "internal" contribution, created by the occurrence of irreversible processes and therefore necessarily positive.

The second possibility is that (0) is an *open* system, one that can exchange both matter and heat with (1). This is by far the more interesting possibility. However, Eqs. (2.24)–(2.26) are now ambiguous, since matter carries with it an associated energy, and the heat transferred is not well defined. It is necessary to rewrite (2.24) in a form which takes explicit account of changes in composition. This form is the Gibbs equation, Eq. (2.19), which will play a major role in our development.

Equation (2.27) can still be written for region (0), but the terms on the right side are no longer identifiable with those of Eq. (2.26). This poses no problem at all. To see more clearly the meaning of $d_i S$, which is central to all further considerations, it is worth considering the entropy changes in both regions of the system, bearing in mind that the irreversible processes relevant to these considerations occur only in (0). The *total* entropy change in the system is evidently given by

$$dS^{\text{total}} = dS^{(0)} + dS^{(1)} \qquad (2.28)$$

However, from Eq. (2.27),

$$dS^{(0)} = d_e S^{(0)} + d_i S^{(0)} \qquad (2.29)$$

$$dS^{(1)} = d_e S^{(1)} = -d_e S^{(0)} \qquad (2.30)$$

since the entropy change in (1) is solely due to exchange with (0).

From Eqs. (2.28), (2.29), and (2.30) it is clear that

$$dS^{\text{total}} = d_tS^{(0)} \tag{2.31}$$

Thus $d_tS^{(0)}$ represents total increase of entropy in the surrounding world due to the processes taking place in (0).

2.3 Dissipation of energy by scalar irreversible processes

Intuitively we feel more at home with the concept of work than with the concept of entropy. It is readily shown that irreversible production of entropy is associated with a loss of free energy, or capability for work. At constant temperature and pressure, the maximum work capability is measured by the Gibbs free energy, G. Under these conditions the changes in G in each region are given by

$$dG^{(0)} = dU^{(0)} + pdV^{(0)} - TdS^{(0)} \tag{2.32}$$

$$dG^{(1)} = dU^{(1)} + pdV^{(1)} - TdS^{(1)} \tag{2.33}$$

Summing these two equations, and remembering that for our system $dV^{(1)} = -dV^{(0)}$, and $dU^{(1)} = -dU^{(0)}$ (since internal energy is conserved in an isolated system),

$$dG^{\text{total}} = dG^{(0)} + dG^{(1)} = -TdS^{\text{total}} \tag{2.34}$$

Since for any real process dS^{total} is necessarily positive, the free energy of the entire system undergoes a decrease. We shall be concerned with the *rate* of this loss or dissipation of free energy, which is given by the dissipation function, Φ:

$$\Phi \equiv T\dot{S}^{\text{total}} = -\dot{G}^{\text{total}} \geq 0 \quad \text{(constant } T \text{ and } p\text{)} \tag{2.35}$$

(The equality to zero refers, of course, only to equilibria or hypothetical reversible processes.) The dissipation function has the dimensions of power and may be measured in watts if convenient.

16   Bioenergetics and linear nonequilibrium thermodynamics

For a system in a stationary state, the evaluation of the dissipation function can be carried out readily by using the Gibbs equation to express the relationship between the different thermodynamic parameters of the surroundings. For example, suppose that the process taking place in region (0) is the enzymatic cleavage of a substrate X into ν identical fragments, giving a product Y:

$$X \rightleftarrows \nu Y \tag{2.36}$$

Now when stationarity is reached, the properties of (0) do not change with time. The rate at which X is taken up balances the rate at which it is converted to Y, and the latter rate just makes up for the loss of Y to the surroundings. Hence the internal energy, entropy, volume, and concentrations associated with (0), or with any small "local" region of (0) if its properties are not uniform, are all constant with the passage of time.[4] The Gibbs equation need therefore be written only for (1):

$$dU^{(1)} = TdS^{(1)} - pdV^{(1)} + \sum_i \mu_i^{(1)} dn_i^{(1)} \tag{2.37}$$

The summation on the right side accounts for changes in composition in (1) resulting from the reaction taking place in (0). The quantities $\mu_i^{(1)}$ and $n_i^{(1)}$ are, respectively, the chemical potential and the number of moles of the ith species in (1). Since $dV^{(1)} = -dV^{(0)} = 0$, and $dU^{(1)} = -dU^{(0)} = 0$ we conclude immediately that

$$TdS^{(1)} = -\sum_i \mu_i^{(1)} dn_i^{(1)} \tag{2.38}$$

Furthermore, since $dS^{(0)} = 0$, we see from Eq. (2.28) that

$$TdS^{\text{total}} = -\sum_i \mu_i^{(1)} dn_i^{(1)} \tag{2.39}$$

It is also worth noting that for the steady state, from Eq. (2.30),

$$dS^{\text{total}} = -d_e S^{(0)} \tag{2.40}$$

that is, the entropy created in region (0) just balances that lost to the surroundings. The dissipation function is obtained by taking the derivative of Eq. (2.39) with respect to time:

$$\Phi = T\dot{S}^{\text{total}} = -\sum_i \mu_i^{(1)} \dot{n}_i^{(1)} \tag{2.41}$$

The rate at which region (0) dissipates free energy can therefore be determined by examining region (1). For the specific enzymatic system we considered above,

$$\Phi = -\mu_X^{(1)} \dot{n}_X^{(1)} - \mu_Y^{(1)} \dot{n}_Y^{(1)} \tag{2.42}$$

The velocity of the reaction, v, may be taken to be the rate of disappearance of X in moles/sec:

$$v = -\dot{n}_X^{(1)} = \frac{1}{\nu}\dot{n}_Y^{(1)} \tag{2.43}$$

Introducing Eq. (2.43) into (2.42) we obtain

$$\Phi = v(\mu_X^{(1)} - \nu\mu_Y^{(1)}) \tag{2.44}$$

The quantity $\mu_X - \nu\mu_Y$ in the above equation is an example of the function of state, first introduced by De Donder, termed the affinity of the reaction. More generally, the affinity is given by

$$A = -\sum_i \nu_i \mu_i \tag{2.45}$$

where ν_i represents the stoichiometric coefficient of the ith species in the reaction, but is taken to have a positive sign if the ith species is formed as a product and to have a negative sign if it is consumed as a reactant. Equation (2.44) can therefore be conveniently rewritten as

$$\Phi = vA \tag{2.46}$$

where A represents the affinity of reaction (2.36) in *region (1)* (measured, for example, in kcal/mole). It should be clear from Eq. (2.41) that if r different independent reactions occurred simultaneously, we would have

$$\Phi = v_1 A_1 + v_2 A_2 + v_3 A_3 + \cdots + v_r A_r \tag{2.47}$$

This is the characteristic form of the dissipation function when reactions only are involved.

Note that we have defined the rate of a given reaction as $v = \dot{n}_i/\nu_i$, where it is obviously immaterial which reactant or product is considered. It is conventional in nonequilibrium thermodynamics to follow De Donder's notation $d\xi = dn_1/\nu_1 = dn_2/\nu_2 = \ldots$, where ξ is called the extent of reaction, the degree of advancement, or the reaction coordinate. Clearly ξ increases by unity each time an equivalent of reaction occurs, and we can write $v = \dot{\xi}$. However, experimentalists customarily choose to regard the rate of reaction as the rate of consumption or production of some easily measured species, say the jth. If we call this measured rate of reaction $v_{(j)}$, then obviously $v_{(j)} = |\nu_j|v$. Since the rate of expenditure of free energy is independent of the convention, the affinity must now be expressed as kcals/mole of species j. Thus $A_{(j)} = A/|\nu_j|$ and $v_{(j)}A_{(j)} = vA$, irrespective of which species j is chosen.

The affinity is related to the change in Gibbs free energy of the system resulting from the reaction. From Eqs. (2.22) and (2.45) it can be seen that for an equivalent of reaction occurring at constant T, p, and μ_i (that is, in a volume sufficiently large that the chemical potentials of all species remain essentially unaltered), the change in Gibbs free energy, ΔG, is given by

$$-\Delta G = A \quad (\text{constant } T, p, \mu_i) \qquad (2.48)$$

Alternatively, the relationship can be formulated in differential form. From Eqs. (2.20) and (2.45), using the definition of ξ given above,

$$-\left(\frac{\partial G}{\partial \xi}\right)_{T,p} = A \qquad (2.49)$$

2.4 Dissipation of energy by vectorial irreversible processes

Equation (2.47) takes into account only the velocities and affinities of chemical reactions, which are scalar processes unassociated with any direction in space. More generally we shall be dealing with vectorial

Equilibrium and nonequilibrium thermodynamics   19

processes as well, since the surroundings need not necessarily be homogeneous, and flows may be occurring across the system. In this case the system will usually be a membrane separating the different regions.

Membrane processes have proved to be a particularly fertile field for the application of nonequilibrium thermodynamics, but because several extensive treatments exist already (Katchalsky and Curran, 1965; Caplan and Mikulecky, 1966; Mikulecky, 1969; Lakshminarayanaiah, 1969), we will present only a bare outline here. To derive the dissipation function, consider the arrangement shown in Fig. 2.3. The membrane (0) is mounted in a suitable chamber and separates the two compartments (1) and (2). The compartments are filled with dilute aqueous solutions containing a single permeant solute, which may or may not be an electrolyte (the concentrations may, of course, differ in the two compartments). In addition, the solutions may contain one or more impermeant solutes and are assumed to be well stirred. The compartments are fitted with vertical tubes of sufficiently wide bore to maintain the pressure head Δp virtually constant over a lengthy period. They are also fitted with electrodes reversible to one of the ions present, for example, Ag/AgCl electrodes.[5] Since the chloride ions in the solutions are in equilibrium with those on the electrodes, the electrodes are considered an integral part of the compartments.

The chamber is immersed in a large constant-temperature air

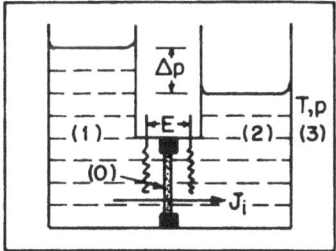

Fig. 2.3. A transport system: irreversible processes are confined to the membrane (0), which is mounted between compartments (1) and (2). Region (3) is enclosed by rigid adiabatic walls. T = temperature; p = pressure; Δp = hydrostatic pressure difference across the membrane; E = electrical potential difference between reversible electrodes (see text); J_i = flux of ith species in the direction shown.

bath (3) enclosed by adiabatic walls. It should be clear that regions (1), (2), and (3) of Fig. 2.3 correspond to region (1) of Fig. 2.2, and the membrane corresponds to region (0). Permeant substances may move between region (1) and region (2) under the influence of concentration and/or electrical driving forces; the electrical potential difference between the electrodes can if desired be set at any chosen value. Compared to the two compartments, the membrane is a region of small capacity and soon reaches a stationary state under given conditions. We now proceed just as we did in considering reaction (2.36). However, since in this case the ith species may be charged, we must take into account its electrochemical potential $\tilde{\mu}_i$ in writing the Gibbs equation, where

$$\tilde{\mu}_i = \mu_i + z_i F \psi \tag{2.50}$$

Here z_i is the charge associated with an ion of species i, F is the faraday, and ψ is the electrical potential; $z_i F \psi$ constitutes the electrical contribution to the partial molar free energy. It should be noted, however, that the development below is perfectly general and applies to an uncharged species on setting $z_i = 0$. We have, therefore,

$$dU^{(1)} = TdS^{(1)} - pdV^{(1)} + \sum_i \tilde{\mu}_i^{(1)} dn_i^{(1)}$$

$$dU^{(2)} = TdS^{(2)} - pdV^{(2)} + \sum_i \tilde{\mu}_i^{(2)} dn_i^{(2)}$$

$$dU^{(3)} = TdS^{(3)} - pdV^{(3)} \tag{2.51}$$

Here, by $dn_{Cl^-}^{(2)}$, for example, we mean the total change in chloride ion in compartment (2), whether in the solution or on the electrode. Thus it is seen that the changes in composition in regions (1) and (2), $dn_i^{(1)}$ and $dn_i^{(2)}$, represent entirely the irreversible flux of species i through the membrane. Adding Eqs. (2.51), and remembering that the dU's and the dV's sum to zero, we obtain

$$TdS^{\text{total}} = -\sum_i \tilde{\mu}_i^{(1)} dn_i^{(1)} - \sum_i \tilde{\mu}_i^{(2)} dn_i^{(2)} \tag{2.52}$$

The flow of the ith species, J_i, through the membrane from side (1) to side (2) is given by

$$J_i = \dot{n}_i^{(2)} = -\dot{n}_i^{(1)} \tag{2.53}$$

and consequently

$$\Phi = T\dot{S}^{\text{total}} = \sum_i J_i \Delta \bar{\mu}_i \tag{2.54}$$

where $\Delta \bar{\mu}_i = \bar{\mu}_i^{(1)} - \bar{\mu}_i^{(2)}$, the difference in electrochemical potential of species i between the two sides.[6] It should be noted that as in Eq. (2.47), the dissipation function has the form of a sum of products of forces and flows. In our case the permeant species are water and a single solute, which may be a salt. Denoting the cation and anion by subscripts 1 and 2, respectively, and water by w,

$$\Phi = J_w \Delta \mu_w + J_1 \Delta \bar{\mu}_1 + J_2 \Delta \bar{\mu}_2 \tag{2.55}$$

2.5 Dissipation of free energy by coupled scalar and vectorial processes

We have considered above the rate of free energy dissipation by scalar and vectorial processes separately. However, in biological systems scalar and vectorial processes are frequently coupled. For example, metabolism of substrates brings about active transport; on the other hand, a transport process may be expected to influence the rate of a coupled metabolic reaction. In such coupled processes, free energy that otherwise would be dissipated is to some extent conserved. Thus in active transport a chemical reaction for which vA is greater than zero may perform electroosmotic work by "driving" a species i against its electrochemical gradient. In this case $J_i \Delta \bar{\mu}_i$ is less than zero. The rate of entropy production is now given by

$$\Phi = J_i \Delta \bar{\mu}_i + vA \tag{2.56}$$

Although by the second law $\Phi > 0$, we see that $\Phi < vA$; in other words, some of the chemical free energy which would have been dissipated in an uncoupled reaction has been conserved. More gener-

22   Bioenergetics and linear nonequilibrium thermodynamics

ally, for the transport process considered in Sec. 2.4, the dissipation function takes the form

$$\Phi = J_w \Delta \mu_w + J_1 \Delta \tilde{\mu}_1 + J_2 \Delta \tilde{\mu}_2 + vA \tag{2.57}$$

2.6 Summary

1. The first and second laws of thermodynamics are presented and discussed for closed systems, making use of the notion of reversibility and introducing the definitions of the internal energy and entropy functions.

2. The two laws are combined, giving internal energy as a function of temperature, pressure, entropy, and volume, and the resulting relation is interpreted in terms of both reversible and irreversible processes. This leads to the definitions of enthalpy, Helmholtz free energy, and Gibbs free energy, all of which are functions of the same four parameters. A simple mnemonic enables one to remember these functions readily.

3. It is shown that the maximum useful work that can be obtained from a process at constant pressure and temperature is equal to the negative change in Gibbs free energy, and this becomes available only in the limiting case of a hypothetical reversible process.

4. The thermodynamic framework thus developed is extended to open systems, which exchange matter with their surroundings. This requires introducing the chemical potential of a species and leads to the Gibbs equation, a generalized form of the combined first and second laws. An important relationship between simultaneous changes in intensive variables, the Gibbs-Duhem equation, is also derived.

5. It is shown that the entropy change in a system which is not at equilibrium with its environment is made up of two contributions: an "exchange" contribution and an "internal" contribution *created* by the occurrence of irreversible processes and therefore necessarily positive. A general procedure is introduced to evaluate the rate of entropy production.

6. The corresponding rate of dissipation of free energy, the dissipation function, is central to our considerations of systems in a stationary state. From the Gibbs equation, the dissipation function is

derived both for scalar and for vectorial irreversible processes. These may be coupled in a single system. In all cases the dissipation function takes the form of a sum of products of thermodynamic forces and flows.

7. The affinity, or driving force, of a chemical reaction is a function of state which arises naturally in this context. It is defined, and its relationship to the change in Gibbs free energy is made clear.

3 Relationships between flows and forces: the Kedem-Katchalsky equations

We saw in the previous chapter that the dissipation function for several different systems could be expressed as a sum of products of conjugate flows and forces (these terms being employed in a generalized sense). This is always possible. Thus for the general case we can write

$$\Phi = J_1 X_1 + J_2 X_2 + J_3 X_3 + \cdots \tag{3.1}$$

Here the J's are thermodynamic flows, such as reaction velocities, rates of transport or muscle contraction, and electrical current; the X's are thermodynamic forces, such as reaction affinities, electrochemical potential differences, and muscular tension, as well as differences in electrical potential and hydrostatic and osmotic pressure.

This dissipation function, if derived from a Gibbs equation in the manner demonstrated in the previous chapter, provides us with a very important piece of information: the number of degrees of freedom in the system. Each term corresponds to a degree of freedom.

In practice it is sometimes found that a given set of forces and flows is experimentally inconvenient. In such cases it often proves possible to transform the original dissipation function so that it becomes a function of forces and flows which can be readily fixed or measured. For example, the representation of the dissipation function given in Eq. (2.55) is of little experimental utility:

$$\Phi = J_w \Delta \mu_w + J_1 \Delta \tilde{\mu}_1 + J_2 \Delta \tilde{\mu}_2$$

None of the flows or forces in this representation is ordinarily measured directly. The transformations which lead to operational forms of this dissipation function for membrane processes were studied in a classic series of papers by Kedem and Katchalsky (1958, 1961, 1963a, b, c), Michaeli and Kedem (1961), and Katchalsky and Kedem (1962).

3.1 The phenomenological equations

Several simple linear relationships between corresponding flows and forces are well known, for example, Ohm's law for flow of electric current, Fick's law for diffusion, Fourier's law for heat flow, and Poiseuille's and Darcy's laws for fluid flow. In each of these cases the flow is given by the product of a conductance coefficient and the force. However, in many other cases we have to consider a multiplicity of interacting flows. For example, the transport of an ion across a biological membrane may be influenced not only by its electrochemical potential gradient, but also by the flows of any other ions which may be present, as well as that of solvent. Despite the complexity of possible interactions, the relationships between forces and flows often remain linear. Nonequilibrium thermodynamics has been usefully applied to such linear processes in a variety of nonliving and living systems. In this approach the simple linear relationship is extended to cover all possible interactions.

To illustrate this, suppose we have a system which gives rise to a three-term dissipation function, indicating the presence of three independent processes. One convenient way of relating the flows and forces would be in terms of the following phenomenological equations, where the flows are written as linear functions of the forces:

$$J_1 = L_{11}X_1 + L_{12}X_2 + L_{13}X_3$$

$$J_2 = L_{21}X_1 + L_{22}X_2 + L_{23}X_3$$

$$J_3 = L_{31}X_1 + L_{32}X_2 + L_{33}X_3 \tag{3.2}$$

It is seen that each flow J_i is related to its "conjugate" force X_i through a "straight" coefficient, L_{ii}. It may also be affected by any other force X_j, the coupling or "cross" coefficient being L_{ij}. On *a priori* grounds this set of linear relations might be expected to hold good only for relatively slow processes close to equilibrium, while faster processes would require the addition of higher-order terms, yielding awkward nonlinear relations.

It might be thought that the region of linearity would be too narrow for a simple linear formulation to be of practical utility. However, the criterion for closeness to equilibrium is empirical, and for vectorial processes linearity is often observed over a surprisingly wide range of magnitudes of the forces. For chemical reactions the criterion is usually much more restrictive. This problem will be discussed in detail in Chapter 6.

A striking advantage of the thermodynamic formulation now presents itself. Rather than having entirely arbitrary values, the phenomenological coefficients of any system, irrespective of the number of degrees of freedom, are subject to important restrictions. First, according to Onsager's law, for a system of flows and forces based on an appropriate dissipation function, the matrix of coefficients is symmetrical, so that

$$L_{ij} = L_{ji} \tag{3.3}$$

This affords a considerable reduction in the number of coefficients to be measured. Second, since the dissipation function can never be negative, it can be shown that the straight coefficients must satisfy the condition

$$L_{ii} > 0 \tag{3.4}$$

and the cross coefficients must satisfy the condition

$$L_{ij}^2 \leq L_{ii}L_{jj} \tag{3.5}$$

Within these constraints the coefficients vary over a wide range of values according to the tightness or looseness of coupling between the processes.

There is an alternative way of formulating the phenomenological equations which is frequently advantageous and which makes use of resistance rather than conductance coefficients. As a consequence of the linearity of Eqs. (3.2), we could equally well have written

$$X_1 = R_{11}J_1 + R_{12}J_2 + R_{13}J_3$$

$$X_2 = R_{21}J_1 + R_{22}J_2 + R_{23}J_3$$

$$X_3 = R_{31}J_1 + R_{32}J_2 + R_{33}J_3 \tag{3.6}$$

The thermodynamic restrictions which apply to this set are completely analogous to those which applied to the previous set,[1] that is,

$$R_{ij} = R_{ji} \tag{3.7}$$

$$R_{ii} > 0 \tag{3.8}$$

$$R_{ij}^2 \leq R_{ii}R_{jj} \tag{3.9}$$

It is necessary at this point to sound a note of caution. The phenomenological coefficients, unfortunately, cannot be considered perfectly constant under all experimental conditions, irrespective of the mode by which the flows are produced. The coefficients are functions of the parameters of state and will be more or less sensitive to variations in the state of the system. However, they are not functions of the forces or the flows, except indirectly in that variations in these may bring about changes in state. It should often prove possible, therefore, to carry out experiments over a range of conditions such that the coefficients do in fact remain sensibly constant. For example, Eqs. (3.2) and (3.6) may fail to characterize ion transport across synthetic membranes driven by large concentration gradients, but they may characterize very well flows induced by equivalent electrical forces.

3.2 Coupling between reactions and flows

Equations (3.2) and (3.6) are very general. Similar equations can be written with any number of degrees of freedom. Depending on the nature of the membrane, these may relate flows and forces of either vectorial or scalar character, provided that the J's and X's are derived from an appropriate dissipation function. The formulation can be applied in particular to the phenomenon of active transport. Kedem (1961) first presented a treatment of this kind in which a scalar chemical reaction was coupled to a vectorial flow. Her formulation was based implicitly on the dissipation function appearing in Eq. (2.57):

$$\Phi = J_w \Delta \mu_w + J_1 \Delta \tilde{\mu}_1 + J_2 \Delta \tilde{\mu}_2 + vA$$

where the subscripts $w, 1$, and 2 refer to water, cation, and anion, respectively. The corresponding phenomenological equations in the resistance formulation are

$$\Delta \mu_w = R_{ww} J_w + R_{w1} J_1 + R_{w2} J_2$$

$$\Delta \tilde{\mu}_1 = R_{1w} J_w + R_{11} J_1 + R_{12} J_2 + R_{1r} v$$

$$\Delta \tilde{\mu}_2 = R_{2w} J_w + R_{21} J_1 + R_{22} J_2 + R_{2r} v$$

$$A = R_{r1} J_1 + R_{r2} J_2 + R_{rr} v \tag{3.10}$$

The reason for choosing an R rather than an L formulation is that in a system in which several flows interact, it is the R cross coefficients that reflect the extent of the interaction directly, as will be shown in Chapter 4. In Eqs. (3.10), nonzero values of the coefficients R_{1r} and R_{2r} indicate coupling between ionic flows and reaction; it is assumed that no direct coupling exists between the flow of water and reaction. Kedem defined the transport of an ion i to be active if the coefficient R_{ir} is different from zero. For example, in certain species the electrical current across the short-circuited frog skin mounted between identical solutions consists essentially of the flow of sodium ions. This observation is consistent with the conclusion that $R_{2r} = R_{12} = 0$, while R_{1r} is nonzero. (For definite knowledge on this point it would be necessary to know the values of R_{1w} and R_{2w}.)

There is an important difference between the coefficients R_{ir} and the other coefficients appearing in Eqs. (3.10), including R_{rr}. The latter are all scalar quantities; however, since both v and A are scalars, while $\Delta\tilde{\mu}_i$ and J_i are vectors, the coefficients R_{ir} and R_{ri} must have a vectorial character.

3.3 The Curie-Prigogine principle

What is the significance of a vectorial coupling coefficient? To understand this, we have to reflect on the nature of coupling between flows and forces of essentially different character. If a reaction within a membrane couples to a flow, as in the short-circuited frog skin mentioned above, it seems evident on intuitive grounds that the direction of the flow must be determined by a property of the membrane. If the membrane were completely isotropic and homogenous, that is, if its equilibrium properties were identical in all directions, it would not be expected that such coupling could occur. In such a membrane there is no obvious reason why consumption of metabolic energy should cause the transport of an ion preferentially in one direction rather than any other. This idea is embodied in the principle originally enunciated by Curie, which in its application to nonequilibrium thermodynamics by Prigogine (1947) and later by others (see, for example, deGroot and Mazur, 1963; Finlayson and Scriven, 1969) indicates that coupling between scalar and vector flows is impossible in isotropic media in the linear regime. However, in anisotropic media such coupling is no longer forbidden. The coupling coefficient must necessarily reflect the anisotropy of the medium and consequently will itself be vectorial.

The Curie-Prigogine principle was originally presented in the context of symmetry considerations concerning cause and effect in crystallographic systems (Curie, 1894). Its interpretation in nonequilibrium thermodynamic terms has been attended by a certain amount of confusion, since as pointed out by Finlayson and Scriven, the drastic restrictions of isotropy *and* linearity which must be invoked to eliminate the possibility of scalar-vector coupling have not always been fully appreciated. In this book our treatment of transport and other processes will generally be confined to the linear regime, and consequently we need only concern ourselves with the isotropy or anisotropy of the medium. The notion of isotropy as used here relates to *local* properties, in other words to the properties of microscopic elements of the medium just sufficiently large to be associated with thermodynamic parameters. If the elements of a system are anisotropic, local scalar-vector coupling is possible in principle. Thus pressure can give rise to electrical polarity in a piezoelectric crystal, and the hydrolysis of ATP can give rise to sodium transport in a cell membrane: anisotropy is known to be a characteristic of the former system and is almost certainly a characteristic of the latter. However, it is also important to consider media that are locally isotropic but not spatially homogenous (DeSimone and Caplan, 1973). For example, a membrane may contain an unsymmetrical distribution of bound enzyme. It is reasonable to expect that such an inhomogeneous distribution of transport and/or reaction parameters, whether continuous or discontinuous, may influence the *global* or overall behavior of the membrane, and it can be shown that "dissymmetry" of this kind may lead to a specific manifestation of scalar-vector coupling even in systems which are locally isotropic. The coupling coefficients in this case are always associated with the global system rather than its local elements, and the phenomenon only occurs under conditions which give rise to "stationary-state" coupling, which will be discussed in Sec. 3.5.

3.4 Stationary states of minimal entropy production

We turn now to a consideration of certain important and characteristic properties of stationary states. If the number of restraints on a system in the steady state is changed, what happens? For example, take the system represented by Eqs. (3.10). If the maximum number of restraints is applied—all four forces, say, being fixed—then the stationary state is fully defined, since no more degrees of freedom are left. If no restraints at all are applied, the forces will all tend to decrease until the system eventually reaches equilibrium. Frequently, however, we impose an intermediate number of restraints. For such situations it has been shown by Prigogine that for linear systems with Onsager symmetry the entropy production gradually assumes the minimal value compatible with the imposed restraints (see, for example, Prigogine, 1961). Thus if some of the forces are fixed, the remainder will reach values in the stationary state such that their conjugate flows become zero.

3.5 Stationary-state coupling

In certain circumstances stationary states can give rise to a special type of interdependence among global flow processes. This interdependence has been termed by Prigogine *stationary-state coupling;* characteristically it manifests itself as a coupling between chemical reaction and diffusional flow. This comes about as a direct consequence of the stationarity condition, which imposes a mutual linear dependence on some of the flows, in effect contracting the dissipation function to fewer terms. A number of striking examples of this kind were described by Hearon (1950) in relation to biological processes; for one simple case, which may be interpreted as a model of the living cell, Prigogine (1961) has analyzed the relationships among the phenomenological equations and shown that a stationary nonequilibrium distribution of matter can arise, determined by the rate of metabolic reaction within the cell. This case involves an open system such as region (0) of Fig. 2.2.

Consider that initially the uncharged species M and O are added to the medium outside the cell. M moves into the cell, where it is transformed into N, some of which flows out of the cell, as shown in Fig. 3.1. The transformation is mediated by the action of an enzyme confined to the interior of the cell. The component O does not take part in any chemical reaction; however, its flow is coupled frictionally to the flow of M. Neither the flow of M nor the flow of O is coupled to the flow of N. After some time has passed, a certain amount

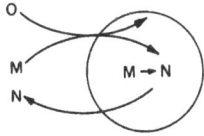

Fig. 3.1. System illustrating the occurrence of a stationary nonequilibrium distribution of matter.

32 Bioenergetics and linear nonequilibrium thermodynamics

of M will have moved into the cell, carrying O with it, and a certain amount of N will have appeared in the external solution. The general phenomenological description of the system in the linear range is as follows:

$$\Delta\mu_M = R_M J_M \quad\quad + R_{MO} J_O$$
$$\Delta\mu_N = \quad\quad R_N J_N$$
$$\Delta\mu_O = R_{MO} J_M \quad\quad + R_O J_O$$
$$A^{in} = \quad\quad\quad\quad\quad\quad R_r v \tag{3.11}$$

Here the J's refer to inward fluxes, $\Delta\mu_i = \mu_i^{ex} - \mu_i^{in}$, and A^{in} refers to the affinity of the reaction in the cell interior. The above equations have been set out so as to illustrate the coupling between certain flows and the lack of coupling between others. Specifically, the metabolic reaction occurring in the interior of the cell is not coupled to any of the flows. With the passage of time the system reaches a state such that the concentrations of M and N, but not necessarily that of O, become constant. In this state of stationarity,

$$v = J_M = -J_N \tag{3.12}$$

Under these conditions the flow of O is effectively coupled to the metabolic reaction as evaluated in the external environment. This is seen as follows.

If we denote by A^{ex} the affinity of the reaction measured externally (where, of course, the requisite enzyme is absent), we have

$$A^{ex} - A^{in} = (\mu_M^{ex} - \mu_N^{ex}) - (\mu_M^{in} - \mu_N^{in})$$
$$= \Delta\mu_M - \Delta\mu_N \tag{3.13}$$

Introducing Eqs. (3.12) into the set of Eqs. (3.11), we see that

$$\Delta\mu_M = R_M v + R_{MO} J_O$$
$$\Delta\mu_N = -R_N v$$
$$\Delta\mu_O = R_{MO} v + R_O J_O$$
$$A^{in} = R_r v \tag{3.14}$$

Combining Eqs. (3.13) and (3.14), we obtain finally

$$\Delta\mu_O = R_{MO}v + R_O J_O$$

$$A^{ex} = (R_r + R_M + R_N)v + R_{MO} J_O \tag{3.15}$$

More simply, Eqs. (3.15) can be written as

$$\Delta\mu_O = R_O J_O + R_{MO} v$$

$$A^{ex} = R_{MO} J_O + R_R v \tag{3.16}$$

where $R_R = R_r + R_M + R_N$. Since the only restraint imposed is the fixing of A^{ex}, eventually the system reaches a state in which $J_O = 0$. In this state there is an accumulation of O within the cell given by

$$\Delta\mu_O = R_{MO} v = \frac{R_{MO}}{R_R} A^{ex} \tag{3.17}$$

The reduction in the number of phenomenological equations required for a functional description of the system seen above is paralleled by a contraction of the dissipation function. Equations (3.11) follow from our standard procedures, taking the view that the cell *membrane* is in a stationary state, since it has a much lower capacity than either the interior or the environment. These equations correspond to the following dissipation function:

$$\Phi = J_M \Delta\mu_M + J_N \Delta\mu_N + J_O \Delta\mu_O + v A^{in} \tag{3.18}$$

The state of stationarity of the *whole cell*, represented by Eqs. (3.12), leads directly to the contracted dissipation function corresponding to Eqs. (3.16):

$$\Phi = J_O \Delta\mu_O + v A^{ex} \tag{3.19}$$

This is seen by combining Eq. (3.18) with Eqs. (3.12) and (3.13).

In the above example, stationary-state coupling occurs between the flow of component O and the reaction. In essence this is a two-compartment system, one compartment of which is inaccessible (except through the membrane) and contains an enzyme. The system is therefore basically unsymmetrical in the sense of Sec. 3.3. We infer from this that to obtain stationary-state coupling between reaction and flow, it is not necessary to have anisotropy, as long as an appropriate unsymmetrical arrangement of isotropic elements is provided.

Active transport in cells such as muscle cells or red cells is not in fact associated with a mechanism of the type considered in the Hearon-Prigogine model discussed above. The coupling is generally held to be a property of the membrane itself, associated with enzymes that are an integral part of the membrane. In the case of membranes such as the frog skin, where both sides of the epithelium are readily accessible, the metabolic reaction clearly occurs within the membrane. A simple model of such a system based on stationary-state coupling was studied experimentally and theoretically by Blumenthal, Caplan, and Kedem (1967). It comprised a series combination of two membranes, one having cation exchange properties, the other anion exchange properties, with a thin layer of an aqueous enzyme solution sealed between them. This system, when supplied externally with a suitable substrate, was able to drive electric current between two identical solutions. In the stationary state its dissipation function takes the contracted form

$$\Phi = IE + vA^{ex} \tag{3.20}$$

where I and E represent electrical current and potential difference, respectively. A further discussion of this model will be presented in Sec. 3.9.

3.6 Transformations of the dissipation function

An important transformation that has proved particularly useful for dilute solutions expresses Φ in terms of flows of volume, salt, and electrical current and the corresponding forces. This form of the dissipation function is developed as follows. If a molecule of the salt under consideration dissociates into ν_1 cations of charge z_1, and ν_2 anions of charge z_2, the condition for electroneutrality is

$$\nu_1 z_1 + \nu_2 z_2 = 0 \tag{3.21}$$

The thermodynamic properties of the salt are related to those of its constituent ions by the relation

$$\mu_s = \nu_1 \tilde{\mu}_1 + \nu_2 \tilde{\mu}_2 \quad \text{or} \quad \Delta\mu_s = \nu_1 \Delta\tilde{\mu}_1 + \nu_2 \Delta\tilde{\mu}_2 \tag{3.22}$$

The electromotive force acting in the system is determined by measuring the electrical potential difference E between two electrodes suitably placed in the two solutions. If these electrodes are reversible to the anion, as in the case of Ag/AgCl electrodes,

$$E = \frac{\Delta\tilde{\mu}_2}{z_2 F} \tag{3.23}$$

It should be noted that the electromotive force E measured by such reversible electrodes is different from the potential difference $\Delta\psi$ that would be measured by reference electrodes such as the calomel electrode with a salt bridge. By referring to the fundamental definition, Eq. (2.50), it is seen that

$$\Delta\psi = E - \frac{\Delta\mu_2^c}{z_2 F} \tag{3.24}$$

where $\Delta\mu_2^c$ is the concentration-dependent part of $\Delta\tilde{\mu}_2$ (the pressure-dependent part of $\Delta\tilde{\mu}_2$, which is usually small, has been neglected). The relationship between the electrical current density and ion flows per unit area is given by

$$I = (z_1 J_1 + z_2 J_2) F \tag{3.25}$$

(It is important to note that in the presence of current flow, the precise application of Eq. (3.23) requires measuring E by means of a separate pair of reversible "sensing" electrodes.) For the case we are considering, the transfer of salt from solution (1) to solution (2) (see Fig. 2.3) may be identified with the flow of cations across the membrane, J_1:

$$J_1 = \nu_1 J_s \tag{3.26}$$

The salt flow cannot in general be associated with J_2, the flow of anions across the membrane, since with $I \neq 0$, some of the anions entering the solution are derived from the current-passing electrode. Hence it is seen that whereas J_1, J_2, J_w, and I represent actual flows across the membrane, J_s must be regarded as a virtual transfer of salt, which may include a component due to processes at the electrodes (Weinstein and Caplan, 1973; Kedem, 1973). Combining Eqs. (3.21), (3.25), and (3.26), the flow of anions is given by

$$J_2 = \nu_2 J_s + \frac{I}{z_2 F} \tag{3.27}$$

Introducing into Eq. (2.55) the values of J_1 from Eq. (3.26), J_2 from Eq. (3.27), and $\Delta \bar{\mu}_2$ from Eq. (3.23) gives, with Eq. (3.22), the result

$$\Phi = J_w \Delta \mu_w + J_s \Delta \mu_s + IE \tag{3.28}$$

This expression, which was first derived by Michaeli and Kedem (1961), has the advantage of incorporating easily measured electrical quantities, but the first two terms are not yet operational. A more convenient form of the dissipation function is derived by substituting appropriate expressions for $\Delta \mu_w$ and $\Delta \mu_s$:

$$\Delta \mu_w = \bar{V}_w (\Delta p - \Delta \pi) = \bar{V}_w (\Delta p - \Delta \pi_i - \Delta \pi_s) \tag{3.29}$$

$$\Delta \mu_s = \Delta \mu_s^c + \bar{V}_s \Delta p \tag{3.30}$$

where $\bar{V}_w$ and $\bar{V}_s$ are the partial molar volumes of water and salt, respectively; $\Delta p = p^{(1)} - p^{(2)}$ is the difference in hydrostatic pressure across the membrane; $\Delta \pi_s$ is the difference in the osmotic pressure due to permeant solute (here salt); $\Delta \pi_i$ is the difference in osmotic pressure attributable to any impermeant solutes which may be present; and $\Delta \mu_s^c$ is the concentration-dependent part of the chemical potential difference of salt. It is convenient to express $\Delta \mu_s^c$ in terms of $\Delta \pi_s$ and a "mean" concentration of salt defined as[2]

$$c_s = \frac{\Delta \pi_s}{\Delta \mu_s^c} = \frac{\Delta \pi_s}{RT \Delta \ln a_s} \tag{3.31}$$

where a represents activity. Introducing Eqs. (3.29)–(3.31) into (3.28) gives

$$\Phi = J_w\overline{V}_w(\Delta p - \Delta\pi) + J_s\overline{V}_s\Delta p + J_s\left(\frac{\Delta\pi_s}{c_s}\right) + IE \qquad (3.32)$$

The need for the experimentally difficult measurement of J_w is now eliminated by the introduction of an expression for the volume associated with the transfer of water and salt between the two solutions:

$$J_v = J_w\overline{V}_w + J_s\overline{V}_s \qquad (3.33)$$

It is noteworthy that as defined here, J_v, like J_s, does not in general represent a rate of material flow across the membrane. Rather, it also is a virtual flow, because under conditions of current passage it incorporates a contribution from electrode processes, in that it contains J_s (Weinstein and Caplan, 1973; Kedem, 1973). Experimentally J_v is evaluated by measuring the change in volume of one or both compartments at opposite surfaces of the membranes. Although this measurement does not reflect the processes at the electrodes, in practice the resultant error is usually insignificant. Special cases necessitating correction of this "observed volume flow" will be considered in Sec. 3.8. Denoting the volume fraction of the solute by $\phi_s = c_s\overline{V}_s$, Eqs. (3.32) and (3.33) combine to give

$$\Phi = J_v(\Delta p - \Delta\pi) + \frac{J_s\Delta\pi_s(1 + \phi_s\Delta\pi/\Delta\pi_s)}{c_s} + IE \qquad (3.34)$$

Finally, the solutions employed in experimental studies are usually sufficiently dilute that $\phi_s\Delta\pi/\Delta\pi_s \ll 1$, so that

$$\begin{aligned}\Phi &= J_v(\Delta p - \Delta\pi) + \frac{J_s\Delta\pi_s}{c_s} + IE \\ &= J_v(\Delta p - \Delta\pi) + J_s\Delta\mu_s^c + IE\end{aligned} \qquad (3.35)$$

The dissipation function of Eq. (3.35) formed the basis of much of Kedem and Katchalsky's extensive treatment of composite membranes (1963b,c). It leads naturally to a convenient set of phenomenological equations. For flows in the linear range we can write (in the L form):

$$J_v = L_{11}(\Delta p - \Delta \pi) + L_{12}\left(\frac{\Delta \pi_s}{c_s}\right) + L_{13}E$$

$$J_s = L_{21}(\Delta p - \Delta \pi) + L_{22}\left(\frac{\Delta \pi_s}{c_s}\right) + L_{23}E$$

$$I = L_{31}(\Delta p - \Delta \pi) + L_{32}\left(\frac{\Delta \pi_s}{c_s}\right) + L_{33}E \tag{3.36}$$

Since Eq. (3.35) is an appropriately derived dissipation function consisting of a sum of products of conjugate flows and forces, Onsager's law tells us that the matrix of the phenomenological Eqs. (3.36) is symmetrical, that is,

$$L_{ij} = L_{ji} \tag{3.37}$$

If it is desired instead to express the forces as functions of the flows, we have

$$(\Delta p - \Delta \pi) = R_{11}J_v + R_{12}J_s + R_{13}I$$

$$\frac{\Delta \pi_s}{c_s} = R_{21}J_v + R_{22}J_s + R_{23}I$$

$$E = R_{31}J_v + R_{32}J_s + R_{33}I \tag{3.38}$$

where the matrix of R coefficients is the inverse of the L matrix, and again

$$R_{ij} = R_{ji} \tag{3.39}$$

Equations (3.36)–(3.39) are relevant to a variety of classical studies of electrokinetic phenomena, since they provide a systematic formalism for describing coupled processes and lead naturally to a variety of symmetry relationships which have long been observed experimentally. As such, they provide striking evidence of the power of the nonequilibrium thermodynamic approach. This is most readily appreciated by considering studies with identical solutions at each surface of the membrane, so that $\Delta\pi = \Delta\pi_s = 0$. Our system now has only two degrees of freedom. In this case

$$\Phi = J_v \Delta p + IE \tag{3.40}$$

and the pertinent phenomenological equations become (in the L form):

$$J_v = L_{11}\Delta p + L_{12}E$$

$$I = L_{12}\Delta p + L_{22}E \tag{3.41}$$

where we have incorporated the Onsager symmetry relation and have written $L_{12} = L_{21}$ for the $L_{13} = L_{31}$ of Eq. (3.36). The phenomenological Eqs. (3.41) embrace the whole of classical electrokinetics. For example, it is clear that the magnitude of the electroosmotic volume flow per unit potential at zero pressure difference, $(J_v/E)_{\Delta p=0}$, and the streaming current per unit pressure difference at short circuit, $(I/\Delta p)_{E=0}$, must be identical. Similarly, we can easily obtain the well-established Saxén relations between force ratios and flow ratios (Miller, 1960; Katchalsky and Curran, 1965)

$$\left(\frac{J_v}{I}\right)_{\Delta p=0} = -\left(\frac{E}{\Delta p}\right)_{I=0} \tag{3.42}$$

$$\left(\frac{J_v}{I}\right)_{E=0} = -\left(\frac{E}{\Delta p}\right)_{J_v=0} \tag{3.43}$$

Various classical models of electrokinetics developed by Helmholtz, von Smoluchowski, and others predict all such relations, but they are not applicable to all types of systems considered (Kedem and Katchalsky, 1958, 1961). Mazur and Overbeek (1951) first demonstrated that the various symmetry relationships observed do not depend on the specific features of any given model but follow quite generally from the linear phenomenological equations of nonequilibrium thermodynamics. Thus any linear model which does not predict these relations is likely to be incorrect.

Often it is useful to study systems in the absence of electrodes. In such cases there can, of course, be no current flow, and the dissipation function of Eq. (3.35) reduces to

$$\Phi = J_v(\Delta p - \Delta \pi) + \frac{J_s \Delta \pi_s}{c_s} \qquad (3.44)$$

a form that emphasizes the relationship between volume flow and salt flow. Sometimes we are interested in processes in which volume flow is associated with a significant separation of salt from water. It then becomes useful to rewrite the above dissipation function in a form that takes this separation into account explicitly. Considering solutions in which $\Delta \pi_i \ll \Delta \pi$, so that $J_v \Delta \pi \simeq J_v \Delta \pi_s$, the introduction of Eq. (3.33) gives

$$\Phi = J_v \Delta p - J_w \overline{V}_w \Delta \pi_s + \left(\frac{J_s}{c_s}\right)(1 - c_s \overline{V}_s)\Delta \pi_s$$

For adequately dilute solutions, $c_s \overline{V}_s \ll c_w \overline{V}_w \simeq 1$, and we have approximately

$$\Phi = J_v \Delta p + \left(\frac{J_s}{c_s} - \frac{J_w}{c_w}\right)\Delta \pi_s = J_v \Delta p + J_D \Delta \pi \qquad (3.45)$$

The expression in parentheses, which is just the velocity of the salt relative to the water, has been termed the diffusional flow. It is clear that the dissipation function expressed by Eq. (3.45) provides a natural basis for the analysis of systems in which mechanical energy derived from volume flow down a hydrostatic pressure gradient is utilized to produce a separation of salt from water in the face of an adverse concentration gradient (Caplan, 1973).

3.7 Transformations of the phenomenological equations: practical phenomenological coefficients

Returning now to a more general consideration of Eqs. (3.36)–(3.39), we see that in each formulation a system with three degrees of freedom is characterized by six independent phenomenological coefficients. In order to determine these experimentally we must impose restrictions such that the pertinent flow or force becomes a function of an appropriate independent variable. For the L formulation the coefficients would be most readily evaluated if it proved possible to control two forces, for example,

$$L_{12} = \left[\frac{J_v}{\Delta \pi_s/c_s}\right]_{\Delta p - \Delta \pi = 0, E = 0} \quad \text{or}$$

$$L_{12} = \left[\frac{\partial J_v}{\partial(\Delta \pi_s/c_s)}\right]_{\Delta p - \Delta \pi, E} \quad (3.46)$$

whereas for the R coefficients it would be desirable to control two flows. Often, however, it is not experimentally convenient to control either two forces or two flows, or suitability for theoretical treatment may dictate other considerations. In such a case it is useful to consider alternative expressions which may be obtained by manipulation of fundamental phenomenological equations without the need for further transformation of the dissipation function. The coefficients of these modified equations will be combinations of the original L_{ij}'s (or R_{ij}'s). As an example of such "practical transport coefficients" we shall consider in detail Kedem and Katchalsky's Set I.

We start with the dissipation function of Eq. (3.35) and consider the corresponding set of phenomenological Eqs. (3.38), expressing the forces as functions of the flows. For practical purposes it is clearly desirable to use relations in which the independent variables are readily controlled experimentally. One convenient set of independent variables is given by J_v, $\Delta \pi_s/c_s$, and I. We therefore recast Eqs. (3.38) in such a way that $\Delta \pi_s/c_s$ becomes an independent variable, while J_s becomes a dependent variable. The appropriate algebraic manipulation leads to

$$(\Delta p - \Delta \pi) = \frac{R_{11}R_{22} - R_{12}R_{21}}{R_{22}} J_v + \frac{R_{12}}{R_{22}} \frac{\Delta \pi_s}{c_s}$$
$$+ \frac{R_{22}R_{13} - R_{12}R_{23}}{R_{22}} I$$

$$J_s = -\frac{R_{21}}{R_{22}} J_v + \frac{1}{R_{22}} \frac{\Delta \pi_s}{c_s} - \frac{R_{23}}{R_{22}} I$$

$$E = \frac{R_{22}R_{31} - R_{21}R_{32}}{R_{22}} J_v + \frac{R_{32}}{R_{22}} \frac{\Delta\pi_s}{c_s}$$

$$+ \frac{R_{22}R_{33} - R_{32}R_{23}}{R_{22}} I \qquad (3.47)$$

Since our independent variables are now mixed forces and flows, the set of composite phenomenological coefficients is no longer symmetrical. However, in this case the loss of symmetry involves only signs.

Table 3.1. Two sets of practical transport coefficients.[a]

Transport coefficients	Set I (vanishing J_v, $\Delta\pi_s$, or I)[b]	Set II (vanishing $(\Delta p - \Delta\pi)$, $\Delta\pi_s$, or I)[b]
Straight:		
Filtration coefficient	$L_p = \left(\dfrac{J_v}{\Delta p - \Delta\pi}\right)_{\Delta\pi_s, I}$	$L_p = \left(\dfrac{J_v}{\Delta p - \Delta\pi}\right)_{\Delta\pi_s, I}$
Solute permeability	$\omega = \left(\dfrac{J_s}{\Delta\pi_s}\right)_{J_v, I}$	$\omega' = \left(\dfrac{J_s}{\Delta\pi_s}\right)_{\Delta p - \Delta\pi, I}$
Electric conductance	$\kappa = (I/E)_{J_v, \Delta\pi_s}$	$\kappa' = (I/E)_{\Delta p - \Delta\pi, \Delta\pi_s}$
Coupling:		
Reflection coefficient	$c_s(1 - \sigma) = -\left(\dfrac{\Delta p - \Delta\pi}{\Delta\pi_s/c_s}\right)_{J_v, I}$ or $\sigma = \left(\dfrac{\Delta p - \Delta\pi_i}{\Delta\pi_s}\right)_{J_v, I}$	$c_s(1 - \sigma) = (J_s/J_v)_{\Delta\pi_s, I}$
Electroosmotic pressure	$P_E = \left(\dfrac{\Delta p - \Delta\pi}{E}\right)_{J_v, \Delta\pi_s}$	Electroosmotic permeability $\beta = (J_v/I)_{\Delta p - \Delta\pi, \Delta\pi_s}$
	$= -(I/J_v)_{\Delta\pi_s, E}$ (streaming current)	$= -\left(\dfrac{E}{\Delta p - \Delta\pi}\right)_{\Delta\pi_s, I}$ (streaming potential)
Transport number	$\tau_1 = \nu_1 z_1 F(J_s/I)_{J_v, \Delta\pi_s}$ $= -\nu_1 z_1 F \left(\dfrac{E}{\Delta\pi_s/c_s}\right)_{J_v, I}$ (membrane potential)	$\tau_1' = \nu_1 z_1 F(J_s/I)_{\Delta p - \Delta\pi, \Delta\pi_s}$ $= -\nu_1 z_1 F \left(\dfrac{E}{\Delta\pi_s/c_s}\right)_{\Delta p - \Delta\pi, I}$ (membrane potential)

a. After Kedem and Katchalsky, 1963a.
b. Subscripts indicate flows or forces kept at zero.

Clearly the complex coefficients of Eqs. (3.47) are too unwieldy for general use. It is therefore useful to replace them by expressions incorporating so-called practical transport coefficients, which may be conveniently evaluated experimentally under conditions in which two of the independent variables $J_v, \Delta\pi_s/c_s$, and I are set equal to zero. (Alternatively it is possible to define the transport coefficients differentially [Michaeli and Kedem, 1961], holding the restrained parameters constant instead of zero.) Such a set of coefficients, presented in Table 3.1, may be immediately identified with six of the coefficients of Eqs. (3.47).[3] The remaining three coefficients may be simply evaluated by realizing that the original R_{ij}'s obey Onsager symmetry, so that

$$\left(\frac{J_s}{J_v}\right)_{\Delta\pi_s, I} = -\left(\frac{\Delta p - \Delta\pi}{\Delta\pi_s/c_s}\right)_{J_v, I} = c_s(1 - \sigma) \quad (3.48)$$

$$-\left(\frac{E}{\Delta\pi_s/c_s}\right)_{J_v, I} = -\left(\frac{J_s}{I}\right)_{J_v, \Delta\pi_s} = \frac{\tau_1}{\nu_1 z_1 F} \quad (3.49)$$

$$\left(\frac{E}{J_v}\right)_{\Delta\pi_s, I} = \left(\frac{\Delta p - \Delta\pi}{I}\right)_{J_v, \Delta\pi_s} = \frac{-\beta}{L_p} \quad (3.50)$$

By these means, Eqs. (3.47) can be usefully rewritten as

$$(\Delta p - \Delta\pi) = \left(\frac{1}{L_p}\right) J_v - c_s(1 - \sigma)\frac{\Delta\pi_s}{c_s} - \left(\frac{\beta}{L_p}\right) I$$

$$J_s = c_s(1 - \sigma)J_v + c_s\omega\frac{\Delta\pi_s}{c_s} + \left(\frac{\tau_1}{\nu_1 z_1 F}\right) I$$

$$E = -\left(\frac{\beta}{L_p}\right) J_v - \left(\frac{\tau_1}{\nu_1 z_1 F}\right)\frac{\Delta\pi_s}{c_s} + \left(\frac{1}{\kappa}\right) I \quad (3.51)$$

The new choice of coefficients is governed in part by the empirical knowledge that certain quantities often have a rather weak concentration dependence. For some purposes it is convenient to rewrite Eqs. (3.51) as flow equations, in the form suggested by Kedem and Katchalsky:

44   Bioenergetics and linear nonequilibrium thermodynamics

$$J_v = L_p(\Delta p - \Delta \pi_i) - \sigma L_p \Delta \pi_s + \beta I$$

$$J_s = c_s(1 - \sigma)J_v + \omega \Delta \pi_s + \left(\frac{\tau_1}{\nu_1 z_1 F}\right) I$$

$$I = \kappa \left(\frac{\beta}{L_p}\right) J_v + \kappa \left(\frac{\tau_1}{\nu_1 z_1 F}\right) \Delta \mu_s^c + \kappa E \qquad \left(\Delta \mu_s^c = \frac{\Delta \pi_s}{c_s}\right) \qquad (3.52)$$

Equations (3.51) and (3.52) are found useful for the treatment of a composite membrane consisting of a series array of elements (Kedem and Katchalsky, 1963c).

A treatment entirely analogous to that above leads to alternative expressions which are readily applied when the restrained variables are chosen from the set $(\Delta p - \Delta \pi)$, $\Delta \pi_s$, or I (Kedem and Katchalsky's Set II, Table 3.1):

$$J_v = L_p(\Delta p - \Delta \pi) + c_s(1 - \sigma)L_p \left(\frac{\Delta \pi_s}{c_s}\right) + \beta I$$

$$J_s = c_s(1 - \sigma)L_p(\Delta p - \Delta \pi) + c_s \omega' \left(\frac{\Delta \pi_s}{c_s}\right) + \left(\frac{\tau'}{\nu_1 z_1 F}\right) I$$

$$E = -\beta(\Delta p - \Delta \pi) - \left(\frac{\tau_1'}{\nu_1 z_1 F}\right)\left(\frac{\Delta \pi_s}{c_s}\right) + \left(\frac{1}{\kappa'}\right) I \qquad (3.53)$$

or, again rewriting in the form of flow equations:

$$J_v = L_p(\Delta p - \Delta \pi_i) - \sigma L_p \Delta \pi_s + \beta I$$

$$J_s = c_s L_p(1 - \sigma)(\Delta p - \Delta \pi) + \omega' \Delta \pi_s + \left(\frac{\tau'}{\nu_1 z_1 F}\right) I$$

$$I = \kappa' \beta (\Delta p - \Delta \pi) + \kappa' \left(\frac{\tau'}{\nu_1 z_1 F}\right) \Delta \mu_s^c + \kappa' E \qquad (3.54)$$

In contrast to the practical coefficients of Set I, those of Set II are useful for the treatment of a composite membrane with a parallel array of elements (Kedem and Katchalsky, 1963b; Weinstein, Bunow, and Caplan, 1972).

As we have emphasized, the practical phenomenological coefficients employed in Eqs. (3.51) and (3.52) were derived from observations which were commonplace long before the advent of nonequilibrium thermodynamics. Nevertheless, combining them in a compact, self-consistent formulation provides a sound basis for the analysis and correlation of a myriad of experimental data and leads to fundamental predictions which should be applicable to a great variety of systems. It is clear that precise characterization of a system requires knowing the number of its degrees of freedom and applying appropriate restraints such that the flow or force of interest becomes a function of a single independent variable. In principle it then becomes possible to evaluate all of the phenomenological coefficients. Although in some cases this will prove impractical, even in such cases it should be possible to state which coefficients have been adequately evaluated and which remain to be determined. Unfortunately, many studies in the biological literature have suffered from attempts to characterize the behavior of a system in terms of incomplete equations, failing to take into account all degrees of freedom. In such cases the experimental observations cannot precisely define the intrinsic characteristics of the system, since they must necessarily reflect also the special experimental conditions employed, in particular the value of the independent variable which was uncontrolled. Hence observations made under different conditions may lead to apparent contradictions. For example, if we consider the determination of a permeability coefficient by measurement of the solute flow induced by a difference of solute concentration across a membrane, Eqs. (3.52) and (3.54) show that in general $\omega = (J_s/\Delta\pi_s)_{J_v,I}$ cannot be expected to equal $\omega' = (J_s/\Delta\pi_s)_{(\Delta p - \Delta\pi),I}$. Furthermore, it is readily seen that

$$\omega' = \omega + c_s(1 - \sigma)^2 L_p \tag{3.55}$$

Thus the formalism serves not only to define suitable conditions for the precise determination of fundamental parameters but also to demonstrate meaningful relationships between them. This may be appreciated here if one examines the significance of the reflection coefficient σ, first introduced by Staverman (1951) to explain the discrepancy between theoretical and experimentally determined values of osmotic pressure. A general feeling for the significance of the reflection coefficient may be obtained by applying any of the Eqs. (3.51)–(3.54) to various experimental situations. A simple example is the case where the permeant solute is a nonelectrolyte, as was considered by Staverman. In this case there is no current flow or electromotive force to be considered, and Eqs. (3.52) reduce to

$$J_v = L_p(\Delta p - \Delta \pi_i) - \sigma L_p \Delta \pi_s \qquad (3.56)$$

$$J_s = c_s(1 - \sigma)J_v + \omega \Delta \pi_s \qquad (3.57)$$

These equations correspond respectively to Eqs. (39) and (41) of Kedem and Katchalsky (1958) for the case of dilute solutions. It is clear that the value of σ must depend on the nature of both the solute and the membrane under study. Considering Eq. (3.57) for the case of volume flow in the absence of a concentration gradient of permeant solute ($\Delta \pi_s = 0$), we see that the quantity $(1 - \sigma)$ is a direct measure of the extent of coupling between solute flow and volume flow: $1 - \sigma = (1/c_s)(J_s/J_v)_{\Delta \pi_s, I=0}$. If the membrane is completely nonselective, $J_s/J_v = c_s$, so that $\sigma = 0$; if the membrane is perfectly selective, permeable only to the solvent, $\sigma = 1$. In most cases σ will lie between 0 and 1, but it is possible for it to lie outside this range (for example, see Weinstein and Caplan, 1968).

Equation (3.56) shows that a solute will influence volume flow only to the extent that its reflection coefficient differs from zero. This equation serves also to elucidate errors in the estimation of osmotic pressure by the use of membrane techniques. If a solution is equilibrated with pure water across an ideally semipermeable membrane, the hydrostatic pressure difference which is developed at equilibrium precisely evaluates the osmotic pressure. Because membranes which are completely impermeable to small solutes require impractically long periods for equilibration, solutions of small solutes have often been studied by the use of slightly permeable "quick" membranes. With this technique the osmotic pressure is evaluated by extrapolating experimental values of Δp to time zero under the presumption that this eliminates the influence of solute leakage. As was clearly pointed out by Staverman, this is not the case. The leakage of solute influences steady-state pressure gradients over and above effects attributable to the change of solution concentration: "the membrane behaves from the start as if not n_i molecules were present in the solution but only $\sigma_i n_i''$ (Staverman, 1951). These considerations are readily appreciated from Eq. (3.56). In the absence of permeant solute ($\Delta\pi_s = 0$), $(\Delta p)_{J_v=0} = \Delta\pi_i$; in other words, the hydrostatic pressure difference at equilibrium equals the osmotic pressure difference, whereas if only a permeant solute is present ($\Delta\pi_i = 0$), $(\Delta p)_{J_v=0} = \sigma\Delta\pi_s$.

3.8 Definition of volume flow in the Kedem-Katchalsky formulation

As has been pointed out, whereas the quantities J_w, J_1, and J_2 represent true material flows across the membrane, the quantities J_s and J_v, as defined respectively by Eqs. (3.26) and (3.33), represent virtual flows, since in the presence of current flow they will include the contribution of electrode processes (Weinstein and Caplan, 1973; Kedem, 1973). The precise evaluation of the practical coefficients of Kedem and Katchalsky's Set I requires accurate measurement or control of J_v, but in practice this quantity is usually approximated by the observed rate of change of volume in either compartment, "J_v^{obsd}". If we neglect second-order effects due to changes in partial molar volumes of anions as they move into or out of the electrodes, it is seen that J_v^{obsd} represents the actual rate of transfer of volume across the membrane, and is therefore given by

$$J_v^{obsd} = J_1\overline{V}_1 + J_2\overline{V}_2 + J_w\overline{V}_w \tag{3.58}$$

If the partial molar volumes of the ions are considered additive,

$$\overline{V}_s = \nu_1\overline{V}_1 + \nu_2\overline{V}_2 \tag{3.59}$$

and with Eqs. (3.21), (3.25), (3.26), (3.33), and (3.58) we obtain the relation between J_v and J_v^{obsd}:

$$J_v = J_v^{obsd} - \overline{V}_2 I/z_2 F \tag{3.60}$$

For the case of Ag/AgCl current-passing electrodes,

$$J_v = J_v^{obsd} + 1.61 \times 10^{-4} I \tag{3.61}$$

where J_v is expressed in cm sec^{-1} and I as A cm^{-2}. While the correction term will often be of little practical importance, in principle it may be of significance. One possible consequence of using J_v^{obsd} to represent J_v is a discrepancy between the values of practical phenomenological coefficients obtained by different experimental means. For example, values of the electroosmotic permeability β obtained appropriately from the measurement of $-(E/(\Delta p - \Delta \pi))_{\Delta\mu_s^c,I}$ may differ appreciably from values based on measurements of flow, since

$$\beta = \left(\frac{J_v}{I}\right)_{(\Delta p - \Delta \pi), \Delta\mu_s^c} = \left(\frac{J_v^{obsd}}{I}\right)_{(\Delta p - \Delta \pi), \Delta\mu_s^c} + 1.61 \times 10^{-4} \quad (3.62)$$

For ion exchange membranes of loose structure that tend to have high values of β, the correction may well be negligible, but in tight membranes it may be significant.

The correction is particularly important in the study of charge-mosaic membranes consisting of alternating anion and cation exchange regions. For such membranes β may be very small, since the electroosmotic flows through the neighboring regions are oppositely directed and tend to cancel. Table 3.2 compares estimated values (β^{est}) and corrected values (β) in such a mosaic membrane and in membranes composed of the elemental anion and cation exchange components making up the mosaic.

Table 3.2. Difference between estimated and true values of the electroosmotic permeability in anion exchange, cation exchange, and charge-mosaic membranes.[a]

Membrane	Solution concentration (M)	$\beta^{est} = J_v^{obsd}/I$ cm^{3}A^{-1}s$^{-1} \times 10^4$	$\beta = J_v/I$ cm^{3}A^{-1}s$^{-1} \times 10^4$
Anion exchange	0.1	− 9.2	− 7.6
	0.01	−11.1	− 9.5
	0.001	−11.7	−10.1
Cation exchange	0.1	10.3	11.9
	0.01	13.0	14.7
	0.001	14.0	15.5
Mosaic	0.1	1.2	2.8
	0.01	1.4	3.1
	0.001	2.6	4.3

a. After Weinstein and Caplan, 1973.

3.9 An active transport model

The model of Blumenthal, Caplan, and Kedem (1967), which was introduced at the conclusion of Sec. 3.5, consisted of two ion exchange membranes in series, one carrying a positive fixed charge, the other a negative fixed charge. A layer of aqueous solution containing a proteolytic enzyme (papain) was sealed between them. This composite membrane was mounted between identical solutions containing a low-molecular-weight substrate for the enzyme. No pressure difference was imposed. In the presence of the enzyme, the substrate hydrolyzed to form a salt, the reaction being essentially

$$XY \rightleftharpoons X^+ + Y^- \tag{3.63}$$

By including product as well as substrate at known concentrations in the reservoirs, the affinity of the reaction in the reservoirs A^{ex} could be calculated using the known value of the equilibrium constant. In this system, transport of a substance across the membrane does not occur but is simulated by "active transport" of electric current. At short circuit the reaction drives a current between identical solutions, while on open circuit a potential difference is developed between identical solutions, providing A^{ex} is nonzero.

Blumenthal, Caplan, and Kedem considered the fluxes of all the permeant species through both component membranes and showed that in the stationary state the dissipation function for the composite membrane contracts to the two terms shown in Eq. (3.20):

$$\Phi = IE + vA^{ex}$$

The same basic phenomenon is involved here as in Eq. (3.19), namely, modification of the local chemical potentials by the reaction. The corresponding phenomenological equations can be written

$$E = R_{11}I + R_{12}v$$

$$A^{ex} = R_{21}I + R_{22}v \tag{3.64}$$

The phenomenological coefficients in Eqs. (3.64) were derived in terms of the kinetic parameters of the reaction and the practical phenomenological coefficients of the component membranes. In particular, the coupling coefficient has the following form:

$$R_{12} = R_{21} = -\frac{\tau_1^\alpha - \tau_1^\beta}{c_s(\omega_s^\alpha + \omega_s^\beta)F} \qquad (3.65)$$

where α and β refer to the two ion exchange membranes, the permeabilities ω_s^α and ω_s^β refer to the salt, and c_s is the mean salt concentration in the ion exchange membranes. The directionality of the composite membrane appears immediately in the difference of the transport numbers.

An important aspect of the physical model studied was that it showed both a rather wide range of linearity and Onsager symmetry. This can be seen in Fig. 3.2(a) and (b), which illustrate, respectively, the dependence of the open-circuit potential difference (emf) on the affinity, and the dependence of the velocity of reaction on the current. Each plot shows two sets of results, each corresponding to a different choice of the anion exchange membrane. The units are chosen so that the slopes of the curves are dimensionless, and it will be apparent that corresponding curves in (a) and (b) are virtually parallel. This is an expression of the Onsager relation

$$\left(\frac{\partial v}{\partial I}\right)_{A^{\text{ex}}} = -\left(\frac{E}{A^{\text{ex}}}\right)_{I=0} \qquad (3.66)$$

The linear relationship between v and I might be regarded as the apparent stoichiometry of the system, but in fact it reflects neither tight coupling between transport and reaction nor linear kinetics. Indeed the straight lines in Fig. 3.2(b) do not pass through the origin just because coupling *is* incomplete. It would be manifestly absurd here to regard the intercept as basal metabolism, which must of course be taken into account in real epithelial membranes (see Chapter 8).

It is important to realize that our ability to describe this composite membrane in terms of stationary-state coupling depends on prior knowledge of its structure; we can consider explicitly events taking place in the enzyme space between the ion exchange membranes. If, however, the structure and internal function of the membrane were unknown, its interior being inaccessible, one would be forced to conclude that R_{12} might represent a *direct* coupling mechanism. It is

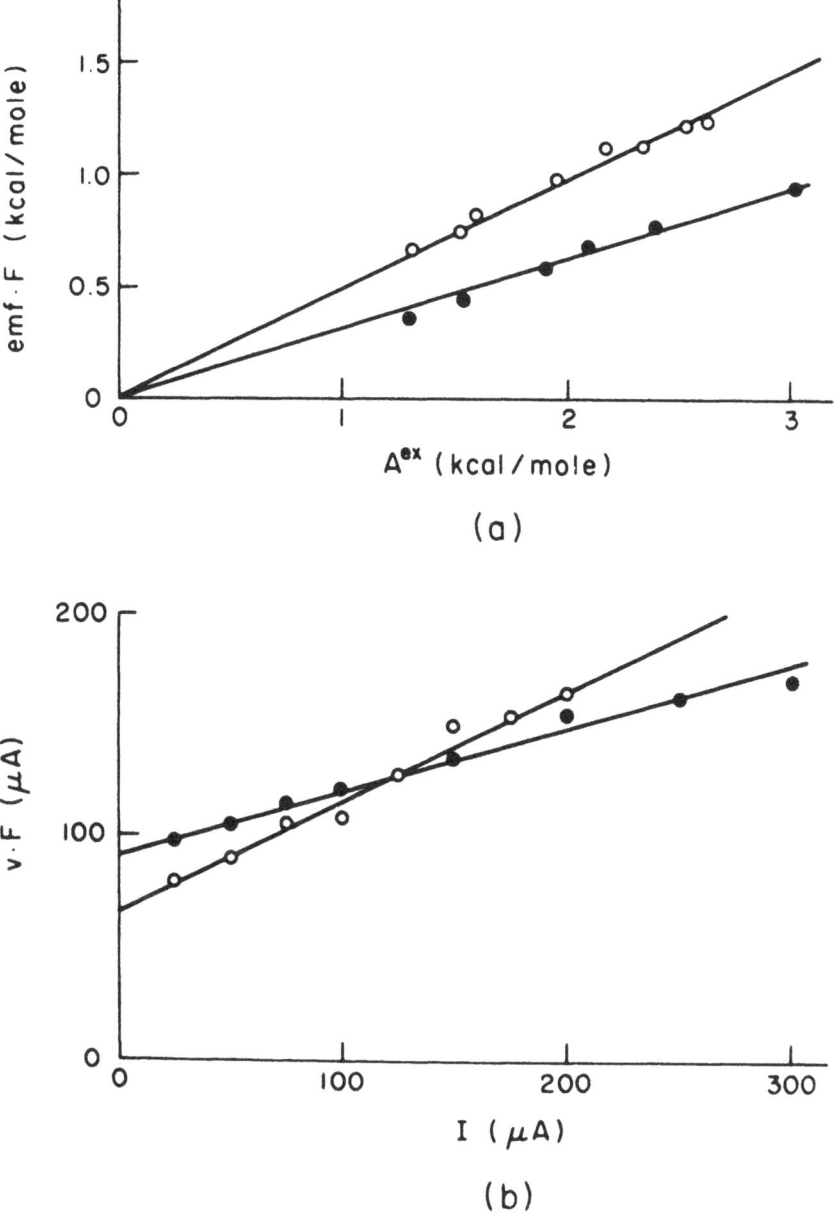

Fig. 3.2. Behavior of a model active transport system (see text). (a) Steady-state electromotive force versus external reaction affinity for two membrane systems. The experiments were performed at zero current. (b) Steady-state reaction flow versus current for two membrane systems. The experiments were performed at an affinity $A^{ex} = 2.6$ kcal/mole. (After Blumenthal, Caplan, and Kedem, 1967.)

thus quite possible that active transport, at least in some cases, will eventually turn out to be stationary-state coupling on a molecular scale.

3.10 Summary

1. The dissipation function, if derived appropriately from a Gibbs equation, indicates the number of degrees of freedom in the system. The corresponding phenomenological equations give the flows as linear functions of the forces (through the use of conductance coefficients) or give the forces as linear functions of the flows (through the use of resistance coefficients). These equations are generalizations of well-known linear relations such as Ohm's law and Fick's law. The range of linearity is generally much greater for vectorial flows than for chemical reactions (a detailed discussion of this issue is given in Chapter 6).

2. The matrices of conductance or resistance coefficients must satisfy two conditions. One ensures that the dissipation function can never be negative. The other is a symmetry condition between all pairs of corresponding cross coefficients; that is, $L_{ij} = L_{ji}$ and $R_{ij} = R_{ji}$ (Onsager reciprocity).

3. Coupling between chemical reaction and transmembrane flow, as in active transport, involves a vectorial coupling coefficient. According to Kedem's definition, active transport occurs if the cross coefficient between reaction and flow in the *resistance* formulation is nonzero.

4. The Curie-Prigogine principle indicates that coupling between scalar and vector flows is impossible in isotropic media in the linear regime—but not in anisotropic media. The vectorial character of the coupling coefficient reflects the anisotropy of the medium.

5. In linear systems with Onsager symmetry, the entropy production assumes the minimal value compatible with the imposed restraints. Such "stationary states of minimal entropy production" can give rise to an interdependence among overall flow processes, which has been termed stationary-state coupling. Examples are discussed.

6. Several important transformations of the dissipation function for membrane transport processes are examined. These give rise to sets of flows and forces which may be experimentally preferable since they can be readily fixed or measured.

7. The transformed forms of the dissipation function can be used to generate transformations of the phenomenological equations, which give rise to practical phenomenological coefficients. Examples of these are the important flow equations of Kedem and Katchalsky.

8. A detailed discussion is given of the coefficients of the Kedem-Katchalsky equations, paying particular attention to the definition of volume flows so as to take into account possible errors due to electrode processes. It is stressed that the well-known reflection coefficient, originally introduced by Staverman, is an indicator of the extent of coupling between solute flow and volume flow.

9. A simple experimental model of active transport based on stationary-state coupling is discussed. In this model a composite membrane incorporating an enzyme, when supplied externally with a suitable substrate, was able to drive electric current between two identical solutions. It exhibited both linearity and symmetry.

4 Effectiveness of energy conversion

Clearly the generality of the treatment of coupled processes in Chapters 2 and 3 permits its application to the analysis of a large variety of biological systems. In particular, it permits a systematic analysis of the effectiveness with which metabolic energy is utilized in the vital processes of active transport and muscular contraction. In Chapter 3 we touched on the definition of active transport and related issues. In this chapter we shall first consider the classical approach to active transport, restricting ourselves for the most part to sodium transport, both because of its fundamental importance and because it has been most intensively studied. We shall point out what we consider to be significant shortcomings of the classical approach, and then shall continue with a general consideration of energy conversion from the viewpoint of nonequilibrium thermodynamics (NET).

4.1 Active transport: background

People have long been aware of the importance of active transport processes in biological systems, as well as their generality and diversity. A casual examination of normal metabolism reveals numerous metabolic products that are formed continuously and that must be eliminated specifically and rapidly, often in the face of adverse concentration gradients. Similarly, normal cellular steady-state content, as well as cellular dynamic function, indicates active transport.

Among the pertinent questions which have been asked concerning active transport processes are: How is active transport unambiguously demonstrated? What is the energy source? What are the mechanisms whereby (scalar) chemical free energy is utilized to produce (vectorial) transport and electroosmotic work? What is the nature of the energetics of the active transport system? This latter question is interesting both in its own right and in relation to the earlier questions.

Attempts to demonstrate the active nature of a transport process, as well as its energy source and mechanism, have often depended on studies of the effects of inhibiting either oxidative or anaerobic metabolism. Such studies have been useful, but there are significant objections. For one, metabolism may affect not only the function of the transport system but also other functions, as well as the structure of a system. Also, the effects of metabolic inhibitors may be either primary or secondary. For example, a primary effect on active sodium transport may result in a secondary effect on passive chloride transport. Thus metabolic inhibition in itself cannot always establish the active nature of a process. Furthermore, even when alteration of metabolic function can differentiate between passive and active transport, it cannot evaluate the effectiveness of utilizing metabolic energy in various states of interest. This is a question of thermodynamics. A thermodynamic definition was stated by Rosenberg (1954), who considered the "demonstration of transport from a lower to a higher potential . . . the only certain criterion of active transport," but this criterion does not incorporate evident cases of active transport in which the electrochemical potential difference may be zero or negative and does not lead naturally to a systematic and comprehensive formulation.

4.2 Classical analysis of energetics: the equivalent circuit model

Most analyses of the energetics of ion transport have been based on that of Ussing, who treated active sodium transport in terms of an equivalent electrical circuit and inquired as to the work required to accomplish the transport of one equivalent of sodium (Ussing and Zerahn, 1951; Ussing, 1960). This was considered to comprise three components: "(a) the work required to overcome the concentration gradient . . . (b) the work required to overcome the potential gradient . . . (c) the work required to overcome the internal sodium resistance of the skin"; this latter quantity was evaluated from the "flux ratio" or ratio of the "unidirectional" fluxes of sodium. The total work was thus evaluated (per mole of sodium) as

$$W = RT \ln \frac{C_i}{C_o} + FE + RT \ln \frac{M_{in}}{M_{out}} \qquad (4.1)$$

where M_{in}/M_{out} represents the flux ratio. Since the skin was treated in terms of an electrical analogue, Ussing spoke also of "the electromotive force of active sodium transport," E_{Na}. This was obtained by dividing the work W by the faraday, F.

Two general methods were employed to evaluate W and E_{Na}: One was to determine the electrochemical potential difference required to reduce the rate of net sodium transport (and hence $\ln M_{in}/M_{out}$) to zero. This was accomplished by manipulation of the concentration difference or electrical potential difference across the membrane or by substitution of impermeant for permeant anions. The second method was to expose the skin to identical solutions at each surface, nullify the spontaneous electrical potential difference, and measure the flux ratio under these "short-circuit" conditions.

As Ussing pointed out, these methods permit the evaluation only of an apparent E_{Na}, since they do not take into account the influence of leak pathways. In addition, on the basis of the work of Hodgkin and Keynes in poisoned squid axons, as well as the possibility of exchange diffusion, it was appreciated that there are fundamental difficulties in attempts to utilize the flux ratio to evaluate energetic parameters. Indeed, it can be shown that in order for the above two techniques to give the same value for E_{Na}, three conditions must be satisfied: (1) absence of leak; (2) absence of isotope interaction; and (3) equality of metabolism at static head and level flow (Kedem and Essig, 1965). (Static head and level flow are two important steady states characterized by zero transmembrane flux and zero transmembrane force, respectively; precise definitions will be given in Sec. 4.6.) Such a combination of circumstances appears quite unlikely. Even if the first two requirements were satisfied, the third is almost certainly not, since the rate of oxidative metabolism in the frog skin, toad skin, and toad bladder has been shown to be clearly dependent on the electrochemical potential difference of sodium across the tissue (see Chapter 8).

Another objection to the above formulation relates to the intrinsically different character of the three components of the "work." The first two terms on the right side of Eq. (4.1) taken together constitute the electrochemical potential difference of sodium and therefore correspond to the rate of performance of electroosmotic work, clearly requiring the expenditure of metabolic energy at an equivalent or greater rate. The third term, on the other hand, represents not a useful conversion of energy but a dissipation of energy, which furthermore cannot be precisely evaluated by measurements performed outside of the system. As such, $RT \ln M_{in}/M_{out}$ might be considered to represent a form of "internal work", as is invoked also in discussions of muscle energetics, where it is presumed to represent dissipation of energy to overcome frictional and viscous resistances. (See Kushmerick, Larsen, and Davies, 1969.) We would question whether it is appropriate to include a dissipative term in an expression for the work accomplished by a process. This is not customary in the thermodynamic analysis of heat engines or other energy-converting systems. In the treatment of such systems it is fundamental to differentiate unambiguously the rate at which energy is supplied and the rate at which energy is recovered.

Another difficulty associated with the classical formulation is that in considering frog skin in terms of an electrical analogue, there is the danger of assuming a constant and stoichiometric relationship between the rates of transport and metabolism under all conditions of operation (Zerahn, 1958; Ussing, 1960). This expectation has occasionally led to the belief that the demonstration of a linear relationship between rates of transport and metabolism constitutes a demonstration of stoichiometry and permits the calculation of a stoichiometric ratio, despite the fact that linearity is not equivalent to proportionality. Also, when variable stoichiometry is found experimentally, as is often the case in studies of oxidative phosphorylation, for example, it is generally attributed to technical imperfections rather than being examined for its intrinsic significance. Complete coupling (stoichiometry) of transport and metabolism is in fact found with perfect electrochemical cells. However, in biological systems partial decoupling would be expected to occur frequently, consequent to either breakdown of metabolic intermediates or dissipation of electrochemical potential gradients by way of leak pathways (Rottenberg, Caplan, and Essig, 1967). In such circumstances an expenditure of metabolic energy would be required to maintain an electrochemical potential difference across a tissue even in the absence of a net flow. This is in apparent conflict with the formulation underlying Eq. (4.1), which implies that in the absence of net flow the rate of performance of work (W × the rate of sodium transport) would equal zero. It might be considered that Eq. (4.1) provides only a minimal value for the necessary rate of expenditure of metabolic energy, but if so the formulation loses some of its value as a means of analyzing the energetics of the transport system with precision.

Another important aspect of the classical analysis of the energetics of active sodium transport is the mode of evaluating the efficiency of the process. This has generally been defined as the quotient of the rate of performance of work, evaluated as described above, and the rate of expenditure of metabolic energy. This latter quantity was taken to be the product of the rate of suprabasal oxygen consumption and the "calorific value" of 1 gram equivalent of oxygen in the oxidation of glucose, as determined from bomb calorimetry. Thus

$$\eta = \frac{W}{Q_{O_2}(-\Delta H)} \qquad (4.2)$$

where the rate of (suprabasal) oxygen consumption Q_{O_2} is expressed in terms of some stoichiometric ratio (moles O_2 per equivalent of Na transported) and $(-\Delta H)$ is the calorific value of glucose. Defining an efficiency function in this manner casts further doubt on the appropriateness of including a dissipative term in the expression for work. Classically one would expect that an increase in dissipation would *decrease* the efficiency of a process, whereas according to Eqs. (4.1) and (4.2) an increase in dissipation of energy within the tissue would *increase* the efficiency. In addition to this objection, one might question evaluating available energy from calorimetry. Such a measurement can evaluate only the enthalpy, under conditions obtaining within the calorimeter, whereas the desired quantity is the Gibbs free energy, under conditions obtaining *in vivo*. Finally, as mentioned above, there has been a widespread tendency to assume the constancy of the ratio of rates of sodium transport and oxygen consumption under all conditions, although in principle, for an incompletely coupled system, this ratio will vary with the thermodynamic forces; furthermore, whatever the degree of coupling and forces, the ratio may possibly vary from one animal to another.

4.3 NET formulation of energetics

Because of the shortcomings of the classical approach to energetics, it is important to develop an appropriate thermodynamic formulation that, it is hoped, will permit a comprehensive and self-consistent treatment of active transport under a wide variety of conditions. Energetic issues are of interest in their own right, but such a formulation should also lead to insights concerning the mechanism of action of factors influencing transport.

Since the systems we are studying are for the most part far from equilibrium, our treatment must be based on nonequilibrium thermodynamics. Various authors have approached problems of biological energy conversion from this point of view (for example, Jardetzky and Snell, 1960). With respect to active transport, Hoshiko and Lindley (1967) have presented a comprehensive linear formalism which facilitates the analysis of active salt and water transport in single salt and bi-ionic systems. For ease of experimental design and analysis, we have limited ourselves to systems with active transport of only one ion, sodium, uncoupled to the flow of other species. In this respect our work is an extension of the linear NET treatment presented originally by Kedem (1961). However, before analyzing the specific characteristics of the sodium active-transport process in these terms, we shall present a more general treatment of two-flow processes, to demonstrate both the relevance of the fundamental thermodynamic considerations presented above and their broad applicability. The general two-flow treatment leads naturally to defining two useful parameters which permit the ready comparison of systems in which the nature of the flows and the forces are different (Kedem and Caplan, 1965). The degree of coupling q is a dimensionless parameter whose absolute value varies from zero for completely uncoupled systems to unity for systems in which flows are related stoichiometrically. For a given degree of coupling and a given ratio of the forces, the parameter Z determines the ratio of the flows; for completely coupled systems, Z represents the stoichiometric ratio.

4.4 Degree of coupling and the relationship of forces and flows

A diagrammatic representation of the type of system we should like to consider is given in Fig. 4.1. Coupling takes place within a "black box" whose mechanism may not in general be understood. This black box usually constitutes the working element of the energy converter, for example, the membrane in an active-transport process. Since there are two processes, the dissipation function will consist of two terms. In general we can select one process that always gives rise to a positive term and hence may be regarded as the energy source or input. The other process generally gives rise to a negative term and constitutes the output. Since the output is usually more accessible than the input in biological systems, it is in a sense proximal to the observer and we shall denote it by 1, while the input is denoted by 2. The dissipation function then takes the form

$$\Phi = J_1 X_1 + J_2 X_2$$
$$ \text{−output} \quad \text{input}$$
$$ \text{power} \quad \text{power} \tag{4.3}$$

Note that the input flow J_2 always takes place in the spontaneous direction; in other words, its direction is in accord with that of its conjugate force X_2. The output flow J_1, however, may take place against its conjugate force X_1 as a consequence of its coupling to J_2. In these cases the output term is negative, indicating that free energy is

Effectiveness of energy conversion   61

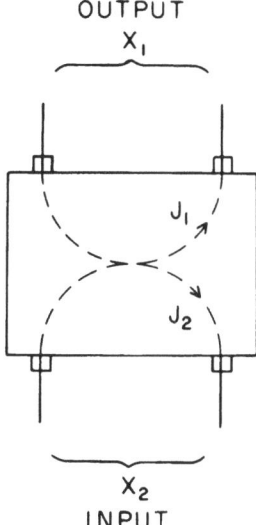

Fig. 4.1. Diagrammatic representation of an energy converter. The clockwise direction of J_2 indicates that the input flow is spontaneous, in accord with its conjugate force, whereas the counterclockwise direction of J_1 indicates that the output flow may take place against its conjugate force as a consequence of its coupling to J_2.

transferred from the working element into the surroundings. Thus part of the free energy expended by—and characteristic of—process 2 is converted into the form characteristic of process 1. Since the dissipation function can never be negative, the output can never exceed the input. It is intuitively clear that effective energy conversion requires tight coupling between the two processes.

The phenomenological flow equations of this system are given by

$$J_1 = L_{11}X_1 + L_{12}X_2 \tag{4.4}$$

$$J_2 = L_{21}X_1 + L_{22}X_2 \tag{4.5}$$

The treatment is much simplified by introducing at this point the Onsager reciprocal relationship between the phenomenological cross coefficients,

$$L_{12} = L_{21} \tag{4.6}$$

Although this relationship has not yet been tested for the case of active transport, it has been shown to apply in the description of a huge variety of physical chemical processes, provided that the flows and forces are appropriately obtained from a proper dissipation function (Miller, 1960). The relationship has been validated also in the synthetic "active-transport" system discussed in Chapter 3 (Blumenthal, Caplan, and Kedem, 1967) and in mitochondrial oxidative phosphorylation (Rottenberg, 1973). Assuming then its general validity, we have

$$J_1 = L_{11}X_1 + L_{12}X_2 \tag{4.7}$$

$$J_2 = L_{12}X_1 + L_{22}X_2 \tag{4.8}$$

It is sometimes more convenient to examine a system by means of a resistance formulation in which forces are expressed as functions of the flows:

$$X_1 = R_{11}J_1 + R_{12}J_2 \tag{4.9}$$

$$X_2 = R_{12}J_1 + R_{22}J_2 \tag{4.10}$$

Equations (4.7) and (4.8) define the ratio of flows J_1/J_2 (which we shall call j) associated with a given ratio of forces X_1/X_2 (which we shall call x). Dividing Eq. (4.7) by Eq. (4.8) and then dividing both the numerator and denominator by $(\sqrt{L_{11}L_{22}})X_2$ gives

$$j = \frac{(\sqrt{L_{11}/L_{22}})x + L_{12}/\sqrt{L_{11}L_{22}}}{(L_{12}/\sqrt{L_{11}L_{22}})x + \sqrt{L_{22}/L_{11}}} \tag{4.11}$$

It is seen that in general the ratio of flows is not constant but varies with the ratio of the forces. The analysis of flow ratios is further clarified by consideration of the parameter $L_{12}/\sqrt{L_{11}L_{22}}$. As $L_{12}/\sqrt{L_{11}L_{22}}$ approaches zero, $j \to (L_{11}/L_{22})x$; that is, each flow becomes proportional to its conjugate force, so that the two flows are independent; on the other hand, as $L_{12}/\sqrt{L_{11}L_{22}}$ approaches ± 1, $j \to \pm\sqrt{L_{11}/L_{22}}$, so that the two flows are in a fixed stoichiometric ratio, whatever the magnitude of the two forces. Accordingly, it is convenient to introduce the definition

$$q = \frac{L_{12}}{\sqrt{L_{11}L_{22}}} \tag{4.12}$$

and to refer to q as the degree of coupling, providing a basis for comparing different types of coupled two-flow systems. It is readily shown that q is also given by

$$q = \frac{-R_{12}}{\sqrt{R_{11}R_{22}}} \tag{4.13}$$

The significance of the sign of q is perhaps clearest if both flows are vectors. For example, the flow of a solute through a membrane may drag another solute along in the same direction ($q > 0$), or by an exchange process may tend to push it back ($q < 0$). If q is positive, L_{12} is positive but R_{12} is negative, and vice versa.[1]

The positive definite character of the dissipation function imposes constraints on q. For our two-flow process, Eqs. (4.7) and (4.8) give

$$\Phi = J_1 X_1 + J_2 X_2$$
$$= L_{11}X_1^2 + 2L_{12}X_1X_2 + L_{22}X_2^2 \geq 0 \tag{4.14}$$

Since this inequality obtains quite generally, we can consider the state in which $X_1 = -1/\sqrt{2L_{11}}$ and $X_2 = 1/\sqrt{2L_{22}}$, giving $1 - L_{12}/\sqrt{L_{11}L_{22}} \geq 0$, whereas by considering the state in which $X_1 = 1/\sqrt{2L_{11}}$ and $X_2 = 1/\sqrt{2L_{22}}$, we see that $1 + L_{12}/\sqrt{L_{11}L_{22}} \geq 0$. Hence

$$-1 \leq q \leq 1 \tag{4.15}$$

The relationship between the flow ratio and the force ratio is further clarified by introducing the definition

$$Z = \sqrt{\frac{L_{11}}{L_{22}}} \tag{4.16}$$

and rewriting Eq. (4.11) in the form

$$j = \frac{Zx + q}{qx + 1/Z} \tag{4.17}$$

As is seen, for given q and x, j is completely determined by Z, which accordingly has acquired the designation "phenomenological stoichiometry" in recent literature (Stucki, 1980); in the limit of complete coupling ($q = \pm 1$) the ratio of flows is fixed and given by Z. In the more general case, Z serves to reduce both the flow and the force ratio to dimensionless numbers. The relationship between the reduced flow ratio j/Z and the reduced force ratio Zx at various values of q is shown in Fig. 4.2. It can be appreciated that although true stoichiometry is observed only if coupling is complete, when unavoidable experimental errors are taken into account an apparently stoichiometric relationship between J_1 and J_2 may be observed over an extensive range even with a moderate extent of uncoupling. Thus in order to carry out an adequate experimental test for uncoupling it may be necessary to set $Z(X_1/X_2)$ at a level adequate to make J_1 near-zero.

The thermodynamic implications of complete coupling are very important. Since in this case the flows are not linearly independent, the dissipation function can be contracted to a single term:

$$\Phi = J_2(X_2 \pm ZX_1) \quad (q = \pm 1) \tag{4.18}$$

If X_1 and X_2 can be varied independently, the composite force $X_2 \pm ZX_1$ can be made as small as desired. Hence Φ may become vanishingly small as compared to input and output. This means that the process approaches reversibility as the rate tends to zero. For example, an electrochemical cell can be completely coupled, provided that there is one homogeneous electrolyte and no side reactions. The dissipation function may then be written as

$$\Phi = IE + vA \tag{4.19}$$

where I and E refer to electrical current and potential, and v and A refer to the rate and affinity of the driving chemical reaction. In the

Effectiveness of energy conversion 65

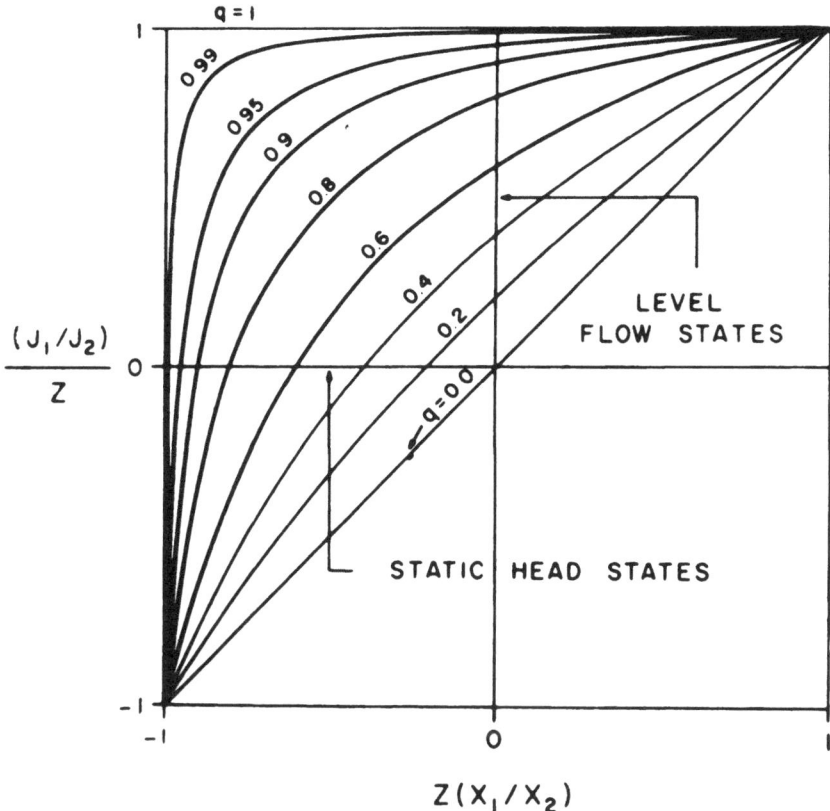

Fig. 4.2. Dependence of the reduced flow ratio on the reduced force ratio for various degrees of coupling. (Adapted from Kedem and Caplan, 1965.)

present notation, $Z = nF$, where n is the number of electrons transferred per mole of reaction, and F is the faraday. Therefore, if coupling is indeed complete,

$$\Phi = v(A + nFE) \qquad (4.20)$$

At reversible equilibrium, $\Phi = v = 0$ and $A = nF(-E)_{I=0}$. The quantity $(-E)_{I=0}$ is the electromotive force of the cell.

4.5 Efficiency of energy conversion

A concept which receives much attention in bioenergetics is that of efficiency. Unfortunately, however, the definition of this quantity differs substantially among different workers, so it is often difficult to compare results obtained in different laboratories. Furthermore, as suggested above, the intuitive notion that biological systems must be efficient in all physiological states has often led to unnatural definitions which are unproductive and internally inconsistent.

Following the concepts of classical thermodynamics, an unambiguous and useful efficiency function is indicated by the dissipation function, Eq. (4.3):

$$\eta = \frac{\text{output}}{\text{input}} = \frac{-J_1 X_1}{J_2 X_2} \tag{4.21}$$

(It is appreciated that η and q are not unique characteristics of the system, since there may be a number of different ways of choosing flows and forces consistent with a given rate of entropy production. In general, however, the choice of input and output is dictated by practical considerations. For complete coupling, $q = 1$ for any choice.)

The states of greatest biological interest are those in which process 2 is spontaneous and process 1 is not. Under these circumstances $J_2 X_2 > 0$, and $J_1 X_1 \leq 0$. In terms of the forces, the condition for such states is that

$$-1 \leq \frac{Z(X_1/X_2)}{q} \leq 0 \tag{4.22}$$

Whenever this condition is satisfied, process 2 "drives" process 1, and

$$0 \leq \eta \leq 1 \tag{4.23}$$

The value of η may readily be obtained as a function of the force ratio and the degree of coupling:

$$\eta = \frac{-q + Z(X_1/X_2)}{q + \frac{1}{Z(X_1/X_2)}} \tag{4.24}$$

Effectiveness of energy conversion 67

Since the efficiency is zero when either J_1 or X_1 is zero, it is obvious that it must pass through a maximum at intermediate values. This maximum value of the efficiency, η_{max}, depends only on the degree of coupling:

$$\eta_{max} = \frac{q^2}{(1 + \sqrt{1 - q^2})^2} \tag{4.25}$$

(A corollary of this result is that if η_{max} is known, q can be determined from $q^2 = 1 - [(1 - \eta_{max})/(1 + \eta_{max})]^2$. Other methods of determining q will be discussed in Chapter 7.) Completely coupled systems can approach $\eta_{max} = 1$ as a limit at reversible equilibrium. However, as pointed out earlier, the rate of the process under these conditions tends toward zero.[2]

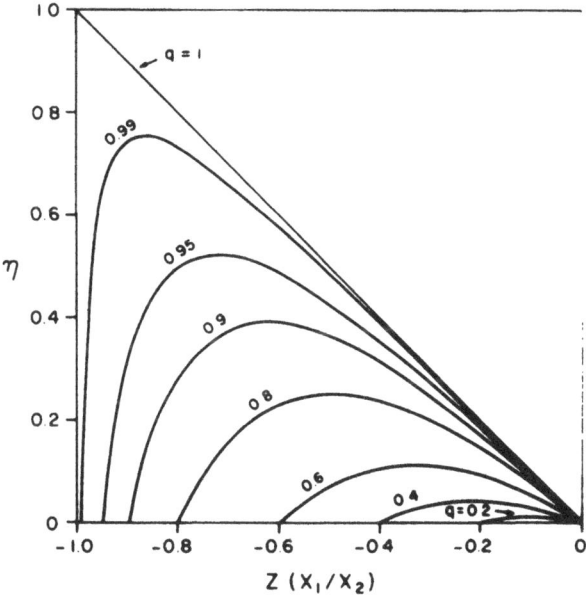

Fig. 4.3. Dependence of the efficiency η on the reduced force ratio for various degrees of coupling. Consideration is limited to the driving region, in which $-1 \leq Z(X_1/X_2)/q \leq 0$. For each value of q, the left-hand limit represents static head and the right-hand limit represents level flow. (Adapted from Kedem and Caplan, 1965.)

68 Bioenergetics and linear nonequilibrium thermodynamics

Figure 4.3 shows the efficiency plotted as a function of the force ratio for various values of q. Note the very rapid fall in η_{max} with decreasing q. Also noteworthy is the fact that, despite the common emphasis on evaluation of efficiency, η is often quite small in regions of physiological interest. Indeed, it is evident that whatever the value of q, the rate of performance of work—and thus the efficiency—is zero when either J_1 or X_1 is zero. Despite the fact that no work is performed under these circumstances, these two states are of sufficient physiological and general importance to merit special attention (Kedem and Caplan, 1965).

4.6 Energy expenditure without performance of work

4.6.1 Static head

If X_2, the input or driving force, is held constant, but no restriction is placed on X_1, the flow J_1 will continue until X_1 achieves a value adequate to bring it to a halt. Thereafter J_1 will remain zero, and thus X_1 will remain constant so long as X_2 and the phenomenological coefficients retain their original values. We shall refer to such stationary states as *static head*. Examples of systems at static head are a fuel cell at open circuit, plant or animal cell membranes maintaining constant concentration gradients by means of active transport, and a muscle in isometric contraction. In an incompletely coupled system, energy must be expended to maintain static head, even though the output is zero.

4.6.2 Level flow

With external reservoirs of finite capacity, X_1 can be maintained constant only if J_1 vanishes or is compensated by an equal flow through the external world. By suitable adjustment of this external flow, one can clamp X_1 at any desired value, as for example in the customary measurement of the short-circuit current in biological systems. Instead of an electric current at zero potential difference, any other force (such as a pressure head) can be held at zero. Although X_1 is often maintained at zero by the use of a compensating device, this may not be necessary. We shall refer to stationary states in which $X_1 = 0$ as *level flow*. Examples of systems at level flow are a fuel cell at short circuit, a kidney proximal tubule transporting salt and water between physiological near-isotonic solutions, and a muscle in unloaded contraction. Clearly, energy must be expended to maintain level flow, even though output is zero.

4.6.3 Effectiveness of energy utilization

A fundamental consideration in the analysis of both living and nonliving systems is the adequacy of energy utilization. The examples above demonstrate that energy may be expended usefully under circumstances where the rate of performance of work and the efficiency are zero. Obviously, therefore, the efficiency function cannot be regarded as a universal criterion for the effectiveness of energy utilization. Indeed, no single criterion appears suitable for this purpose; rather, the conditions of operation of a system determine the appropriate criteria.

Efficiency is clearly of interest when a system converts one form of energy into another to an appreciable extent, for example, when a muscle raises a weight. Another important example occurs in the loop of Henle, where chloride is transported against large differences of electrochemical potential at a rate which permits water flow in the collecting tubule to concentrate the urine. Near static head and level flow, however, where the extent of energy conversion becomes small, the function of a system is not energy conversion and therefore efficiency is of no significance. Nevertheless, it is necessary to evaluate the effectiveness of energy utilization under such circumstances. Thus, in studies of muscle contraction, for example, it has been suggested that comparisons of different tissues or different states might be usefully carried out in terms of an "isometric efficiency," defined as the tension developed per unit rate of consumption of high-energy phosphate (Awan and Goldspink, 1972). However, the free energy change per unit rate of reaction might well differ from one tissue to another. We would suggest therefore that it is more appropriate to relate the output force (or flow) to the rate of energy expenditure (Essig and Caplan, 1968).

4.6.4 Efficacy of force and flow

Near static head the function of a system is apparently to maintain an electrochemical potential difference, a tension, or some other appropriate force. In this case a parameter of interest is the force developed per given rate of expenditure of metabolic energy. This quantity, the *efficacy of force*, is denoted by ϵ_{X_1}:

$$\epsilon_{X_1} = \frac{-X_1}{J_2 X_2} \qquad (4.26)$$

For fixed X_2 the force efficacy increases monotonically as the magnitude of $(-X_1)$ increases, reaching its maximal value (in the driving region) at static head.

Near level flow the function of the system appears to be rapid transport of material. Here it is useful to consider the rate of transport per given rate of expenditure of metabolic energy. This quantity, the *efficacy of flow*, is denoted by ϵ_{J_1}:

$$\epsilon_{J_1} = \frac{J_1}{J_2 X_2} \qquad (4.27)$$

For fixed X_2 the flow efficacy increases monotonically as the magnitude of X_1 decreases (unless $q^2 = 1$), reaching its maximal value at level flow.

These considerations show that for systems of fixed input force whose function is not conversion of energy, but maintenance of an output force or flow, energy is utilized most effectively at static head or level flow, respectively.

4.7 The degree of coupling in nonisothermal systems: heat engines and the Carnot cycle

For completeness, we shall relate the entropic efficiency as defined above to the thermal efficiency of a heat engine. In the thermodynamics of heat engines, the reversible Carnot machine is the standard of comparison. We may follow the working substance through a series of steps, each of which approaches reversibility in the ideal case, and waste of free energy can be attributed to lack of reversibility in a certain step. This subdivision, however, is not possible in a thermoelectric element such as a thermocouple, where the essentially irreversible heat flow is coupled to the flow of electric current. In this sense the thermocouple is a closer analogue, in general, of biological energy-converting systems.

Consider now a heat engine working in cycles between T_1 and T_0. Suppose we choose a time scale long enough that the time of one cycle in some condition of steady operation is vanishingly small. We can then consider the flow of heat J_q entering the engine at T_1 to be virtually steady. Suppose further that the engine is coupled to some process taking place at T_0 which permits a steady or virtually steady withdrawal of power; for example, the piston may be coupled to a d.c. generator. The entropy production in this system is given by

$$d_i S/dt = \frac{IE}{T_0} + \frac{J_q \Delta T}{T_1 T_0} \qquad (4.28)$$

where I represents electric current, E represents electrical potential difference, and $\Delta T = T_1 - T_0$. The same expression would apply in the case of a thermocouple (Kedem and Caplan, 1965). The entropic efficiency is constructed just as before:

$$\eta = -\frac{IE}{J_q(\Delta T/T_1)} \qquad (4.29)$$

The thermal efficiency η_t is defined as the ratio (output)/(heat entering the system):

$$\eta_t = \left(\frac{\Delta T}{T_1}\right) \eta \qquad (4.30)$$

where the proportionality factor is the efficiency of a Carnot cycle operating between the same temperature limits. Hence the thermal efficiency of this heat engine cannot exceed the efficiency of a Carnot cycle. It can approach the Carnot efficiency only if the degree of coupling is unity, and as it does so the rate at which it operates tends toward zero.

The definition of efficiency arising out of entropy production is general and includes cyclic processes as a special case. Indeed, for nonisothermal systems it represents a normalization of the thermal efficiency. The additional insight it gives stems from the dependence of η on q. In the above example, even when the heat engine is operated infinitely slowly, Carnot efficiency can only be achieved if it is completely coupled to the output process. There must be no losses between the engine and the terminals of the generator. If there are losses, the degree of coupling is not complete, but we can calculate the maximum efficiency and the finite operating velocity at which it will be achieved. The fraction of the input dissipated by friction still decreases with decreasing rate. On the other hand, the energy dissipated by losses such as a thermal leak between the reservoirs is a larger fraction of the total energy expended at low rates. The maximum efficiency represents the optimal compromise between these two types of wastage.

4.8 Summary

1. The classical treatment of active transport in terms of an equivalent electrical circuit is subjected to critical review. In particular, it is shown that there are theoretical and practical difficulties associated with the concepts of work, efficiency, electromotive force of active ion transport, and stoichiometry, as used in this approach.

2. The nonequilibrium thermodynamic formulation of energetics is outlined for the case of a two-flow system. The degree of coupling (q) and the phenomenological stoichiometry (Z) are defined and are used to relate ratios of forces to ratios of flows. The implications of complete coupling ($q = 1$) are discussed.

3. It is shown that an unambiguous and useful definition of efficiency arises from the dissipation function (or more generally, for nonisothermal systems, from the entropy production function). Attention is drawn to two important physiological states in which energy expenditure occurs without performance of work, and hence the efficiency is zero. These are static head (output flow zero) and level flow (output force zero).

4. The series of steady states between static head and level flow are characterized by efficiencies greater than zero and less than unity. Within this region of operation (the "driving region"), the efficiency must obviously pass through a maximum. The maximum value of the efficiency depends only on the degree of coupling.

5. Only completely coupled systems can approach an efficiency of unity as a limit. This limit corresponds to reversible equilibrium, and the rate of the process under these conditions tends toward zero.

6. Near static head or level flow, the efficiency of the system may be very low, but its function in static head is maintenance of an electrochemical potential difference, a tension, or some other appropriate force, while in level flow the function is rapid transport of material. The parameters of interest in these states are the efficacy of force, that is, the force developed per given rate of expenditure of metabolic energy, and the efficacy of flow, or the rate of transport per given rate of expenditure of metabolic energy.

7. It is shown that the definition of efficiency arising out of the entropy production is general and includes cyclic processes as a special case. For nonisothermal systems it represents a normalization of the thermal efficiency with respect to the efficiency of a Carnot cycle operating between the same temperature limits. Carnot efficiency can only be achieved at an infinitely slow rate of operation if there is complete coupling to the output process.

5 The diagram method

In this chapter we briefly consider an important alternative approach that both complements and extends the formalism we have been employing. This is the "diagram method" for analyzing the steady-state kinetics of macromolecular energy-transducing systems, introduced by T. L. Hill and developed extensively by Hill and coworkers over the past decade. Since the technique has recently been described in great detail in an excellent monograph (Hill, 1977), we will not attempt to outline the quantitative methodology rigorously here. Instead, we will endeavor to give the reader a somewhat qualitative description, sufficient to indicate the power and utility of the method and its ability to provide insight into the nature of the coupling process.

5.1 States and cycles

To begin, we will consider a simple example of the diagram method. Figure 5.1a shows a hypothetical model of coupled transport (symport) discussed in detail by Hill (1977). What is actually being con-

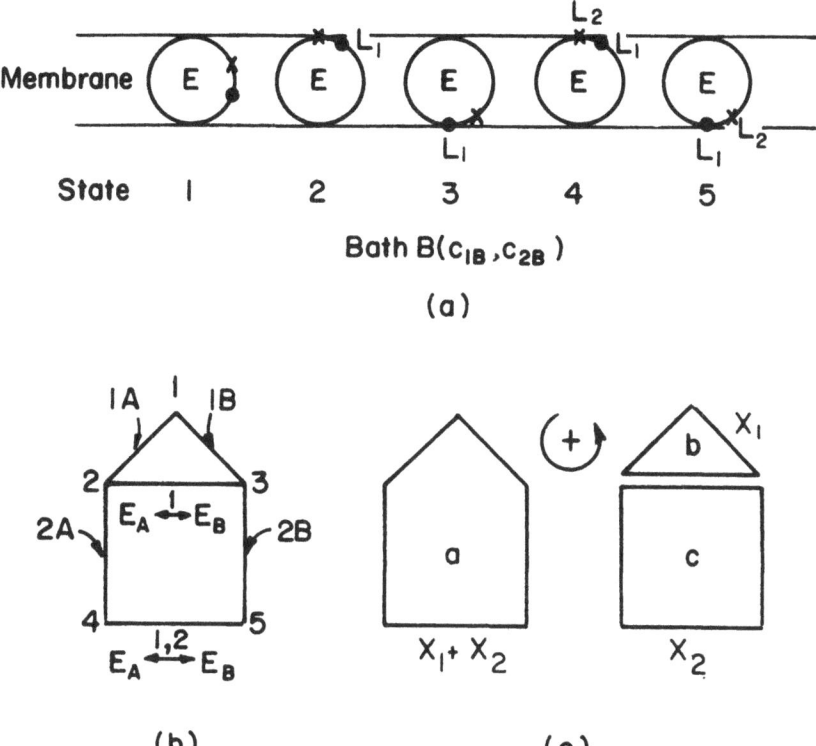

Fig. 5.1. (a) An illustrative model for membrane transport of two ligands L_1 and L_2 between two baths A and B: ●, L_1 site; X, L_2 site; E, protein. (b) Diagram corresponding to (a). (c) Cycles (the forces operating in each cycle are indicated). (After Hill, 1977.)

sidered here is an ensemble, that is, a large number, N, of equivalent and independent units, where one transporting protein molecule constitutes a unit. Each unit may exist in any one of the discrete states enumerated. The two baths A and B contain both ligands L_1 and L_2 at the concentrations indicated; however, the L_2 site on the protein E is activated only if L_1 is already bound. In this state E can undergo a conformational change ($2 \rightleftarrows 3$ or $4 \rightleftarrows 5$), in effect switching the bath to which the binding sites are accessible. The diagram for this system is shown in Fig. 5.1b. Arrows entering the diagram indicate binding; for example, 1A indicates binding of L_1 from

A; desorption arrows are omitted. The conformational changes in E are indicated by horizontal arrows. As the system moves according to the indicated transitions around the diagram, completing cycles of the three possible types indicated in Fig. 5.1c, the net effect is to transport L_1 and L_2 from one bath to the other. At equilibrium, since $c_{1A} = c_{1B}$ and $c_{2A} = c_{2B}$, there is no net transport of either ligand. But in the steady state, with unequal bath concentrations, a sufficient concentration difference in one ligand can cause a net flux in the other ligand against its own concentration gradient (note that the more general case of charged ligands is readily handled in essentially the same terms). An explicit derivation of the forces acting in each cycle will be presented later.

A kinetic description of the system can evidently be given in terms of first-order (or pseudo-first-order) rate constants for each transition; it is seen that each line in the diagram corresponds to two rate constants, one for the forward transition and one for the reverse. The power of the method now emerges in the calculation of steady-state transition fluxes (between any pair of states) and state probabilities (relative occupancies). It is shown that the number of *independent*, nonzero, steady-state transition fluxes for a diagram is given by the number of lines in the diagram minus one less than the number of states. In order to obtain the fluxes and probabilities as explicit functions of all the rate constants, it is necessary in principle to solve a substantial number of linear algebraic equations (in complex models this may involve a great deal of tedious labor). Instead of doing this, one can find and write down the solutions directly by means of a graphical algorithm of a type originally developed by King and Altman (1956). For this purpose it is necessary to construct a complete set of *partial diagrams*, each of which contains the maximum number of lines that can be included without forming any cycle, or closed path. In the present case there are eleven such partial diagrams. The next step is to derive a set of *directional diagrams* from the partial diagrams, one set for each state, making fifty-five directional diagrams in all for the present example. This is done by introducing arrows into the partial diagrams to make all connected paths, in effect, flow toward a given state. The directional diagrams corresponding to state 2 of the model system under consideration are illustrated in Fig. 5.2. Denoting the first-order rate constant for the transition $i \to j$ as α_{ij}, one can assign an algebraic value to each direc-

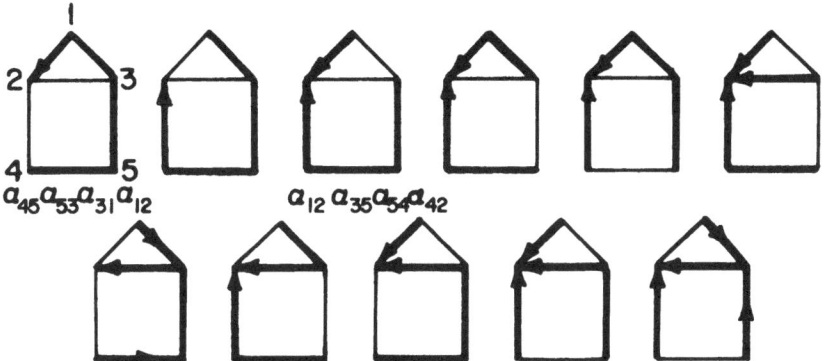

Fig. 5.2. Directional diagrams for state 2 of the model shown in Fig. 5.1; the algebraic values of the first and third directional diagrams are given. (After Hill, 1977.)

tional diagram as indicated. Then the steady-state probability of the ith state, p_i^∞, is given by

$$p_i^\infty = \text{(sum of directional diagrams of state } i)/\Sigma \tag{5.1}$$

where

$$\Sigma \equiv \text{sum of directional diagrams of } \textit{all} \text{ states} \tag{5.2}$$

Thus, for state 2,

$$p_2^\infty = \frac{(\alpha_{45}\alpha_{53}\alpha_{31}\alpha_{12} + \alpha_{13}\alpha_{35}\alpha_{54}\alpha_{42} + 9 \text{ other terms})}{(\alpha_{45}\alpha_{53}\alpha_{31}\alpha_{12} + \alpha_{13}\alpha_{35}\alpha_{54}\alpha_{42} + 53 \text{ other terms})} \tag{5.3}$$

The net mean steady-state transition flux between states i and j, in the direction $i \to j$, is denoted J_{ij}^∞. It can now be written down immediately:

$$J_{ij}^\infty = N(\alpha_{ij}p_i^\infty - \alpha_{ji}p_j^\infty) \tag{5.4}$$

From the criterion given earlier, it is obvious that if a diagram consists of only a single cycle, there is only one independent transition flux; in other words, for such a system the steady-state transi-

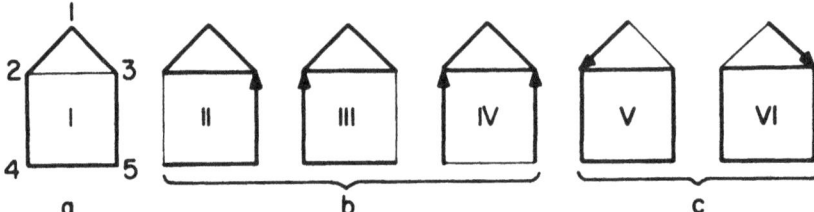

Fig. 5.3. Six flux diagrams for the model of Fig. 5.1, classified according to cycle. (After Hill, 1977.)

tion fluxes between consecutive pairs of states around the cycle are equal. This gives rise to the notion of a "cycle flux," which may be generalized to multicycle diagrams. Indeed, cycle fluxes are more informative than transition fluxes and, as will be seen, are fundamental to nonequilibrium thermodynamics. Moreover, multicycle diagrams readily clarify the concept of nonintegral and variable stoichiometry. If we focus again on Fig. 5.1, it is evident that cycle a transports L_1 and L_2 between baths A and B, cycle b transports L_1, and cycle c transports L_2. A *cyclic diagram* is obtained from a directional diagram by adding one arrow to complete a cycle of arrows. Such diagrams occur in pairs that differ only in the direction of traversal of the cycle, the algebraic difference between a pair of cyclic diagrams being defined as a *flux diagram*. Flux diagrams contain one cycle only (omitting arrows since both directions are traversed) and generally include one or more directed "streams" flowing into the cycle. Figure 5.3 illustrates all possible flux diagrams for the model we have been considering. More explicitly, if any line $i-j$ is included in the cycle of a flux diagram, we consider the $j \rightarrow i$ flux diagram to be defined as follows (in terms of algebraic values):

$$(j \rightarrow i \text{ flux diagram}) = (j \rightarrow i \text{ cyclic diagram})$$
$$- (i \rightarrow j \text{ cyclic diagram}) \quad (5.5)$$

For example, for flux diagram II of Fig. 5.3 we have

$$\text{II}_{12} = (\alpha_{12}\alpha_{23}\alpha_{31} - \alpha_{21}\alpha_{32}\alpha_{13})\alpha_{45}\alpha_{53}$$

The counterclockwise cycle direction is conventionally considered to generate a positive flux diagram, as indicated in Fig. 5.1. It can be shown that the transition fluxes are given by

$$J_{ji}^\infty = (N \times \text{sum of } j \to i \text{ flux diagrams})/\Sigma \tag{5.6}$$

Different flux diagrams containing the same cycle may be collected and summed, providing the same directionality convention holds for all terms. The sum is also referred to as a cycle. Thus, for Fig. 5.3,

$$\text{cycle } a = \text{I}, \quad \text{cycle } b = \text{II} + \text{III} + \text{IV},$$

$$\text{cycle } c = \text{V} + \text{VI} \tag{5.7}$$

The *cycle fluxes* J_κ, where $\kappa = a,b,c$, are given by

$$J_a = N(\text{cycle } a)/\Sigma, \quad J_b = N(\text{cycle } b)/\Sigma,$$

$$J_c = N(\text{cycle } c)/\Sigma \tag{5.8}$$

These are defined only for the steady state. From Eq. (5.6) and generalizing Eqs. (5.7) and (5.8), we have

$$J_{ji}^\infty = \text{sum of } j \to i \text{ cycle fluxes} \tag{5.9}$$

For the example of Fig. 5.1,

$$J_{12}^\infty = J_a + J_b, \quad J_{23}^\infty = J_b - J_c, \quad J_{24}^\infty = J_a + J_c \tag{5.10}$$

Also,

$$J_{31}^\infty = J_{12}^\infty, \quad J_{24}^\infty = J_{45}^\infty = J_{53}^\infty \tag{5.11}$$

Note that there are three cycles and hence three independent cycle fluxes.

To proceed further it is necessary to be specific about the algebraic expressions. If we define $\Pi_{a+} = \alpha_{12}\alpha_{24}\alpha_{45}\alpha_{53}\alpha_{31}$, $\Pi_{b-} = \alpha_{21}\alpha_{32}\alpha_{13}$, and so on, we see from Eqs. (5.7) and (5.8) that in general,

$$J_\kappa = N(\Pi_{\kappa+} - \Pi_{\kappa-})\Sigma_\kappa/\Sigma \tag{5.12}$$

80   Bioenergetics and linear nonequilibrium thermodynamics

where Σ_κ is a sum, over all flux diagrams belonging to κ, of the "appendages" feeding into the cycle. From Fig. 5.3,

$$\Sigma_a = 1, \quad \Sigma_b = \alpha_{45}\alpha_{53} + \alpha_{54}\alpha_{42} + \alpha_{42}\alpha_{53},$$

$$\Sigma_c = \alpha_{12} + \alpha_{13} \tag{5.13}$$

It can be shown that it is always possible to split cycle fluxes into *separate unidirectional* cycle fluxes, for example:

$$J_{\kappa^+} = N\Pi_{\kappa^+}\Sigma_\kappa/\Sigma, \quad J_{\kappa^-} = N\Pi_{\kappa^-}\Sigma_\kappa/\Sigma \tag{5.14}$$

Before we consider the forces acting in the system, it is necessary to emphasize a general assumption that the rate constants for binding of the ligands are pseudo-first-order, that is, dependent on the concentration (or activity) of the bath from which the ligand binds. Thus $\alpha_{12} = \alpha_1^* c_{1A}$ and $\alpha_{13} = \alpha_1^* c_{1B}$, where α_1^* is the second-order rate constant for binding of L_1. Similarly, $\alpha_{24} = \alpha_2^* c_{2A}$, $\alpha_{35} = \alpha_2^* c_{2B}$. Accordingly, the chemical potential differences are given by[1]

$$X_1 \equiv \Delta\mu_1 = RT \ln\left(\frac{c_{1A}}{c_{1B}}\right) = RT \ln\left(\frac{\alpha_{12}}{\alpha_{13}}\right)$$

$$X_2 \equiv \Delta\mu_2 = RT \ln\left(\frac{c_{2A}}{c_{2B}}\right) = RT \ln\left(\frac{\alpha_{24}}{\alpha_{35}}\right) \tag{5.15}$$

It should be noted that the simplifying assumptions introduced above ($\alpha_{12}^* = \alpha_{13}^* = \alpha_1^*$ and $\alpha_{24}^* = \alpha_{35}^* = \alpha_2^*$) are not at all necessary for the following argument. Now at *equilibrium*, for cycle b (Fig. 5.1c),

$$\Pi_{b^+} = \Pi_{b^-} \quad \text{or} \quad \alpha_{12}\alpha_{23}\alpha_{31} = \alpha_{21}\alpha_{32}\alpha_{13} \tag{5.16}$$

But since $c_{1A} = c_{1B}$ and hence $\alpha_{12} = \alpha_{13}$ at equilibrium, we arrive at a perfectly general relationship between the rate constants:

$$\alpha_{23}\alpha_{31} = \alpha_{21}\alpha_{32} \tag{5.17}$$

Thus, at the steady state,

$$\Pi_{b^+} - \Pi_{b^-} = \left(\frac{c_{1A}}{c_{1B}} - 1\right)\Pi_{b^-} \tag{5.18}$$

Here, using Eqs. (5.14) and (5.15),

$$\frac{\Pi_{b^+}}{\Pi_{b^-}} = \frac{J_{b^+}}{J_{b^-}} = \frac{c_{1A}}{c_{1B}} = e^{X_1/RT} \qquad (5.19)$$

Similarly, for cycle c,

$$\Pi_{c^+} - \Pi_{c^-} = \left(\frac{c_{2A}}{c_{2B}} - 1\right)\Pi_{c^-} \qquad (5.20)$$

$$\frac{\Pi_{c^+}}{\Pi_{c^-}} = \frac{J_{c^+}}{J_{c^-}} = \frac{c_{2A}}{c_{2B}} = e^{X_2/RT} \qquad (5.21)$$

For cycle a we find

$$\Pi_{a^+} - \Pi_{a^-} = \left(\frac{c_{1A}c_{2A}}{c_{1B}c_{2B}} - 1\right)\Pi_{a^-} \qquad (5.22)$$

$$\frac{\Pi_{a^+}}{\Pi_{a^-}} = \frac{J_{a^+}}{J_{a^-}} = \frac{c_{1A}c_{2A}}{c_{1B}c_{2B}} = e^{(X_1+X_2)/RT} \qquad (5.23)$$

The steady-state fluxes of ligands 1 and 2 between the baths and across the membrane will be denoted J_1^∞ and J_2^∞, respectively. Evidently $J_1^\infty = J_{12}^\infty$ and $J_2^\infty = J_{24}^\infty$. From Eqs. (5.10) and (5.12), making use of (5.18), (5.20), and (5.22), and introducing for convenience the definition

$$\kappa = N\Pi_{\kappa^-}\Sigma_\kappa/\Sigma = J_{\kappa^-} \qquad (\kappa = a, b, c) \qquad (5.24)$$

(compare Eq. 5.14), as well as the identity

$$e^{(X_1+X_2)/RT} - 1 = (e^{X_1/RT} - 1)(e^{X_2/RT} - 1)$$
$$+ (e^{X_1/RT} - 1) + (e^{X_2/RT} - 1)$$

we have

$$J_1^\infty = (a + b)(e^{X_1/RT} - 1) + a(e^{X_2/RT} - 1)$$
$$+ a(e^{X_1/RT} - 1)(e^{X_2/RT} - 1) \qquad (5.25)$$

$$J_2^\infty = a(e^{X_1/RT} - 1) + (a + c)(e^{X_2/RT} - 1)$$
$$+ a(e^{X_1/RT} - 1)(e^{X_2/RT} - 1) \tag{5.26}$$

It will often be convenient to choose one bath, say bath B, as the *reference bath*, restricting all variations in X_1 and X_2 to changes in bath A. In this case we would like the coefficients κ to be independent of variations in bath A. Indeed, this consideration has already entered into the above formulation, except that Σ contains α_{12} and α_{24}, which are proportional to c_{1A} and c_{2A}, respectively, and Σ_c contains α_{12} (see Eq. 5.13). Assuming that Σ will be comparatively insensitive to variations in bath A, we use a substitution for c which is readily derived from Eqs. (5.13), (5.15), and (5.24):

$$c = [(e^{X_1/RT} - 1) + 2]c' \qquad (c' = N\alpha_{13}\Pi_c^-/\Sigma) \tag{5.27}$$

Then Eq. (5.26) becomes

$$J_2^\infty = a(e^{X_1/RT} - 1) + (a + 2c')(e^{X_2/RT} - 1)$$
$$+ (a + c')(e^{X_1/RT} - 1)(e^{X_2/RT} - 1) \tag{5.28}$$

It is of interest that this type of transformation is also possible when $\alpha_{12}^* \neq \alpha_{13}^*$. In this case the substitution is

$$c = [(e^{X_1/RT} - 1) + (1 + \gamma)]c'$$

$$\left(\gamma = \frac{\alpha_{13}^*}{\alpha_{12}^*}, \quad c' = N\alpha_{12}^* c_{1B} \Pi_c^- / \Sigma\right)$$

and the second term on the right side of Eq. (5.28) takes the form $[a + c'(1 + \gamma)](e^{X_2/RT} - 1)$. Equations (5.25) and (5.28) give the ligand fluxes at an arbitrary steady state, not necessarily near equilibrium, as functions of the forces. If variations are restricted to bath A, the coefficients will be constant to the extent that Σ is constant (alternatively, the equations may be multiplied through by Σ, giving rise to constant coefficients relating the forces to "flows" containing a variable factor). It will be noted that the cross coefficients of the "single-force" terms obey a reciprocal relation. (Notice that they refer to properties of the given steady state, *not* of the equilibrium state.) It is therefore clear that on expanding the exponential terms for near-equilibrium conditions—X_1 and X_2 small—the usual phenomenological equations are recovered in the limit; for example, from Eqs. (5.25) and (5.26),

$$J_1^\infty = (a+b)X_1/RT + aX_2/RT \qquad (5.29)$$

$$J_2^\infty = aX_1/RT + (a+c)X_2/RT \qquad (5.30)$$

(Somewhat further from equilibrium we find equations of a type which will be discussed in Chapter 13: compare Eqs. 13.33 and 13.34.) The net steady-state rate of free energy dissipation in the ensemble + bath A + bath B is given, as usual, by

$$\Phi = J_1^\infty X_1 + J_2^\infty X_2 \qquad (5.31)$$

Comparing Eqs. (5.29) and (5.30) with Eqs. (5.24) and (5.14), we see that the coefficients are simply combinations of unidirectional cycle fluxes. More generally, it can be said that whenever a pair of cycles in a diagram share a common line, they represent coupled processes. If the common line is eliminated a larger cycle is formed, embracing both of the previous cycles; this larger cycle then represents a hypothetical "completely coupled" contribution. (Of course, if two forces indeed act in a single cycle, the corresponding processes must be fully coupled.) The larger cycle may itself be coupled to additional cycles, and so on. The phenomenological coefficients for equations such as (5.29) and (5.30) may be written down as follows: for the "straight" coefficient of any process, unidirectional cycle fluxes are summed covering the cycle relating to the given process and all possible larger cycles including it; for the "cross" coefficient between two processes, unidirectional cycle fluxes are summed for all larger cycles which include *both* elementary cycles relating to the given processes (if negative coupling is involved, the summation has a negative sign).

Instead of writing the dissipation function in terms of the products of operational fluxes and forces, as in Eq. (5.31), one may write it as a sum of products of cycle fluxes and forces. Introducing Eqs. (5.10) into (5.31), we obtain

$$\Phi = J_a X_a + J_b X_b + J_c X_c \qquad (5.32)$$

where

$$X_a = X_1 + X_2, \quad X_b = X_1, \quad X_c = X_2 \tag{5.33}$$

Here the quantity X_κ associated with each cycle flux J_κ is the sum of all the thermodynamic forces operative in cycle κ. Note that it is possible for a cycle to have zero net force. As we have seen (Eqs. 5.19, 5.21, and 5.23) X_κ is given by

$$e^{X_\kappa/RT} = \frac{\Pi_\kappa^+}{\Pi_\kappa^-} \tag{5.34}$$

and J_κ is given by Eq. (5.12). From this it follows that J_κ and X_κ will always have the same sign, and in Eq. (5.32) *each term* in the sum is positive or zero.

5.2 Free energy levels

Hill and coworkers (Hill, 1977) also introduced the explicit use of free energy levels in describing the states of a macromolecular unit or system, and this is of considerable utility in understanding the nature of the forces operating in the system. It is important to realize here that although in general states are not in equilibrium with *each other*, each state is in equilibrium *internally*. (Should relatively slow transitions occur between more elementary "internal" substates, so that some state turns out not to be in internal equilibrium, it is necessary to subdivide that state into substates for which internal equilibrium does hold.) Three kinds of free energy change are defined for any given transition ij: standard, basic, and gross. These are referred to as free energy level differences between states i and j; for convenience, free energy levels are associated with individual states, although only the differences are significant.

On statistical mechanical grounds it is shown that for an ensemble of condensed systems of the kind we have been considering, the chemical potential of an arbitrary component in state j is given by

$$\mu_j = G_j + RT \ln p_j \tag{5.35}$$

where p_j, as before, is the probability or occupancy fraction of the jth state. If all systems are in state j, $p_j = 1$ and $\mu_j = G_j$. Thus G_j may be thought of either as the molar Gibbs free energy of a single isolated system in state j or as the molar Gibbs free energy per system, if the whole ensemble is in state j. Equation (5.35) indicates that G_j is also the standard Gibbs free energy of state j, the standard state referred to being $p_j = 1$. The standard free energy G_j is an intrinsic equilibrium property of state j alone, while the chemical potential μ_j is clearly a property of the whole ensemble (since it involves p_j). As usual, standard free energies can be related to equilibrium constants. To do this, we first consider transitions in either direction between states i and j, chosen so that no binding or release of ligand is involved. If we imagine a hypothetical equilibrium of the system (or even of the two states alone with other transitions blocked), then from the principle of detailed balancing

$$\alpha_{ij} p_i^e = \alpha_{ji} p_j^e \tag{5.36}$$

where e indicates an equilibrium value. Furthermore,

$$\mu_i = G_i + RT \ln p_i^e = \mu_j = G_j + RT \ln p_j^e \tag{5.37}$$

and hence the equilibrium constant K_{ij} is given by

$$K_{ij} \equiv \frac{\alpha_{ij}}{\alpha_{ji}} = \exp[-(G_j - G_i)/RT] \tag{5.38}$$

Equation (5.38) is always valid, since the G's and α's are intrinsic properties of states i and j of each system. We next consider the possibility that a ligand L is bound in the transition $i \to j$. Again, equilibrium between the states leads to Eq. (5.36), but now we are dealing with the pseudo-first-order rate constant $\alpha_{ij} = \alpha_{ij}^* c_L$, where c_L is the concentration of L in solution. The equilibrium condition is

$$\mu_i + \mu_L = G_i + RT \ln p_i^e + \mu_L = \mu_j$$
$$= G_j + RT \ln p_j^e \tag{5.39}$$

and therefore

$$K_{ij} \equiv \frac{\alpha_{ij}}{\alpha_{ji}} = \exp\{-[G_j - (G_i + \mu_L)]/RT\} \tag{5.40}$$

Alternatively,

$$\frac{K_{ij}}{c_L} = K_{ij}^* \equiv \frac{\alpha_{ij}^*}{\alpha_{ji}} = \exp\{-[G_j - (G_i + \mu_L^0)]/RT\} \quad (5.41)$$

where the "second-order" equilibrium constant K_{ij}^* is invariant, that is, independent of c_L, and we have introduced the standard free energy (chemical potential) μ_L^0 of L via the relation

$$\mu_L = \mu_L^0 + RT \ln c_L \quad (5.42)$$

The term *basic free energy change* will be used in what follows to indicate the kind of free energy change directly related to *first-order* rate constants that appears in Eqs. (5.38) and (5.40). If a ligand is involved in the process at concentration c_L, μ_L must be taken into account. It should be noted that basic free energy changes are time-independent intrinsic properties of each individual system of the ensemble. The *standard free energy* changes which correspond to these basic free energy changes are seen in Eqs. (5.38) and (5.41). If the process is purely isomeric (Eq. 5.38), the standard and basic changes are the same.

While basic and standard free energy changes are characteristic properties of individual macromolecular systems plus their surrounding bath or baths (and hence independent of boundary conditions and time), a third free energy change of importance relates to the ensemble as a whole at an arbitrary time t. If it has a composition specified by $p_1, p_2, \ldots, p_n$, then the free energy changes in the entire ensemble plus its surrounding baths (associated with the processes $i \to j$ in Eqs. 5.38 and 5.40) are

$$\mu_j - \mu_i = G_j - G_i + RT \ln \left(\frac{p_j}{p_i}\right) \quad (5.43)$$

$$\mu_j - (\mu_i + \mu_L) = G_j - (G_i + \mu_L) + RT \ln \left(\frac{p_j}{p_i}\right) \quad (5.44)$$

These will be designated *gross free energy* changes. They are essentially the conventional Gibbs free energy changes that determine equilibrium in the ensemble plus baths. Following Hill, a convenient notation and sign convention for basic and gross free energy changes for the present argument is the following. For the process $i \to j$, whether it is an isomeric change or involves binding and release of a ligand, we write

$$\Delta G'_{ij} = G'_i - G'_j$$

$$\Delta \mu'_{ij} = \mu'_i - \mu'_j \qquad (5.45)$$

for the basic and gross free energy changes, respectively (the quantities on the right may be considered the basic and gross free energy levels of the states). The primes indicate that ligand or substrate chemical potentials have been taken into account where necessary. Then

$$\frac{\alpha_{ij}}{\alpha_{ji}} = \exp(\Delta G'_{ij}/RT) = K_{ij} \qquad (5.46)$$

$$\Delta \mu'_{ij}(t) = \Delta G'_{ij} + RT \ln \left[\frac{p_i(t)}{p_j(t)}\right] \qquad (5.47)$$

The product of equations such as (5.46) taken successively around any cycle κ, for example with states numbered in counterclockwise order $1, 2, \ldots, m$, is (compare Eq. 5.34):

$$\frac{\Pi_\kappa^+}{\Pi_\kappa^-} = e^{X_\kappa/RT} = K_{12} K_{23} \cdots K_{m1} \qquad (5.48)$$

The total thermodynamic force in the cycle, X_κ, is therefore given by the sum of the successive basic free energy changes around the cycle:

$$X_\kappa = \Delta G'_{12} + \Delta G'_{23} + \cdots + \Delta G'_{m1} \qquad (5.49)$$

Since the probability terms cancel if Eq. (5.47) is summed around the cycle at some arbitrary t, we also have

$$X_\kappa = \Delta \mu'_{12}(t) + \Delta \mu'_{23}(t) + \cdots + \Delta \mu'_{m1}(t) \qquad (5.50)$$

Thus $\Delta G'_{ij}$ is the "drive" associated with the transition ij of the cycle.

The subdivision represented by Eq. (5.49) is invariant and characterizes each individual system in the ensemble, while that represented by Eq. (5.50) is a time-dependent property of the entire ensemble.

An important conclusion may be drawn from the above considerations which relates to the direction of spontaneous transition. For example, consider an ensemble of N systems with an arbitrary kinetic diagram, and at any time t (not necessarily in the steady state) let p_i be the probability of state i. For any transition ij, the net mean transition flux $i \rightarrow j$ is given by (compare Eq. 5.4):

$$J_{ij}(t) = N[\alpha_{ij} p_i(t) - \alpha_{ji} p_j(t)] \tag{5.51}$$

From Eqs. (5.46) and (5.47),

$$\frac{\alpha_{ij} p_i(t)}{\alpha_{ji} p_j(t)} = e^{\Delta \mu'_{ij}(t)/RT} \tag{5.52}$$

By comparing the above two equations we see that the transition flux J_{ij} for any transition ij always has the same sign as the gross free energy level difference $\Delta \mu'_{ij}$ (at any time t). If i has the higher gross free energy level, the net mean flux will be in the direction $i \rightarrow j$ (stochastic exceptions may occur in single systems or small groups of systems). Hence, net positive flux (or reaction) always occurs in a downhill direction with respect to a set of gross free energy levels. This is *not* true of the basic free energy levels. It also follows that the dissipation function for the ensemble plus baths can be written

$$\Phi = \sum_{ij} J_{ij}(t) \Delta \mu'_{ij}(t) \tag{5.53}$$

where the sum is over all lines in the diagram, each term being positive.

The above approach leads to a consideration which cannot be emphasized too strongly: in studying ion pumps and similar energy-transducing mechanisms, it is fruitless to search for the crucial step or steps in the enzymatic cycle at which free energy transfer between ligands actually occurs. As has been pointed out forcefully by Hill and Eisenberg (1981), free energy transfer between the small molecules is an indivisible property of the *entire* cycle.

We will illustrate the use of these concepts by discussing their ap-

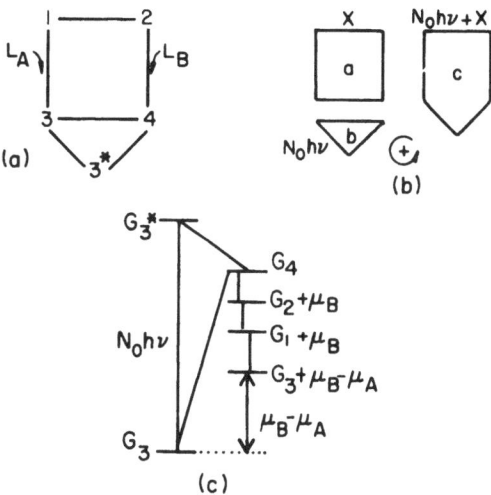

Fig. 5.4. A model that combines transport of a ligand L with light absorption. State 3^2 is an excited state. (a) Diagram. (b) Cycles. (c) Basic free energy levels for macromolecule + ligand. (After Hill, 1977.)

plication to a simple model of phototranslocation in the well-known purple membrane, given by Hill (1977). (The light-driven proton pump in purple membrane will be examined in more detail in Chapter 13.) This model is described in Fig. 5.4. The thermodynamic force corresponding to ligand transport is negative: $c_B > c_A$, $\mu_B > \mu_A$, $X = \mu_A - \mu_B < 0$. To begin with, consider only cycle a. The equations of the type (5.38) or (5.40) are

$$\frac{\alpha_{13}}{\alpha_{31}} = \exp\{[(G_1 + \mu_A) - G_3]/RT\} = K_{13} \qquad (5.54)$$

$$\frac{\alpha_{34}}{\alpha_{43}} = \exp[(G_3 - G_4)/RT] = K_{34} \qquad (5.55)$$

$$\frac{\alpha_{42}}{\alpha_{24}} = \exp\{[G_4 - (G_2 + \mu_B)]/RT\} = K_{42} \qquad (5.56)$$

$$\frac{\alpha_{21}}{\alpha_{12}} = \exp[(G_2 - G_1)/RT] = K_{21} \qquad (5.57)$$

Multiplying these together, we obtain

$$\frac{\Pi_+}{\Pi_-} = e^{(\mu_A - \mu_B)/RT} = e^{X/RT} = K_{13}K_{34}K_{42}K_{21} \tag{5.58}$$

The difference in basic free energy between state 3 and state 4 is seen, from Eq. (5.55), to be $G_3 - G_4$. Hence we may assign a basic free energy level G_3 to state 3, and a basic free energy level G_4 to state 4. Since the transition $3 \to 4$ may be regarded as carrying bound ligand from one side of the membrane to the other, we shall assume that in the present example the quantity $(G_3 - G_4)$ is negative, and therefore the basic free energy level G_3 lies below, perhaps well below, G_4. This is indicated in Fig. 5.4c. By reading off the basic free energy differences in Eqs. (5.56), (5.57), and (5.54) in succession, we complete one counterclockwise circuit around cycle a and are led to assign basic free energy levels as shown to states 2 and 1. On returning to state 3, a molecule of ligand has been transferred against the chemical potential difference $\mu_B - \mu_A$, and therefore the basic free energy level has increased by this amount. This new level for state 3 now forms the baseline for a second set of levels corresponding to a second cycle, and so on *ad infinitum*. Actually, if light played no role in this system, it would spontaneously cycle backward, or clockwise around cycle a, producing on the average a net negative flux J_a, that is, a net flux of ligand from bath B to bath A. This flux might be very small because of the uphill transitions $3 \to 1$, $1 \to 2$, $2 \to 4$ before the major drop in basic free energy associated with $4 \to 3$. It will be clear that for such systems free energy levels are repeated indefinitely both above and below those shown, at appropriate intervals, one set for each cycle. From a stochastic point of view it can be said that at the individual transition level, such a system performs a biased one-dimensional walk on the free energy levels. The transition probabilities are the first-order rate constants, and the bias is in favor of downward transitions as determined by the basic free energy differences.

We now suppose that there exists an excited state of state 3, 3^*, reached from state 3 by absorption of a photon having an energy $h\nu$. A transition between 3^* and 4 is possible. This expands the kinetic diagram to include cycles b and c, as shown in Fig. 5.4b. We further suppose that the energy level G_{3^*} is above G_4, as shown in Fig. 5.4c.

In fact the energy level diagram assumes that $G_{3^*} - G_3 = N_0 h\nu$, which is a useful approximation. (Since our discussion is on a molar basis, we consider the energy change per einstein[2] absorbed.) On exposure of the ensemble to steady radiation of frequency ν and of sufficient intensity, cycle c will operate in the positive direction, dominating the situation. Thus each complete circuit of cycle c will transport a molecule of L from bath A to bath B, against its chemical potential gradient, at the expense of part of the photon energy $h\nu$. Cycle b utilizes a photon but accomplishes no transport and is therefore wasteful. The cycle fluxes J_b and J_c are positive, but J_a is negative. Since light energy is consumed at the rate $N_0 h\nu(J_b + J_c)$, the efficiency of the light energy to free energy transduction is

$$\eta = \frac{-X(J_a + J_c)}{N_0 h\nu(J_b + J_c)} \tag{5.59}$$

Clearly, cycles a and b both tend to reduce η.

When conditions are such that $\eta > 0$, the one-dimensional walk on the free energy levels (discussed above in relation to cycle a alone) becomes an upward walk for the dominant cycle c. Under experimental conditions, however, the stationary state under study is frequently static head. In this case, if no other leakage pathways are to be taken into consideration, $J_a + J_c = 0$; that is, clockwise cycling of cycle a occurs at the same rate as counterclockwise cycling of cycle c. In this case the free energy $\mu_B - \mu_A$ gained in a traversal of cycle c is lost (simultaneously) in a traversal of cycle a.

The twelve first-order rate constants corresponding to Fig. 5.4a may be used to calculate steady-state probabilities of the states, cycle fluxes, and operational fluxes as before. However, the formalism breaks down when it is concerned with the relationship of $N_0 h\nu$ to basic free energy levels, rate constants and, more generally, nonequilibrium thermodynamics. To some extent photon absorption may be treated as though it were analogous to ligand binding, this being a strictly formal notion enabling one to show absorption of an einstein in a cycle as a "basic free energy" drop of magnitude $N_0 h\nu$. But in other respects this analogy is not close, because of the essentially "nonthermodynamic" character of a beam of photons of frequency ν at some arbitrary intensity. To be specific, the fundamental relation (5.46) breaks down for the transitions 33* in cycles b and c; the ratio

92 Bioenergetics and linear nonequilibrium thermodynamics

$\alpha_{33^*}/\alpha_{3^*3}$ has nothing to do with detailed balance at equilibrium. Consequently, although the kinetic result

$$\frac{\Pi_{\kappa^+}}{\Pi_{\kappa^-}} = \frac{J_{\kappa^+}}{J_{\kappa^-}} \qquad (\kappa = b, c) \tag{5.60}$$

holds as usual, Π_{b^+}/Π_{b^-} is *not* equal to $e^{N_0h\nu/RT}$ and Π_{c^+}/Π_{c^-} is *not* equal to $e^{(N_0h\nu+X)/RT}$. Indeed, both Π ratios depend on the beam intensity, whereas the quantities $N_0h\nu$ and X do not. Furthermore, the use of nonequilibrium thermodynamics, which implies the validity of the reciprocal relations, is excluded: since the "thermodynamic force" $N_0h\nu$ is constant for a monochromatic beam of photons, one cannot meaningfully consider a gradual approach to equilibrium as a limiting steady state. Again, gross free energy levels can be defined as before, but neither Eq. (5.52) nor Eq. (5.53) is applicable if these equations refer to, or in the latter case include, photon transitions. For these transitions J_{ij} need not have the same sign as $\Delta\mu'_{ij}$, and the gross levels are accordingly of diminished interest.

5.3 Stoichiometry and coupling

The diagram method permits us to see in a rather clear way the nature of the phenomenological stoichiometry Z and the degree of coupling q. In Chapter 4 we found that these quantities are of considerable utility in characterizing the performance of linear energy converters. We can now reinterpret them in terms of the unidirectional cycle fluxes $a, b,$ and c for systems of the type described in Fig. 5.1. To do this, we rewrite Eqs. (5.25) and (5.26) in the following way:

$$J'_1 = (a + b)(e^{X'_1/RT} - 1) + a(e^{X_2/RT} - 1) \\ + a(e^{X'_1/RT} - 1)(e^{X_2/RT} - 1) \tag{5.61}$$

$$J_2 = a(e^{X'_1/RT} - 1) + (a + c)(e^{X_2/RT} - 1) \\ + a(e^{X'_1/RT} - 1)(e^{X_2/RT} - 1) \tag{5.62}$$

Here we drop the superscript ∞, it being understood that we are dealing only with steady states. The reason for writing J'_1 and X'_1 is that in the general case, where n moles of ligand L_1 may bind to the protein E per mole of L_2 bound, the description of the flows must remain consistent with a summation of the forces as required in cycle a. This requirement may give rise to an awkward choice of units, which is readily avoided by making use of the transformation to "natural" quantities J_1 and X_1, where

$$J_1 = nJ_1' \tag{5.63}$$

$$X_1 = \frac{X_1'}{n} \tag{5.64}$$

The integer (or ratio of integers) n will be designated the "mechanistic stoichiometry." Substituting (5.63) and (5.64) in (5.61) and (5.62), we obtain

$$J_1 = n(a + b)(e^{nX_1/RT} - 1) + na(e^{X_2/RT} - 1)$$
$$+ na(e^{nX_1/RT} - 1)(e^{X_2/RT} - 1) \tag{5.65}$$

$$J_2 = a(e^{nX_1/RT} - 1) + (a + c)(e^{X_2/RT} - 1)$$
$$+ a(e^{nX_1/RT} - 1)(e^{X_2/RT} - 1) \tag{5.66}$$

Notice that the force acting in cycle a (compare Eq. 5.23) is now $(nX_1 + X_2)$, which will be zero either when $X_1 = X_2 = 0$, or when $X_1 = -X_2/n$. The substitution giving rise to Eq. (5.28) may, of course, be made here as well. In the linear range corresponding to Eqs. (5.29) and (5.30), Eqs. (5.65) and (5.66) give

$$J_1 = n^2(a + b)X_1/RT + naX_2/RT \tag{5.67}$$

$$J_2 = naX_1/RT + (a + c)X_2/RT \tag{5.68}$$

Hence we can write, for models of this kind,

$$Z = n\sqrt{\frac{(a + b)}{(a + c)}} \tag{5.69}$$

$$q = \frac{1}{\sqrt{(1 + b/a)(1 + c/a)}} \tag{5.70}$$

The degree of coupling is now seen to depend explicitly on the relative magnitudes of the cycle fluxes b and c as compared with a. When b and c are both zero we have complete coupling, and the phenomenological stoichiometry is identical to the mechanistic stoichiometry. In the general case $Z \neq n$, except in the fortuitous circumstance that $b = c$. Such a possibility will only rarely arise in practice. Indeed, it is readily shown from Eqs. (5.69) and (5.70) that for this type of model, at any given degree of coupling q, the quantity Z/n may lie anywhere within the range $1/q$ to q. At sufficiently high degrees of coupling the approximation $Z \simeq n$ will therefore be close.

5.4 Usefulness of diagram formulations

The diagram method is an approach of considerable utility in conceptualizing complex kinetic systems. While it can be used as a computational tool, its greatest utility probably is in clarifying the important interactions in a multicyclic kinetic system, as well as the nature of the degree of coupling and the phenomenological stoichiometry. In this regard it is clearly related to the formalism of network thermodynamics (Oster, Perelson, and Katchalsky, 1973; Mikulecky, 1977; Mikulecky and Thomas, 1978). Both give rise to linear nonequilibrium thermodynamics in the near-equilibrium limit, and it is to be expected that the circuit diagrams of network thermodynamics are isomorphic with the kinetic diagrams discussed here. However, in a very important sense circuit diagrams are at least one step further removed from the microscopic real world than kinetic diagrams; therefore they are perhaps less congenial to work with from the point of view of molecular biology, although not from the point of view of electrophysiology, where the problems are generally posed at a higher level of organization. The two formulations serve similar purposes and appear to be complementary.

5.5 Summary

1. As an important alternative approach to problems of free energy transduction, we draw attention to the "diagram method"

developed by Hill and coworkers. This is particularly appropriate for a discussion of complex and/or nonlinear systems. A good example is the photocycle of bacteriorhodopsin, owing to its multicyclic character.

2. A simple example (the cotransport of two ligands by a membrane protein) is used to illustrate the relationship between states and cycles in the diagram method. A brief account is given of the use of partial diagrams, directional diagrams, cyclic diagrams, and flux diagrams in the calculation of the steady-state probabilities of states, the transition fluxes, and the cycle fluxes.

3. The origin of reciprocity as a property of the steady state is discussed, especially in regard to the derivation of the phenomenological equations of nonequilibrium thermodynamics in the near-equilibrium range.

4. The use of free energy levels is discussed. Free energy level diagrams are drawn and interpreted for a simplified model of bacteriorhodopsin. The essential "nonthermodynamic" character of this particular system is shown to exclude the use of nonequilibrium thermodynamics, in particular the reciprocal relations.

5. It is misleading to claim that free energy is exchanged between the small molecules of the system in a particular part of the transducing cycle. The free energy changes of the small molecules and of the enzyme are inseparable in individual transitions. They can be identified only at the complete cycle level.

6. The phenomenological stoichiometry Z and the degree of coupling q are derived in terms of the unidirectional cycle fluxes of the simple model for cotransport. The mechanistic stoichiometry n of the system is identified and shown to be equal to Z only if $q = 1$ (except in one unlikely case of incomplete coupling). However, when q is close to unity, Z does not differ greatly from n.

7. The relationship of the diagram method to network thermodynamics is discussed briefly.

6 Possible conditions for linearity and symmetry of coupled processes far from equilibrium

We have seen in earlier chapters that the formulation of useful relationships between forces and flows is perhaps the most important problem in the theoretical and experimental analysis of biological reactions and transport processes. We need relationships which permit the self-consistent analysis of data under diverse conditions. Thus, for uncoupled flows, Fick's law and Ohm's law for solute flow and electrical current are of great utility, as is also the Michaelis-Menten description of the kinetics of enzymatic reactions. This utility in large part derives from the ease with which linear relationships can be handled. It was the attempt to incorporate the influence of coupling of flows that led to the formulations of linear NET.

Many workers continue to feel, however, that at least for biology, Onsager's relations are of greater theoretical than practical value, since biological processes are commonly far removed from equilibrium. In particular, doubts are expressed with respect to complex processes involving chemical reactions. Even for a simple uncoupled reaction, the rate is generally found to be a highly nonlinear function of the conjugate force, the thermodynamic reaction affinity A, unless $A \ll RT$ (Prigogine, 1955; deGroot and Mazur, 1962). It has been suggested that since biological systems may be characterized by processes which take place in many steps, each of which is nearly reversible, linear phenomenological relationships between rates and affinities may apply even if the overall affinity $A \gg RT$ (Prigogine, 1955). This point of view overlooks the fact that the phenomenological coefficients of the elemental reactions will in general be sensitive to the concentrations of reaction intermediates, which will vary with change of A and hence rule out linearity over a wide range.

Nevertheless, it is a remarkable experimental finding that extensive linearities between flows and forces characterize transepithelial active Na^+ and H^+ transport and are equally characteristic of mitochondrial oxidative phosphorylation. This has been demonstrated repeatedly: specific examples will be presented in Chapters 8 and 13. Consequently, efforts have been made to analyze these processes by means of NET (Caplan and Essig, 1977). These efforts are, of course, based on the assumption that the linearity observed is *thermodynamic linearity*—a linear dependence of the flows on the thermodynamic forces that extrapolates smoothly to equilibrium. This is to be contrasted with what we may call *kinetic linearity*, a region of approximate linearity, narrow or broad, exhibited by some generally nonlinear kinetic scheme (examples are dealt with in Chapter 13). The observation of kinetic linearity in the behavior of a given far-from-equilibrium system may strongly suggest the presence of thermodynamic linearity, and it may even be associated with a kind of quasi-thermodynamic linearity (to be discussed below). This possibility can neither be excluded nor guaranteed on *a priori* grounds; in some systems it is subject to experimental test. If thermodynamic linearity is indeed present, then measurements in far-from-equilibrium states reflect the properties of the corresponding near-equilibrium states to which they can be extrapolated, even though such states may not be realizable in practice.

Thermodynamic linearity remote from equilibrium may arise for various reasons, not the least being an intrinsic linearity of the system itself. In the present chapter we put forward two possible interpretations of some of the observed linearities based on kinetic considerations, using models that are admittedly highly simplified. However, these interpretations suggest that circumstances can and do exist in which it is legitimate to apply NET to the analysis of linear behavior in a coupled system. If that is so, studies in amphibian epithelia, for example, permit evaluation by nondestructive means of the affinity of an oxidative reaction driving transepithelial active transport, thereby circumventing uncertainties concerning tissue compartmentalization, standard free energies, and activity coefficients.

6.1 Proper pathways: uncoupled processes

The key to the first approach resides in the fact that in general a set of phenomenological equations provides an incomplete description of the processes under study, since a given thermodynamic force may be induced in an infinite number of ways. Accordingly, as emphasized by Sauer (1973), the rate of a process cannot be predicted solely from the value of the force, without knowledge also of the "reference state." For example, the flow of a solute across a membrane depends not only on its electrochemical potential difference, but also on its thermodynamic state on both sides of the membrane. Thus it is not to be expected that the flow of an uncharged solute from region I to region II will be the same when the concentration $c^I = 2$ and $c^{II} = 1$ as when $c^I = 4$ and $c^{II} = 2$, despite the fact that in the two cases, disregarding activity coefficients, the chemical potential difference $\Delta\mu = RT \ln (c^I/c^{II})$ is the same. An analogous consideration applies to chemical reactions where, for example, for a first-order reaction $S \to P$ with a single substrate and a single product, doubling the substrate and product concentrations will double the reaction rate, despite constancy of the affinity.

The implications of this point of view may be made more precise by analyzing specific examples that show how constancy of phenomenological coefficients may be assured by appropriate constraint of the means employed to vary the force. We consider first the uncoupled flow of a solute across a membrane, which may be analyzed by means of the Nernst-Planck equation (again ignoring activity coefficients):

$$J = -uC \left(\frac{RT}{C} \frac{dC}{dx} + zF \frac{d\psi}{dx}\right) = -uC \left(\frac{d\bar{\mu}}{dx}\right) \tag{6.1}$$

where u represents mobility, C concentration, x position within the

Possible conditions for linearity 99

membrane, z charge, ψ electrical potential, and $\bar{\mu}$ electrochemical potential; R, T, and F have their usual significance. In the steady state, when J is everywhere constant,

$$\int_0^{\Delta x} \frac{J dx}{uC} = J \int \frac{dx}{uC} \equiv JR = -\Delta \bar{\mu} \qquad (6.2)$$

so that

$$JR = X \quad \text{or} \quad J = LX \qquad (6.3)$$

Here, since $X \equiv -\Delta \bar{\mu}$ is the appropriate conjugate force for J, we have a solution in the form of the equations of NET. This solution is of limited value, however, without knowledge of the nature of the phenomenological coefficient L (or R). This will, of course, reflect the nature of the membrane, but also the means employed to vary the force X. Thus if a homogeneous thin membrane is exposed at each surface to the same concentration c of the species of interest, flow being induced solely by an electrical potential difference, it is seen that L is constant with variation of X, being given by

$$L = \frac{\beta u c}{\Delta x} \qquad (\Delta c = 0) \qquad (6.4)$$

where β represents the solvent-membrane partition coefficient. If, on the other hand, X consists only of the chemical potential difference, determined by the bath solute concentrations,

$$L = \frac{\beta u \bar{c}}{\Delta x} \qquad (\Delta \psi = 0) \qquad (6.5)$$

where $\bar{c}$ represents the logarithmic mean bath concentration, $\Delta c / \Delta \ln c$. In this case L will in general vary with perturbation of Δc, but if some value $\bar{c}$ is selected and the concentrations are then constrained to the locus $\Delta c = (\bar{c}) \Delta \ln c$, again L will be invariant on perturbation of X (Fig. 6.1a). The reader will recall encountering the logarithmic mean concentration in Chapter 3, where it plays an essential role in the linearization introduced by Kedem and Katchalsky. If X is varied by perturbing both the bath concentrations

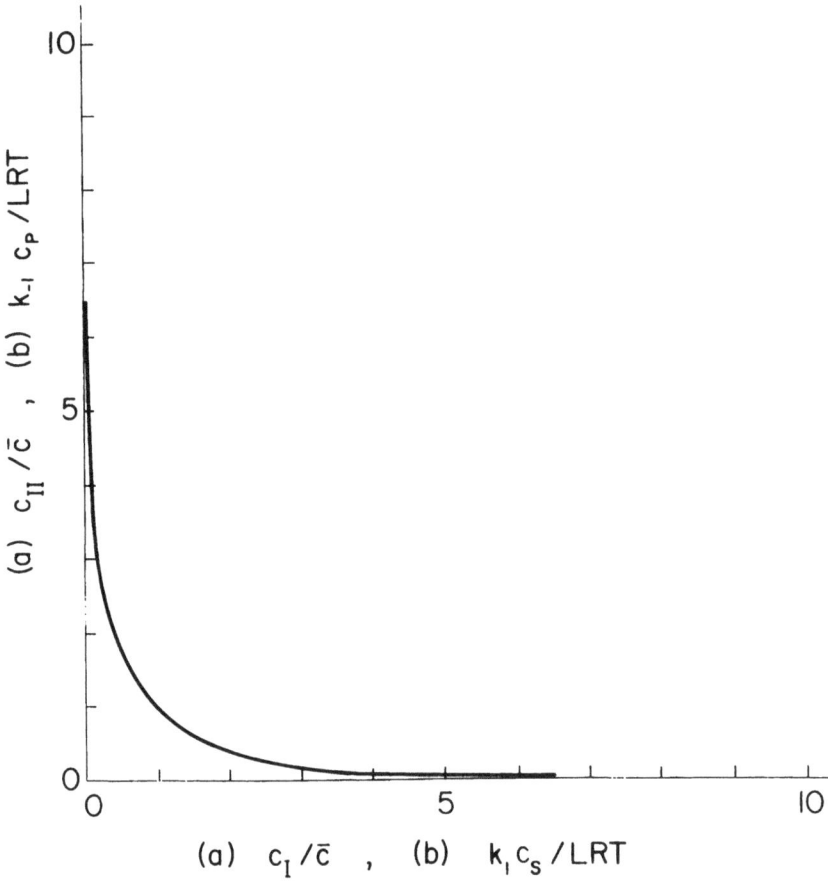

Fig. 6.1. Loci of constant L (proper pathways). (a) Solute concentrations for linearity of flow induced by a chemical potential difference across a homogeneous membrane. For any given logarithmic mean bath concentration $\bar{c}$, restricting the bath concentrations c^I and c^{II} to the indicated locus will assure that $J = LX$, where L is given by Eq. (6.5). (b) Reactant and product concentrations for linearity of reaction rate for the first-order reaction of Eq. (6.7). For any given values of k_1, k_{-1}, and $L' = RTL$, restricting c_S and c_P to the indicated locus will assure that $J = LA$. (Generalization to more complex cases is straightforward.)

Possible conditions for linearity

and $\Delta\psi$, L becomes more complex. For a membrane with a constant electrical field, as is commonly invoked in the Goldman-Hodgkin-Katz treatment,

$$L = \frac{\beta u c^{II}}{\Delta x} \cdot \frac{(-zF\Delta\psi/RT)(e^{X/RT} - 1)}{(e^{-zF\Delta\psi/RT} - 1)(X/RT)} \tag{6.6}$$

Here again it is simple to maintain L constant, since in principle at each value of X, two undetermined parameters exist that will be given by solving two simultaneous equations. For example, suppose L is evaluated in some stationary state according to Eq. (6.3), by measuring J and X. It may be helpful for this consideration to rewrite Eq. (6.6) in the following way:

$$\frac{\beta u c^{II}}{\Delta x} \cdot \left(\frac{-zF\Delta\psi/RT}{e^{-zF\Delta\psi/RT} - 1}\right) = L\left(\frac{X/RT}{e^{X/RT} - 1}\right) \tag{6.6a}$$

Then for any other new value of X, the right side of Eq. (6.6a) is fully determined. If a value is chosen for $\Delta\psi$, Eq. (6.6a) gives the appropriate value of c^{II}, while the expression for $\Delta\tilde{\mu}$ gives the requisite concentration ratio c^{II}/c^{I}, and hence the absolute values of both concentrations are obtained. In the general case, of course, it may often be impossible to predict the appropriate constraint on X on *a priori* grounds, but here experiment may well define the locus which will provide a single-valued L that is characteristic of the system under study.

Similarly for chemical reactions, given adequate knowledge of the kinetics of the system, it is possible to define conditions for constancy of L. For example, for the above-mentioned first-order reaction, the rate is given by

$$J = k_1 c_S - k_{-1} c_P = k_{-1} c_P (e^{A/RT} - 1) \tag{6.7}$$

Suppose now that in a given steady state far from equilibrium, we choose to describe the reaction by the relation

$$J = LA = L'A/RT \tag{6.8}$$

where $L' = RTL$, and proceed to evaluate L' (by measuring J and A).

Then

$$k_{-1}c_P = L' \left(\frac{A/RT}{e^{A/RT} - 1}\right) \quad (6.9)$$

(compare Eq. 6.6a; note that if u is replaced by the diffusion coefficient $RTu = D$, L becomes L'). It is seen that in other stationary states corresponding to different values of A, the same value of L' will continue to describe the reaction rate according to Eq. (6.8), providing the concentrations are chosen appropriately. Again, two undetermined parameters are obtained by solving two simultaneous equations. In this case Eq. (6.9) determines c_P at any specified value of A, and the expression for A determines the concentration ratio c_P/c_S. This procedure therefore maintains constancy of L by restricting c_P and c_S to an appropriate locus. (See Fig. 6.1b, and note that along this locus $c_S + c_P$ is not conserved.) An obvious conclusion is that as one approaches equilibrium under this restriction, that is, as A tends to zero, $k_{-1}c_P$ tends to the value L'. It is readily shown that this procedure may be used equally well in more complex reactions. With the exception of two species (substrate and product), all concentrations may be fixed arbitrarily; the remaining two concentrations which enable Eq. (6.8) to be satisfied at any given value of A, when L' has been predetermined, are obtained as before. For example, consider a reaction involving multiple substrates and products

$$\nu_1 S_1 + \nu_2 S_2 + \ldots + \nu_j S_j \to \nu_{j+1} P_{j+1} + \ldots + \nu_n P_n$$

where the ν's are stoichiometric coefficients. Then $J = k_1 \pi_S - k_{-1} \pi_P = LA$, where π_S is the product $c_1^{\nu_1} c_2^{\nu_2} \ldots$ for the substrate concentrations, and π_P the corresponding product for the product concentrations. The analogy with Eqs. (6.7)–(6.9) is complete.

Henceforth we shall designate loci in state space (such as those in Fig. 6.1) which for any reason give rise to thermodynamic linearity as "proper pathways."

6.2 Proper pathways: coupled processes

Although the artificiality of the means employed above to vary the forces may appear to make some of the examples irrelevant for practical experiments, their consideration facilitates understanding of the treatment of coupled flows presented below. It is essential to appreciate that we are *not* seeking an algorithm which will impart linearity, but rather showing that linearity is possible and that when it occurs for whatever reason, it may be of great help to the experimenter. Thus, as will be discussed, even kinetic linearity may be accompanied by near-reciprocity.

The relationship between flows and forces for two coupled processes can be written *a priori* as

$$J_1 = L_{11}X_1 + L_{12}X_2$$

$$J_2 = L_{21}X_1 + L_{22}X_2 \qquad (6.10)$$

where it is required only that (1) at equilibrium both flows are zero, and (2) each flow depends on both forces. Since we have not yet set limits on the means employed to vary X_1 and X_2, it is not to be expected that the flows will be single-valued for any given values of the forces, nor that any L coefficient need be constant. We can extend the concepts discussed above for uncoupled flows by using the formalism introduced in Chapter 5. The kinetic models treated there have considerable generality, presenting a level of detail somewhere between a molecular description and a phenomenological description, and in fact encompassing both. Figure 6.2 represents a hypothetical model of this kind in which metabolism of ATP could result in the transport of a sodium ion across a membrane against an electrochemical potential gradient. (We emphasize that this simple model is presented only to facilitate examination of the general principles involved.) In contrast to the case of overall observable fluxes, the rate of net flux around any component cycle will be determined by a single composite rate coefficient. From Eqs. (5.12) and (5.18)–(5.24), the steady-state cycle fluxes are seen to take the form:[1]

$$J_a = a(e^{(X_1+X_2)/RT} - 1) \qquad (6.11)$$

104 Bioenergetics and linear nonequilibrium thermodynamics

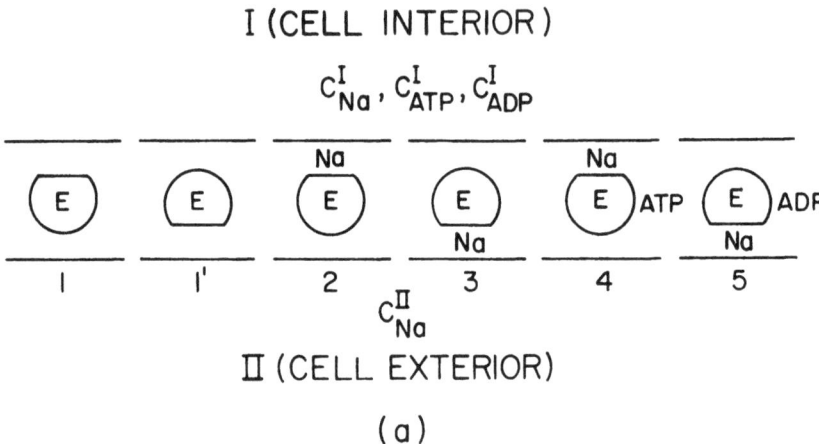

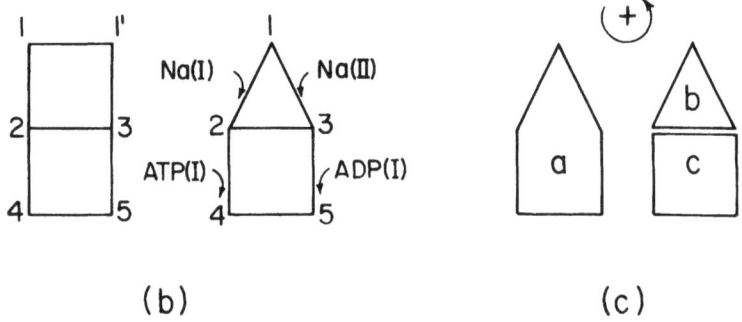

Fig. 6.2. Illustrative model for active sodium transport. (a) Hypothetical scheme for coupled transport of sodium and metabolism of ATP. E is a transporting enzyme such as Na,K ATPase; for simplicity we omit consideration of an associated cyclic flux of potassium, although this may well be necessary in a more realistic model. Six states of the enzyme are indicated. When combined with Na, the enzyme can interact with intracellular ATP and transport Na against an electrochemical potential gradient from the cell interior to the exterior, simultaneously converting ATP to ADP and P_i. P_i is assumed constant inside the cell and is not explicitly considered. Two sources of uncoupling are incorporated in this model, namely back flux of sodium without net generation of ATP and splitting of ATP in the absence of transport. In the toad urinary bladder, although back transport through the active pathway has been demonstrated (Chen and Walser, 1975; Wolff and Essig, 1977; Dawson and Al-Awqati, 1978), back leakage into the cell independent of metabolism is not demonstrable (Canessa, Labarca, and Leaf, 1976). Thus

$$J_b = b(e^{X_1/RT} - 1) \tag{6.12}$$

$$J_c = c(e^{X_2/RT} - 1) \tag{6.13}$$

where X_1 and X_2 represent $-\Delta\bar{\mu}_{Na}$ and A, respectively. The quantities a, b, and c are functions of individual rate constants and external concentrations, and hence change on manipulation of the forces. It will be observed that these equations are analogous to Eq. (6.7). (Note that, in general, electrostatic effects appear not only in the thermodynamic forces but also in the rate constants. An extension of this model incorporating electrostatic effects explicitly will be considered below; however, so far as the present considerations are concerned, no new principles emerge.)

The far-from-equilibrium steady-state fluxes conjugated to X_1 and X_2 are then given by[2]

$$\begin{aligned}J_1 = J_b + J_a &= b(e^{X_1/RT} - 1) \\ &+ a(e^{(X_1+X_2)/RT} - 1)\end{aligned} \tag{6.14}$$

$$\begin{aligned}J_2 = J_a + J_c &= a(e^{(X_1+X_2)/RT} - 1) \\ &+ c(e^{X_2/RT} - 1)\end{aligned} \tag{6.15}$$

As usual, the above relations can be expanded to give

$$\begin{aligned}J_1 = (b + a)(e^{X_1/RT} - 1) &+ a(e^{X_2/RT} - 1) \\ &+ a(e^{X_1/RT} - 1)(e^{X_2/RT} - 1)\end{aligned} \tag{6.16}$$

$$\begin{aligned}J_2 = a(e^{X_1/RT} - 1) &+ (a + c)(e^{X_2/RT} - 1) \\ &+ a(e^{X_1/RT} - 1)(e^{X_2/RT} - 1)\end{aligned} \tag{6.17}$$

---

the rate constant for the transition from state 3 to state 2 may be very low. (b) Diagrams corresponding to (a), showing possible transitions between the states. Because of rapid transitions between conformations of the free enzyme, states 1 and 1' are combined, leading to the contracted diagram at the right. The arrows represent the direction of the transition of the indicated binding. (c) Cycles corresponding to the contracted diagram in (b). Cycle a represents complete coupling between transport and metabolism, while cycles b and c represent the sources of uncoupling. All three cycles occur concurrently.

Consider a reference steady state far from equilibrium, characterized by given values of X_1 and X_2. We wish to investigate the existence of proper pathways in the neighborhood of this state along which the components of X_1 and X_2 are varied in such a manner as to result in linearity of the flows in the forces.[3] In analogy to the treatment above of uncoupled solute transport and chemical reaction, we express the cycle fluxes in the form of the equations of NET. Accordingly, we introduce the phenomenological coefficients L_a, L_b, and L_c. Dealing first with the sodium flux, we write the equation used to evaluate L_b:

$$J_b = L_b X_1 = L_b' X_1 / RT \qquad (6.18)$$

We now require that all variations of X_1 refer to a proper pathway specified by the procedure described above so as to maintain constancy of L_b. Turning next to the cycle flux J_a, we write the equation used to evaluate L_a:

$$J_a = L_a(X_1 + X_2) = L_a'(X_1 + X_2)/RT \qquad (6.19)$$

We now require that all variations in X_1 be along the proper pathway already specified so as to maintain constancy of L_b (Eq. 6.18). This leaves two undetermined parameters in X_2 to be manipulated so as to maintain constancy of L_a. Variations of X_1 and X_2 must be limited to a range sufficiently small that the values of L_a and L_b do not deviate appreciably from those in the reference steady state.

Finally we consider the equation used to evaluate L_c:

$$J_c = L_c X_2 = L_c' X_2 / RT \qquad (6.20)$$

It cannot, of course, be assumed *a priori* that the proper pathway for variation of X_2 so as to maintain constancy of L_c in Eq. (6.20) will always correspond to its proper pathway so as to maintain constancy of L_a in Eq. (6.19). However, if this is the case, we have

$$J_1 = (L_b' + L_a')X_1/RT + L_a' X_2/RT \qquad \text{(proper pathways)} \qquad (6.21)$$

$$J_2 = L_a' X_1 / RT + (L_a' + L_c')X_2/RT \qquad \text{(proper pathways)} \qquad (6.22)$$

The linearity of both J_1 and J_2 in X_1 observed in the experimental studies of active transport in epithelia suggests that it is plausible to assume the correspondence of proper pathways, since transport and metabolism in these tissues appear not to be completely coupled (Caplan and Essig, 1977), and thus correspondence is not automatic. In any case the equations will be a good approximation in highly coupled systems, irrespective of the uniqueness of the proper pathway for X_2, since the cycle fluxes J_b and J_c play an increasingly minor role as the degree of coupling increases. It should be noted that Eqs. (6.21) and (6.22) correspond to the near-equilibrium range of Eqs. (6.14) and (6.15), where $a \simeq L'_a$, $b \simeq L'_b$, and $c \simeq L'_c$. In analogy with Eq. (6.10), Eqs. (6.21) and (6.22) may be rewritten more compactly, using bars to indicate that variations along proper pathways only are considered:

$$J_1 = \bar{L}_{11}\bar{X}_1 + \bar{L}_{12}\bar{X}_2 \qquad (6.23)$$

$$J_2 = \bar{L}_{21}\bar{X}_1 + \bar{L}_{22}\bar{X}_2 \qquad (6.24)$$

Here the $\bar{L}$'s indicate constant coefficients, and each flow is now a single-valued function of each force. As can be seen from Eqs. (6.21) and (6.22), $\bar{L}_{12} = \bar{L}_{21}$.

The feasibility of the above approach may be evaluated by more detailed analysis of the model of Fig. 6.2. As shown in Chapter 5, the explicit form of the kinetic coefficients in Eqs. (6.11)–(6.13) is given by $\kappa = N\Pi_{\kappa-}\Sigma_{\kappa}/\Sigma$ ($\kappa = a, b, c$) where N is the (large) number of transport molecules in the membrane, $\Pi_{\kappa-}$ is the product of rate coefficients in cycle κ taken in the negative or clockwise direction, Σ_{κ} is a sum of products of rate constants of processes feeding into cycle κ, and Σ is a sum over all possible states of products of rate constants of processes feeding into a given state. In the model under consideration, all the rate constants α_{ij} (where α_{ij} refers to the transition $i \rightarrow j$) are true first-order rate constants except $\alpha_{12} = c^I_{Na}\alpha^*_{12}$, $\alpha_{13} = c^{II}_{Na}\alpha^*_{13}$, $\alpha_{24} = c^I_{ATP}\alpha^*_{24}$, and $\alpha_{35} = c^I_{ADP}\alpha^*_{35}$, where the α^*_{ij}'s are second-order rate constants. In these terms the coefficient c is given by

$$c = \frac{N}{\Sigma}(\alpha_{42}\alpha_{54}\alpha^*_{35}\alpha_{23})\alpha^*_{12}\left(\frac{c^I_{Na}}{c^{II}_{Na}} + \frac{\alpha^*_{13}}{\alpha^*_{12}}\right)c^I_{ADP}c^{II}_{Na}$$

$$= \left(\frac{C}{\Sigma}\right)(e^{X_1/RT} + \gamma)c^I_{ADP}c^{II}_{Na} \qquad \left(\gamma = \frac{\alpha^*_{13}}{\alpha^*_{12}}\right) \qquad (6.25)$$

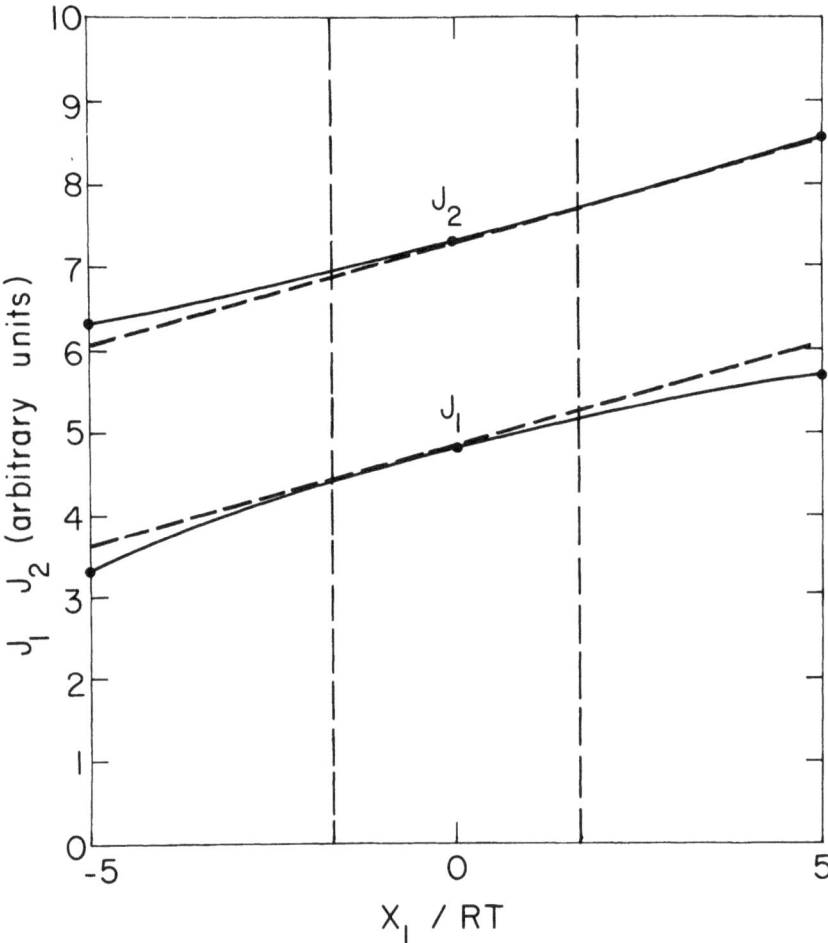

Fig. 6.3. Flow-force relationships for the model of Fig. 6.2. The values of the kinetic coefficients have been chosen in the relative proportions $A = 10\,\Sigma$, $B = 0.1\,\Sigma$, $C = 5\,\Sigma$ so as to give a value of q of approximately 0.8 (more exactly, $q = 0.82$). The other parameter values used are as follows: $X_2 = 20\,RT$, $c_{Na}^{II} = 0.1$ M, $c_{ADP}^{I} = 1$ mM, $\gamma = 0.005$. The dashed lines represent predictions calculated from the phenomenological equations (Eqs. 6.21 and 6.22), using reference steady-state values of the phenomenological coefficients obtained from Eqs. (6.26), (6.29), and (6.30). The solid lines represent the behavior based on the kinetic equations (Eqs. 6.14 and 6.15), using values of a, b, and c calculated from Eqs. (6.27), (6.28), and (6.25), respectively. In evaluating a, b, and c for stationary states perturbed from the reference steady state, c_{Na}^{II} was evaluated from Eq. (6.30), and a value of $c_{ADP}^{I} c_{Na}^{II}$ was chosen

Introducing Eqs. (6.13) and (6.20) into Eq. (6.25), we obtain the constraint for a proper pathway,

$$c_{ADP}^{I} c_{Na}^{II} = \frac{L_c'}{(C/\Sigma)(e^{X_1/RT} + \gamma)} \cdot \frac{X_2/RT}{e^{X_2/RT} - 1} \quad (6.26)$$

Analogously, from consideration of cycles a and b,

$$a = \frac{N}{\Sigma}(\alpha_{21}\alpha_{42}\alpha_{54}\alpha_{35}^*\alpha_{13}^*)c_{ADP}^{I} c_{Na}^{II} = \left(\frac{A}{\Sigma}\right) c_{ADP}^{I} c_{Na}^{II} \quad (6.27)$$

$$b = \frac{N}{\Sigma}(\alpha_{21}\alpha_{32}\alpha_{13}^*)(\alpha_{45}\alpha_{53} + \alpha_{54}\alpha_{42} + \alpha_{42}\alpha_{53})c_{Na}^{II}$$

$$= \left(\frac{B}{\Sigma}\right) c_{Na}^{II} \quad (6.28)$$

and hence, using Eqs. (6.11) and (6.19), and (6.12) and (6.18), respectively,

$$c_{ADP}^{I} c_{Na}^{II} = \frac{L_a'}{(A/\Sigma)} \cdot \frac{(X_1 + X_2)/RT}{e^{(X_1+X_2)/RT} - 1} \quad (6.29)$$

$$c_{Na}^{II} = \frac{L_b'}{(B/\Sigma)} \cdot \frac{X_1/RT}{e^{X_1/RT} - 1} \quad (6.30)$$

---

between the values given by Eqs. (6.26) and (6.29). No attempt was made to optimize this choice with respect to linearity, although in principle this could be done. In this calculation we have considered Σ constant with variation of X_1, since the variations of those of its 55 terms which are concentration-dependent tend to be compensatory over a sufficiently small range. Over a wider range, explicit consideration of the variation in Σ is necessary. To the extent that Σ is not constant, this has no influence on the ratio between values of $c_{ADP}^{I} c_{Na}^{II}$ calculated from Eqs. (6.26) and (6.29), and therefore no influence on the deviation of this ratio from unity, which results in the deviation of the solid from the dashed curves. No explicit account has been taken here of the effect of electric fields on the rate constants (see text). However, the total range of perturbations of X_1 studied corresponds to $\sim \pm 125$ mV and the contracted range (between dashed vertical lines) to $\sim \pm 40$ mV. The contracted range corresponds more closely to current experimental practice than the full range.

For any given reference steady state, L'_a, L'_b, and L'_c are well defined and determined by Eqs. (6.18)–(6.20). In principle, from a knowledge of all the rate constants one would know the values of A, B, C, γ, and Σ, and hence, as shown above, it is possible to evaluate the concentrations c^{II}_{Na} and c^{I}_{ADP} associated with perturbations from the reference steady state along proper pathways. The question then arises whether the concentrations evaluated from Eqs. (6.26), (6.29), and (6.30) are consistent, as is necessary for cycles a, b, and c to follow proper pathways simultaneously. Specifically, values of the product $c^{I}_{ADP} c^{II}_{Na}$ evaluated from Eqs (6.26) and (6.29), that is, from cycles c and a, must agree when the forces are varied. The range of agreement has been tested in the neighborhood of a reference steady state at level flow ($X_1 = 0$) where X_2 has been chosen to have the value 20 RT (about 12–13 kcal/mole ATP) (Essig and Caplan, 1981). The concentrations chosen for the reference steady state were $c^{II}_{Na} = 0.1$ M, $c^{I}_{ADP} = 1$ mM. The quantity γ was assigned the value 0.005 on grounds that the apparent affinity of the sodium pump for external sodium is about 160 times less than the apparent affinity for internal sodium (Garay and Garrahan, 1973). Computations were made for several degrees of coupling q of the system. If indeed the values of the product $c^{I}_{ADP} c^{II}_{NA}$ calculated from Eqs. (6.26) and (6.29) agreed perfectly, the calculated flows evaluated from the phenomenological equations (Eqs. 6.21 and 6.22) would be identical with those evaluated from the kinetic equations (Eqs. 6.14 and 6.15); if not, the flows would diverge. Figure 6.3 shows the result obtained for $q = 0.82$, a value close to those obtained experimentally (Caplan and Essig, 1977). The dashed lines and the solid lines are derived from the phenomenological equations and the kinetic equations, respectively. Deviations from linearity were found to increase at $q = 0.7$ and to decrease slightly at $q = 0.9$. The range of linearity at all values of q tested was substantially less extensive in the neighborhood of static head ($J_1 = 0$).

For highly coupled systems, the case discussed above may essentially reduce to the case of a single uncoupled flow. Evidently, for processes which are coupled completely, linear dependencies on conjugate forces are necessarily associated with linear dependencies on nonconjugate forces, as can be seen if we consider cycle a alone, assuming cycles b and c to play a negligible role. On relaxation of the completeness of coupling, we may expect departures from linearity not to be abrupt. Further studies will be necessary to analyze the extent to which completeness of coupling may be relaxed without inducing appreciable nonlinearity. However, in the neighborhood of static head ($X_1 \simeq -X_2$), one would clearly expect linearity when q is very close to unity.

6.3 The multidimensional inflection point

Rothschild et al. (1980) have demonstrated the existence in certain circumstances of a multidimensional inflection point (MIP) in the force-flow space of a system of enzyme-mediated reactions, in other words, a steady state in the vicinity of which linear relationships between steady-state flows and their conjugate thermodynamic forces occur over a considerable range. This range of kinetic linearity may be very far from equilibrium. A set of sufficient conditions can be used to test any given first-order or pseudo-first-order discrete-state kinetic mechanism for the possible presence of an MIP; as pointed out by Rothschild et al., these conditions are not overly restrictive. It was also found that while, in general, reciprocity is not obtained at the MIP, it can exist in specific cases, the most obvious being coincidence of the MIP with equilibrium. The more interesting possibility of reciprocity away from equilibrium was exemplified by a four-state model of two coupled processes, such as the facilitated exchange of two similar ions across a membrane: for a particular (symmetrical) assignment of the rate constants, which in addition imposed a severe constraint on the reactant concentrations, the two fluxes became equal at the MIP. This situation was found to be characterized by reciprocity. We shall interpret this observation below; for the moment it should be noted that in terms of the Hill diagram for the system, equality of the fluxes at a particular steady state in a sense simulates complete coupling, since the net mean "leakage" transition flux must be zero. While a proof of the existence of the MIP is beyond the scope of our present considerations, it is the purpose of this section to show that all completely coupled systems exhibit reciprocity at the MIP, however far from equilibrium, and highly coupled systems such as are often encountered in biology approximate reciprocity rather closely (Caplan, 1981). This is a unique characteristic of the MIP, since it is well established that both linearity and reciprocity of chemical reactions, even completely coupled systems of reactions, are invariably restricted to the immediate vicinity of equilibrium. Thus the Jacobian matrix characterizing perturbations about a steady state away from equilibrium is not symmetrical (Oster, Perelson, and Katchalsky, 1973; Oster and Perelson, 1974; Mikulecky, 1977; Bunow, 1978), and correspondingly the excess entropy production (Glansdorff and Prigogine, 1971; Nicolis and Prigogine, 1977; Stucki, 1978) is not necessarily positive-definite. In some systems, however, as has been discussed above, an exception to this may be found or constructed that rests on the nonexclusive dependence of a given thermodynamic force on any single set of state parameters. In these cases perturbations are constrained to occur along proper pathways derived by appropriate manipulation of all the components of the forces.

As before, we consider here an ensemble of N identical enzyme molecules or molecular complexes without cooperativity (in most systems of interest these will be membrane proteins). This macromolecular array couples together M processes (reactions, vectorial flows). The Hill diagram for such a system may take various forms, but in general it consists of a set of cycles and subcycles corresponding to cyclic fluxes that relate to the different processes taken individually and in all possible combinations in which coupling can occur. For example, the diagram depicted in Fig. 6.4, where $M = 3$, exemplifies the central features to be demonstrated. Denoting the cycle flux in cycle κ as J_κ ($\kappa = a, b, \ldots, h$), the steady-state thermodynamic flows are given by (see Chapter 5):

$$J_1 = J_a + J_b + J_f$$

$$J_2 = J_a + J_b + J_c + J_g$$

$$J_3 = J_a + J_c + J_h \tag{6.31}$$

If the mechanistic (reaction-mechanism-determined) stoichiometric ratios in each cycle in which coupling occurs are unity, which is the simplest case, Eqs. (6.31) yield

$$J_1 = a(e^{(X_1+X_2+X_3)/RT} - 1) + b(e^{(X_1+X_2)/RT} - 1)$$
$$+ f(e^{X_1/RT} - 1) \tag{6.32}$$

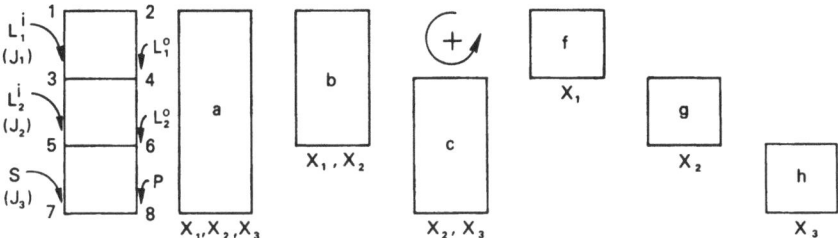

Fig. 6.4. Hill diagram and component cycles for a representative coupled three-flow membrane system. The transporting enzyme has eight states, with transitions between them as shown. The steady-state thermodynamic fluxes (J_1, J_2, J_3) refer to the processes indicated. J_1 and J_2 represent transport. L_1 and L_2 are ligands binding or unbinding at the inner (i) or outer (o) membrane surface; the direction of the binding transition only is specified. J_3 represents a reaction involving attachment of substrate (S) and release of product (P) at one surface, usually the inner. The thermodynamic forces (X_1, X_2, X_3) appear below the cycles in which they act. The cycles are taken to be positive in the counterclockwise direction.

and corresponding expressions for J_2 and J_3, where X_1 is the thermodynamic force conjugate to J_1, and so on. Here the kinetic coefficient κ preceding the exponential term associated with cycle κ includes the multiplicative factor N/Σ (usually written explicitly), where Σ is the sum of directional diagrams of all states. Each κ (that is, $a, b, \ldots, h$) is of course a function of concentrations as well as rate constants; the detailed structure of the function for the present model will be considered below.

We neglect electrical effects to begin with and assume that the system satisfies the criteria of Rothschild et al. (1980) for the existence of an MIP, namely: (1) each reactant whose concentration (activity) is to be varied influences the transition rates for leaving one state only, (2) the kinetics of the transition involving the given reactant are of fixed order with respect to that reactant, and (3) for each possible combination of reactants whose concentrations are varied, at least one directional diagram is present containing only that combination and no others. The first criterion immediately eliminates from consideration autocatalytic systems such as the well-known Brusselator (Glansdorff and Prigogine, 1971; Nicolis and Prigogine, 1977), but for many biological energy transducers these criteria may well be satisfied. Generalizing Eqs. (6.31), we focus for convenience on the *i*th and *j*th coupled processes, which are now presumed to be related by a mechanistic stoichiometry n (for simplicity these processes, considered individually, are taken to be unimolecular with respect to the concentrations varied). We expand these two flows as functions of their conjugate forces in a Taylor series about some reference steady state, assuming all other forces to be fixed. This gives, to first-order terms, the finite differences

$$\delta J_i = \left[\frac{\partial}{\partial X_i} n \sum_{\kappa \in \{\kappa\}_i} J_\kappa\right] \delta X_i$$

$$+ \left[\frac{\partial}{\partial X_j} n \sum_{\kappa \in \{\kappa\}_i} J_\kappa\right] \delta X_j + \ldots$$

$$\delta J_j = \left[\frac{\partial}{\partial X_i} \sum_{\kappa \in \{\kappa\}_j} J_\kappa\right] \delta X_i$$

$$+ \left[\frac{\partial}{\partial X_j} \sum_{\kappa \in \{\kappa\}_j} J_\kappa\right] \delta X_j + \ldots \quad (6.33)$$

where $\{\kappa\}_i$ and $\{\kappa\}_j$ are the sets of cycles associated with J_i and J_j, respectively, and the coefficients [ ] are evaluated at the reference state. Rothschild et al. have shown that expansion of the flows in a Taylor series about an MIP gives expressions linear in $\ln(c_i)$ and $\ln(c_j)$ up to third order in either one if c_i and c_j are the reactant concentrations varied. Thus if Eqs. (6.33) represent an expansion about an MIP, with $X_i(c_i)$ and $X_j(c_j)$ varied through c_i and c_j while all other concentrations remain at the values defining the MIP, they correspond to the Rothschild et al. expansion and are linear within the same degree of approximation.[4] An attractive feature of equations such as Eqs. (6.33), from our point of view, is that the proper conjugate forces appear explicitly. Note that it is not necessary that the reference steady state be an inflection point except with respect to c_i and c_j.

Consider the two cross coefficients in Eqs. (6.33). Examination of Fig. 6.4 and Eqs. (6.31) and (6.32) makes it clear that X_i acts in all cycles in the set $\{\kappa\}_i$, and hence X_j influences J_i *directly* (that is, as part of one or more of the forces thermodynamically conjugate to the cycle fluxes) through a subset of cycles $\{\kappa\}_{ij}$ in which both X_i and X_j act.

Conversely, X_i influences J_j directly (in the above sense) through the subset of cycles $\{\kappa\}_{ji} \subset \{\kappa\}_j$, where, clearly, $\{\kappa\}_{ji} \equiv \{\kappa\}_{ij}$. Consequences of this type of symmetry at an arbitrary steady state have been commented on previously (Hill, 1977). We now show that for any cycle κ in the subset $\{\kappa\}_{ij}$, the following relation is satisfied at the MIP:

$$\frac{\partial}{\partial X_j} n[\kappa(e^{(nX_i + X_j + \Sigma X_k)/RT} - 1)] = \frac{\partial}{\partial X_i}[\kappa(e^{(nX_i + X_j + \Sigma X_k)/RT} - 1)] \quad (6.34)$$

where X_k ($k \neq i, j$) denotes any other (constant) force that may be acting in the cycle. To prove this, it is only necessary to show that at the MIP,

$$n(\partial\kappa/\partial X_j) = (\partial\kappa/\partial X_i) \quad (6.35)$$

As discussed previously, the structure of κ is given by $\kappa = N\Pi_{\kappa^-}\Sigma_\kappa/\Sigma$, where Π_{κ^-} is the negative (clockwise) product of rate constants around cycle κ, and Σ_κ is a sum of appendages feeding into the cycle analogous to a sum of directional diagrams feeding into a state. Since X_i and X_j both act in cycle κ, Σ_κ must be constant by criterion (1) for the existence of an MIP. With regard to Π_{κ^-} we consider three cases:

Case 1: the variables c_i and c_j are both substrate concentrations, that is, both govern binding steps (in the positive direction of cycling) at some point in the cycle. Then Π_{κ^-} is constant. Since the system satisfies criteria (2) and (3), Σ has the structure (Rothschild et al., 1980):

$$\Sigma = f + f_1 c_i^n + f_2 c_j + f_{12} c_i^n c_j \quad (6.36)$$

where the f's are constants. Equation (6.36) may be rewritten

$$\Sigma = f + f_1' e^{nX_i/RT} + f_2' e^{X_j/RT} + f_{12}' e^{nX_i/RT} e^{X_j/RT} \quad (6.37)$$

Therefore,

$$n(\partial\kappa/\partial X_j) = -(n\kappa/RT)[(f_2 c_j + f_{12} c_i^n c_j)/\Sigma]$$
$$= -(n\kappa/RT)(\sigma_j/\Sigma) \quad (6.38)$$

$$(\partial \kappa / \partial X_i) = -(n\kappa/RT)[(f_1 c_i^n + f_{12} c_i^n c_j)/\Sigma]$$
$$= -(n\kappa/RT)(\sigma_i/\Sigma) \qquad (6.39)$$

where σ_i and σ_j represent the contributions to Σ of terms involving c_i and c_j, respectively. But at the MIP, c_i and c_j are constrained by the relations $c_i^n = \sqrt{ff_2/f_1 f_{12}}$ and $c_j = \sqrt{ff_1/f_2 f_{12}}$, and hence $f_1 c_i^n = f_2 c_j$ and $f = f_{12} c_i^n c_j$ (Rothschild et al., 1980). From this it is readily seen that at the MIP,

$$\sigma_i = \sigma_j = \Sigma/2 \qquad (6.40)$$

and Eq. (6.35) follows, that is:

$$n(\partial \kappa / \partial X_j) = (\partial \kappa / \partial X_i) = -n\kappa/2RT \qquad (6.41)$$

Case 2: the variables c_i and c_j are both product concentrations, that is, both are external concentrations at unbinding steps (in the positive direction of cycling) at some point in the cycle. Then $\Pi_{\kappa^-} = \Pi_{\kappa^-}^* c_i^n c_j$, where $\Pi_{\kappa^-}^*$ is constant. Dividing the expression for κ both above and below by $c_i^n c_j$, and proceeding essentially as before, Eq. (6.41) is again found to hold at the MIP.

Case 3: only one of the variables c_i and c_j is a product concentration in the above sense. If it is c_i, we have $\Pi_{\kappa^-} = \Pi_{\kappa^-}^* c_i^n$, where again $\Pi_{\kappa^-}^*$ is constant. Dividing the expression for κ both above and below by c_i^n, proceeding as before, and applying the constraints of the MIP, Eq. (6.41) is obtained.

If J_i and J_j are completely coupled, $\{\kappa\}_{ij}$ is the only set of cycles to contribute terms to the cross coefficients of Eqs. (6.33). In this case the Jacobian of the system is symmetrical within the degree of approximation of linearity. If J_i and J_j are incompletely but nevertheless highly coupled, the terms contributed by $\{\kappa\}_{ij}$ dominate, and symmetry of the Jacobian remains a useful approximation, the precision of which depends on the degree of coupling. This is illustrated in the next section by means of a simple model that involves a charged ligand and hence requires the introduction of electrical terms. A constraint on any such model is that the electrical factors must enter in such a way as to satisfy the three criteria for existence of an MIP.

6.4 Linearity and symmetry in a model electrogenic ion pump

A simple model of active ion transport, having properties consistent with the existence of an MIP, is shown in Fig. 6.5. This is obviously an expanded version of the models given in Figs. 5.1 or 6.2 and might represent, for example, an ATP-driven proton pump, with $n = 3$ or 4 (depending on the system) and P_i at a constant level. Since similar models have already been described extensively, we outline only the relevant electrochemical characteristics. Consider first the full diagram on the left of Fig. 6.5. With regard to the transitions involving the charged ligand, we assign the entire effect of the phase-boundary potentials $(\psi_i - \psi_i')$ and $(\psi_o' - \psi_o)$ to the rates of un-

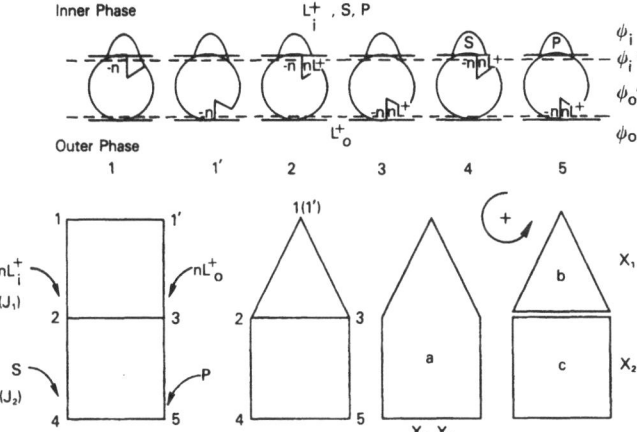

Fig. 6.5. Explicit model for an electrogenic cation pump. The univalent cation L^+ binds to an internal or external site on the transporting enzyme bearing a charge $-n$. Binding is assumed to occur in successive steps with transient intermediates which need not be specified (Hill, 1977). Six states of the enzyme are shown: each may represent the "reduction" of several states in sufficiently rapid equilibrium (Hill, 1977). Binding of the substrate S and release of the product P take place at the internal surface. The electrical potentials are ψ_i and ψ_o at the inner and outer bulk phases, respectively, and ψ_i' and ψ_o' at planes close to the inner and outer membrane surfaces that correspond, as indicated, to the positions of the binding sites. J_1 and J_2, the transport and reaction flows, refer to the processes shown, their conjugate forces being X_1 and X_2.

binding, taking binding to be essentially a diffusion-controlled process. This is a convenient simplifying approximation but also a reasonable one, since it is the rate constants for unbinding that depend primarily on the depths of the binding free energy wells from which L^+ must escape in order to return to either bath (Hill, 1977). Furthermore, we suppose binding and unbinding to occur at or very close to the surfaces of the membrane, so that the phase-boundary potentials are determined by the properties of the surfaces and may be assumed essentially constant. The magnitudes of the phase-boundary potentials do not affect the following argument, but for simplicity we will consider them equal and opposite, so that $(\psi'_i - \psi'_o) = (\psi_i - \psi_o) = \Delta\psi$. We therefore have the following relations for the rate constants α:

$$\alpha_{12} = \alpha_{12}^* c_{L_i^+}^n \qquad \alpha_{1'3} = \alpha_{1'3}^* c_{L_o^+}^n \qquad (6.42)$$

where α_{12}^* and $\alpha_{1'3}^*$ are $(n + 1)$th-order rate constants, and

$$\alpha_{21} = \alpha_{21}(0) e^{n\Delta\psi_b F/RT}$$

$$\alpha_{31'} = \alpha_{31'}(0) e^{n\Delta\psi_b F/RT} \qquad (6.43)$$

where F is the faraday constant (molar charge), $\Delta\psi_b = (\psi'_i - \psi_i) = (\psi'_o - \psi_o)$, and $\alpha_{21}(0)$ and $\alpha_{31'}(0)$ are the values of these rate constants in the absence of the relevant potential difference. If the diagram is reduced as shown in Fig. 6.5, assuming states 1 and 1' to be in rapid equilibrium, we may drop the primed subscripts in Eqs. (6.42) and (6.43), but α_{12}^* and α_{13}^* are no longer independent of electrostatic effects since they are now functions of $\alpha_{11'}$ and $\alpha_{1'1}$:

$$\alpha_{12}^* = \frac{\alpha_{12}^0 \alpha_{1'1}}{\alpha_{11'} + \alpha_{1'1}} \qquad \alpha_{13}^* = \frac{\alpha_{13}^0 \alpha_{11'}}{\alpha_{11'} + \alpha_{1'1}} \qquad (6.44)$$

where α_{12}^0 and α_{13}^0 represent the original values of α_{12}^* and $\alpha_{1'3}^*$ prior to reduction. Considering the equilibrium of the transition 11', we have

$$\frac{\alpha_{11'}(\Delta\psi)}{\alpha_{1'1}(\Delta\psi)} = \left[\frac{\alpha_{11'}(0)}{\alpha_{1'1}(0)}\right] e^{-n\Delta\psi F/RT} \qquad (6.45)$$

Hence,

$$\alpha_{12}^* = \alpha_{12}^0 \alpha_{1'1}(0) e^{n\Delta\psi F/RT}/D \qquad \alpha_{13}^* = \alpha_{13}^0 \alpha_{11'}(0)/D \qquad (6.46)$$

where

$$D = \alpha_{11'}(0) + \alpha_{1'1}(0) e^{n\Delta\psi F/RT} \qquad (6.47)$$

These rate constants are readily shown to be consistent with detailed balancing of cycles a and b at equilibrium.

If we calculate the kinetic coefficients κ ($\kappa = a, b, c$) for this model we find

$$a = (A/D\Sigma) c_P c_{L_i^+}^n$$

$$b = (B/D\Sigma) c_{L_i^+}^n$$

$$c = (C/D\Sigma) c_P c_{L_i^+}^n (e^{nX_1/RT} + \gamma) = c'(e^{nX_1/RT} + \gamma) \qquad (6.48)$$

where A, B, C, and γ are constants. The quantity γ, which we have seen before, is of interest. It is given by

$$\gamma = \frac{\alpha_{13}^0 \alpha_{11'}(0)}{\alpha_{12}^0 \alpha_{1'1}(0)} \qquad (6.49)$$

and thus is related to the ratio of external to internal binding constants for the ligand L^+. Examination of Eqs. (6.42) and (6.46) shows that α_{12} includes the factor $c_{L_i^+}^n e^{n\Delta\psi F/RT}/D$, and α_{13} includes the factor $c_{L_i^+}^n/D$. The quantities in these factors never appear in any other combination in directional diagrams. It is also easily verified that no directional diagram contains the product $\alpha_{12}\alpha_{13}$, so the product of the two factors cannot appear in Σ. If we choose $c_{L_i^+}^n$ and either c_S or c_P as variable parameters, criterion (3) for the existence of an MIP is satisfied, and we can write, for example,

$$\Sigma = f + f_1 c_{L_i^+}^n e^{n\Delta\psi F/RT}/D + f_2 c_S + f_{12} c_S c_{L_i^+}^n e^{n\Delta\psi F/RT}/D \qquad (6.50)$$

where $f = f' + f'' c_{L_i^+}^n/D$ and $f_2 = f_2' + f_2'' c_{L_i^+}^n/D$. This suggests the possibility of varying $e^{n\Delta\psi F/RT}$ rather than $c_{L_i^+}^n$ while still satisfying criterion (3). To verify this, consider the denominator $D\Sigma$ in Eqs. (6.48). We may conveniently define it as Σ', since after regrouping we find the required bilinear form:

$$\Sigma' \equiv D\Sigma = \phi + \phi_1 e^{n\Delta\psi F/RT} + \phi_2 c_S + \phi_{12} c_S e^{n\Delta\psi F/RT} \quad (6.51)$$

The numerators of the cycle fluxes corresponding to Eqs. (6.48) are equally readily seen to be linear functions of $e^{n\Delta\psi F/RT}$. Thus we have shown that for this model an MIP exists when one of the variable parameters is $\Delta\psi$, and the symmetry considerations discussed earlier apply. In effect, the quantity $e^{n\Delta\psi F/RT}$ plays the role of a concentration, but it should be noted that Σ' no longer contains the combination $c_{Lt}^n e^{n\Delta\psi F/RT}$ exclusively.

The flow equations for this model are

$$J_1 = na(e^{(nX_1+X_2)/RT} - 1) + nb(e^{nX_1/RT} - 1)$$

$$J_2 = a(e^{(nX_1+X_2)/RT} - 1) + c'(e^{nX_1/RT} + \gamma)(e^{X_2/RT} - 1) \quad (6.52)$$

which once again may be conveniently expanded in the form

$$J_1 = n(a + b)(e^{nX_1/RT} - 1) + na(e^{X_2/RT} - 1)$$
$$\quad + na(e^{nX_1/RT} - 1)(e^{X_2/RT} - 1)$$

$$J_2 = a(e^{nX_1/RT} - 1) + [a + c'(1 + \gamma)](e^{X_2/RT} - 1)$$
$$\quad + (a + c')(e^{nX_1/RT} - 1)(e^{X_2/RT} - 1) \quad (6.53)$$

Linearizing Eqs. (6.53) near equilibrium, we find for the degree of coupling

$$q \equiv \sqrt{\left(\frac{\partial J_1}{\partial J_2}\right)_{X_1} \left(\frac{\partial J_2}{\partial J_1}\right)_{X_2}} = \sqrt{\left(1 + \frac{B}{Ac_P}\right)\left(1 + \frac{C(1+\gamma)}{A}\right)} \quad (6.54)$$

and for the phenomenological stoichiometry

$$Z \equiv \sqrt{\frac{(\partial J_1/\partial J_2)_{X_1}}{(\partial J_2/\partial J_1)_{X_2}}} = n\sqrt{\frac{\left(1 + \dfrac{B}{Ac_P}\right)}{\left(1 + \dfrac{C(1+\gamma)}{A}\right)}} \quad (6.55)$$

where c_P may be replaced by Kc_S, K being the equilibrium constant of the reaction, if c_S is to be maintained constant. The equilibrium values of q and Z no longer have their usual significance when linearity breaks down, but in a sense they continue to characterize the system. For example, we have seen that in the linear range near equilibrium, static head (the stationary state with $J_1 = 0$) is given by $-X_1/X_2 = q/Z$. If we define a reduced phenomenological stoichiometry $\zeta = Z/n$ ($q \leq \zeta \leq 1/q$ from Eqs. 6.54 and 6.55), the static head relation is $-nX_1/X_2 = q/\zeta$. For comparison, the static head relation in the nonlinear range is, from Eqs. (6.52), (6.54), and (6.55),

$$\frac{e^{-nX_1/RT} - 1}{e^{X_2/RT} - 1} = \frac{q}{\zeta} \tag{6.56}$$

From this it follows that, if X_2 is not too small,

$$\frac{-nX_1}{X_2} = 1 + \left(\frac{RT}{X_2}\right) \ln\left(\frac{q}{\zeta}\right) \tag{6.57}$$

Thus for highly coupled systems and large values of X_2, $-nX_1/X_2 \simeq 1$ at static head. This result will be used below. At level flow ($X_1 = 0$), the relation $J_1/J_2 = qZ$ holds true whatever the magnitude of X_2.

Far from equilibrium, q may still be defined by the left-hand side of Eq. (6.54). To evaluate it at the MIP, it is simplest to write Eqs. (6.33) in the following form, first introduced by Rottenberg (1973) and since used extensively by others (Westerhoff and van Dam, 1979):

$$J_1 = L_{11}X_1 + L_{12}X_2 + K_1$$

$$J_2 = L_{21}X_1 + L_{22}X_2 + K_2 \tag{6.58}$$

Here K_1 and K_2 are constants, that is, $K_1 = J_1^0 - L_{11}X_1^0 - L_{12}X_2^0$ and $K_2 = J_2^0 - L_{21}X_1^0 - L_{22}X_1^0$, where X_1^0, X_2^0 and J_1^0, J_2^0 denote the values of the forces and flows at the reference MIP. (A further discussion of the significance of K_1 and K_2 is given in the appendix to this chapter.) We have then $q_{\text{MIP}} = \sqrt{L_{12}L_{21}/L_{11}L_{22}}$. Evaluating the coefficients as discussed above, we find

$$L_{11} = (n^2a/2RT)(e^{(nX_1^0 + X_2^0)/RT} + 1) + (n^2b/2RT)(e^{nX_1^0/RT} + 1)$$

$$L_{12} = (na/2RT)(e^{(nX_1^0+X_2^0)/RT} + 1) + (nb/2RT)(e^{nX_1^0/RT} - 1)$$

$$L_{21} = (na/2RT)(e^{(nX_1^0+X_2^0)/RT} + 1)$$
$$+ (nc'/2RT)(e^{nX_1^0/RT} - \gamma)(e^{X_2^0/RT} - 1)$$

$$L_{22} = (a/2RT)(e^{(nX_1^0+X_2^0)/RT} + 1)$$
$$+ (c'/2RT)(e^{nX_1^0/RT} + \gamma)(e^{X_2^0/RT} + 1) \quad (6.59)$$

Clearly q_{MIP}^2 cannot exceed 1. Equations (6.59) enable us to find an approximate relationship between the ratio of cross coefficients L_{21}/L_{12} and q. For a highly coupled pump operating under physiological conditions, $X_2^0 \gg RT$, $-nX_1^0 \gg RT$. Note that limits on b/a and c'/a may be established using Eq. (6.54); for example, if $q = 0.95$ near equilibrium, these quantities could vary (in opposite senses) between 0.1 and zero. For a high value of q_{MIP}, c'/a must approach the lower limit unless γ is very small. It turns out that if $\alpha_{32} \ll \alpha_{31}$, and if $\alpha_{23} \ll \alpha_{21}$, which will generally be the case in a highly coupled system, then for varying $\Delta\psi$ and c_S, one finds at the MIP, to a close approximation,

$$c_S^0 = \sqrt{\frac{\phi\phi_1}{\phi_2\phi_{12}}} \approx \frac{B}{\gamma KC} \quad (6.60)$$

(compare Eq. 6.51). To a slightly lesser degree of approximation a similar relation holds when $c_{L_t}^{n+}$ varies instead of $\Delta\psi$. It follows that the "uncoupled" contributions to L_{12} and L_{21} in Eqs. (6.59) are essentially equal in this model. With the parameters given in Table 6.1, this has been found to hold true over a wide range of concentrations. Even if α_{23} and α_{32} are together increased by a factor of 10^7, so that the equilibrium value of q at $c_P = 10^{-3}$ M (which is usually much lower than q_{MIP}) falls to $\sim 10^{-3}$, L_{21}/L_{12} departs from unity by only $\sim 0.6\%$. At higher degrees of coupling, even though q_{MIP} may fall appreciably below unity, L_{21}/L_{12} is imperceptibly different from 1. (The MIP is remarkably insensitive to changes in ligand or reactant concentrations.) It should be noted that circumstances may arise when $q_{MIP} \approx 1$, even though cycle a plays a negligible role. In this case, results inconsistent with complete coupling will be found at static head and level flow.

Table 6.1. An arbitrarily chosen but self-consistent set of parameters used in simulating an ion pump. The first-order rate constants α_{ij} are in sec^{-1}. For explanation, see text

Parameter	Value	Parameter	Value
$\alpha_{1'1}(0)$	10^{12}	α_{24}^*	20
$\alpha_{11'}(0)$	5×10^9	α_{42}	0.2
$\alpha_{21}(0)$	5×10^2	α_{35}^*	2
$\alpha_{31}(0)$	5×10^2	α_{53}	20
α_{12}^0	10^2	α_{45}	50
α_{13}^0	10^2	α_{54}	0.1
α_{23}	10^{-11}	K	10^8
α_{32}	2×10^{-9}	γ	5×10^{-3}
		$\Delta\psi_b$	-130 mV

Near equilibrium, $q = 0.816$ at $c_P = 10^{-4}$ M, 0.976 at $c_P = 10^{-3}$ M.

6.5 Proper pathways at the multidimensional inflection point

It has been shown above that at least in the case of one simple model, an ion pump may possess an MIP when one of the variable parameters is the electrical potential difference across the membrane. It has also been shown that in the vicinity of the MIP a highly coupled system will be characterized by approximate reciprocity, such that (for the model examined) the "reciprocity ratio" L_{21}/L_{12} is essentially unity even if the *equilibrium* value of q is as low as 10^{-3}. There is no reason to think these results may not be more general, and indeed studies in mitochondria (Rottenberg, 1973; Rottenberg and Gutman, 1977; Westerhoff and van Dam, 1979; Stucki, 1980b) suggest that they are.

Stucki (1980b) has demonstrated in mitochondria that variation of the phosphate potential (X_{phos}) while maintaining the oxidation potential (X_{ox}) constant yields linear flow–force relationships, such that dividing the intercept of (J_{phos} versus X_{phos}) by the slope of (J_{ox} versus X_{phos}) results in an estimate of X_{ox} correct to within a few percent. If Eqs. (6.58) are applicable to this case, then not only should q_{MIP} be high (which apparently it is), but K_1 should be negligibly small. The latter condition is automatically achieved when the MIP is sufficiently close to static head. If static head is included in the linear range, we have, from Eqs. (6.58), the static head relation

$$\frac{-nX_1}{X_2} = \frac{nL_{12}}{L_{11}} + \frac{nK_1}{L_{11}X_2} \qquad (6.61)$$

Since this result is general for values of X_2 close to X_2^0, we may conclude that for a highly coupled system, by analogy with Eq. (6.59), $nL_{12}/L_{11} \simeq 1$, and by analogy with Eq. (6.57), $nK_1/L_{11}X_2$ ($\simeq K_1/L_{12}X_2$) must be negligibly small. This provides a kinetic basis for the type of experimental approach just described, which has been used extensively for determining the affinity of the reaction driving active sodium transport in epithelial membranes where extensive ranges of linearity are found (see Chapter 8). Although no independent estimation of the driving force in epithelia is at present possible, it is known that the sodium pump operates close to static head. Maximum efficiency may well be reached within the linear range, although far from equilibrium q_{MIP} does not bear a simple relation to maximum efficiency (unless for some reason K_2 may also be neglected). However, to a limited extent in the vicinity of static head, kinetic linearity in effect simulates thermodynamic linearity at the MIP. The system behaves as if it were on a proper pathway so far as affinity determinations of the input process are concerned.

6.6 Concluding remarks

We commented earlier on a special case discussed by Rothschild et al. (1980), in relation to a four-state model—the steady state in which both flows are identical at the MIP. In our terms this represents the condition $J_1 = nJ_2$, which if applied to Eqs. (6.53) yields a relation between the forces:

$$b(e^{nX_1^0/RT} - 1) = c'(e^{nX_1^0/RT} + \gamma)(e^{X_2^0/RT} - 1) \qquad (6.62)$$

Discarding trivial solutions, it is readily seen that X_1^0 and X_2^0 must have the same sign for this relation to hold, which would mean that the system is in a nonphysiological range. For $X_2^0 \gg RT$, $nX_1^0 \gg RT$, and γ of the order of 1 or less, Eqs. (6.59) and (6.62) give $L_{12} = L_{21}$.

It is important to realize that the MIP is not a unique inflection point; other conditions may exist where J_1 and J_2 simultaneously pass through an inflection point on variation of X_1 but not on variation of X_2, and vice versa. In this case there is no reciprocity. Consequently, although linear behavior is undoubtedly found under essentially physiological conditions during experiments *in vitro*, and hence may be assumed to occur *in vivo* as well, the question arises: have we reason to suppose that this behavior reflects the presence of an MIP? A possible answer may lie in stability considerations. Examination of Eqs. (6.46) and (6.47) shows that the kinetics are far from first order in $e^{n\Delta\psi F/RT}$, even in this simple model. A stability analysis is outside the scope of this chapter, but in general there may be a physiological advantage in the near-linearity and reciprocity conferred on a highly coupled energy transducer at the MIP, since local asymptotic stability is guaranteed by these conditions (Nicolis and Prigogine, 1977; Stucki, 1978). Although it is exceedingly unlikely that the whole machinery that can be purchased by exact linearity and reciprocity—such as minimum dissipation and the evolution criterion (Glansdorff and Prigogine, 1971)—applies to the systems described here, the existence of near-linearity and reciprocity may automatically provide a proper pathway and thus enable one to obtain experimental information that, at least in the case of epithelial membranes, is currently inaccessible by any other means.

In principle the above formulations, where applicable, permit us to characterize a system under a wide variety of circumstances. If a system is sufficiently well understood that we know how to vary both X_1 and X_2 along proper pathways, there may of course be little additional benefit to be gained from quantifying the phenomenological coefficients beyond an estimate of the effective degree of coupling in the reference state. On the other hand, frequently it is not possible to vary both forces independently in the manner desired, particularly in biological systems in which one or the other force may be experimentally inaccessible for perturbation by specific, nondestructive means. For this reason it is important to observe that if one of the forces, say X_1, can be varied experimentally along a proper pathway (as described by Eqs. 6.23 and 6.24) while X_2 remains constant, the response of the flows to perturbation of $\tilde{X}_1$ will permit a complete thermodynamic characterization of the system. This can be seen by realizing that along a proper pathway the slopes $(\partial J_1/\partial \tilde{X}_1)_{\tilde{X}_2} = \tilde{L}_{11}$ and $(\partial J_2/\partial \tilde{X}_1)_{\tilde{X}_2} = \tilde{L}_{21}$. Onsager reciprocity then gives $\bar{L}_{12} = \bar{L}_{21}$, so that $(J_1[\tilde{X}_1 = 0])_{\tilde{x}_2}/(\partial J_2/\partial \tilde{X}_1)_{\tilde{x}_2} = \tilde{X}_2$; finally $(J_2[\tilde{X}_1 = 0])_{\tilde{x}_2}/\tilde{X}_2 = \bar{L}_{22}$. For example, in Fig. 6.3 the points where the curves calculated from the kinetic equations (solid lines) intersect the ordinates at $X_1/RT = -5, 0$, and 5 are indicated by dots, which are analogous to experimental points. Evaluating the best straight lines through these points for both J_1 and J_2 by least squares permits an estimate of X_2, which exceeds the true value by only 4%. In practice this type of analysis has been applied experimentally to frog abdominal skin and toad urinary bladder, tissues which carry out vigorous transepithelial active sodium transport driven by oxidative metabolism. A detailed discussion will be given in Chapter 8.

A very interesting final consideration, briefly referred to in the introduction, is that proper pathways may be followed because the system is *intrinsically* highly linear (in the thermodynamic sense) over some physiological range. This perhaps could be achieved by a very general form of "thermodynamic buffering" of the enzymes involved (Stucki, 1980a), which is certainly not comprehended in the kinetic schemes presented here. Through analysis of a model of the type presented in Fig. 6.2, however, it can be shown that in general, providing the degree of coupling is not extremely low, the maximum efficiency of the system suffers a drastic reduction as operation is moved from the domain of thermodynamic linearity to the domain of nonlinearity (Stucki, Compiani, and Caplan, 1982). This suggests that mechanisms exhibiting intrinsic or "built-in" linearity would have an energetic advantage and may well have emerged as a consequence of evolutionary pressure.

6.7 Summary

1. Conventional phenomenological equations of nonequilibrium thermodynamics constitute an incomplete description of the processes under study, since a given thermodynamic force may be induced in an infinite number of ways. In general, therefore, both uncoupled and coupled flows are nonlinear functions of the forces, and the Onsager reciprocal relations are obeyed only very near equilibrium.

2. It is shown that in the case of uncoupled transport and reaction processes the forces can be constrained to "proper pathways" so

that linear behavior is observed. For this purpose a distinction is made between "thermodynamic" linearity, which implies a linear dependence of the flows on the thermodynamic forces extrapolating smoothly to equilibrium, and "kinetic" linearity, which is not necessarily characterized by this property. Proper pathways are associated with thermodynamic linearity.

3. If the forces of two coupled processes can be simultaneously constrained to proper pathways such that each flow is a linear function of each force, the phenomenological cross coefficients are equal far from equilibrium. The nature of such proper pathways is investigated in terms of a simple model of a sodium active transport system.

4. It has been demonstrated that coupled enzymatic processes may possess, for a particular choice of the state variables, a multidimensional inflection point (MIP) in thermodynamic force-flow space. The conditions for reciprocity in the kinetically linear region near such a reference state, which may be far from equilibrium, are investigated. It is shown by examining the associated Hill diagrams that all cycles in which a given pair of forces act contribute a corresponding pair of symmetrical terms to the Jacobian matrix characterizing perturbations about this stationary state. To the extent that these cycles dominate—that is, to the extent that the system is highly coupled—reciprocity or near-reciprocity will be obeyed.

5. Since local asymptotic stability is guaranteed by local symmetry, the observation of linear behavior in many highly coupled biological energy-transducing systems may well reflect operation at or near a multidimensional inflection point. If so, certain applications of linear nonequilibrium thermodynamics may be justified; that is, the kinetic linearity may to some extent simulate thermodynamic linearity and give rise to a proper pathway.

6. The properties of the MIP are illustrated by a simple kinetic model for active ion transport in which electrical forces are explicitly taken into account. In this model it is shown that even "uncoupled" cycles contribute symmetrical terms to the Jacobian over an extremely wide range.

7. Where consideration in terms of proper pathways is appropriate, it permits a complete thermodynamic characterization of a system even when only one of the two forces can be controlled experimentally while the other remains constant.

8. It is suggested that "built-in" linearity may well be energetically advantageous and may have emerged as a consequence of evolutionary pressure.

Appendix: Linear phenomenological coefficients and equilibrium unidirectional cycle fluxes: behavior of biochemical systems near static head

It has been pointed out by Hill (1982a, b) that the phenomenological coefficients in the linear flux–force equations for an arbitrary diagram describing a system in a steady state *near equilibrium* have a simple physical interpretation: the L_{ij}'s are essentially sums of the equilibrium unidirectional cycle fluxes. Those cycles that include process i contribute to L_{ii}, while those that include both process i and process j contribute to L_{ij}. For example, from Eqs. (5.12), (5.14), and (5.34) we have in general (as we have seen earlier)

$$J_\kappa = J_{\kappa^+} - J_{\kappa^-} = J_{\kappa^-}\left(\frac{J_{\kappa^+}}{J_{\kappa^-}} - 1\right) = J_{\kappa^-}(e^{X_\kappa/RT} - 1) \quad (6.A1)$$

Near equilibrium this gives

$$J_\kappa = J^e_{\kappa\pm}(X_\kappa/RT) \quad (|X_\kappa| \ll RT) \quad (6.A2)$$

Since the two equilibrium unidirectional cycle fluxes for cycle κ are equal, that is,

$$J^e_{\kappa^+} = J^e_{\kappa^-} = J^e_{\kappa\pm} \quad (6.A3)$$

$J^e_{\kappa\pm}$ denotes either of the two unidirectional fluxes. Thus Eqs. (5.67) and (5.68) can be written, in view of Eq. (5.24),

$$J_1 = n^2(J^e_{a\pm} + J^e_{b\pm})X_1/RT + nJ^e_{a\pm}X_2/RT \quad (6.A4)$$

$$J_2 = nJ^e_{a\pm}X_1/RT + (J^e_{a\pm} + J^e_{c\pm})X_2/RT \quad (6.A5)$$

We see that

$$L_{11} = n^2(J^e_{a\pm} + J^e_{b\pm})/RT$$

$$L_{12} = nJ^e_{a\pm}/RT$$

$$L_{22} = (J^e_{a\pm} + J^e_{c\pm})/RT \tag{6.A6}$$

Of much greater interest to us is the situation near static head. Since biochemical systems are usually highly coupled, there will be one dominant cycle that operates close to equilibrium at static head, while the remaining "leakage" cycles, however limited their role, operate very far from equilibrium. To consider a steady state in this region from the same viewpoint, we start with Eqs. (6.14) and (6.15), but we take stoichiometry explicitly into account by replacing J_1 by J_1/n, and X_1 by nX_1 (compare Eqs. 5.63 and 5.64). These equations then transform directly into Eqs. (6.52). We have

$$J_1 = nJ_b + nJ_a \tag{6.A7}$$

$$J_2 = J_a + J_c \tag{6.A8}$$

Introducing the unidirectional cycle fluxes, and taking into consideration that cycle a is close to equilibrium, we obtain

$$J_1 = n(J_{b^+} - J_{b^-}) + nJ^e_{a\pm}X_a/RT \tag{6.A9}$$

$$J_2 = J^e_{a\pm}X_a/RT + (J_{c^+} - J_{c^-}) \tag{6.A10}$$

Now, although X_a ($= nX_1 + X_2$) is very small, $X_b(= nX_1)$ is large and negative, while X_c ($= X_2$) is large and positive. By comparison with Eq. (6.A1), we see that in this case J_{b^+} and J_{c^-} may be neglected, and Eqs. (6.A9) and (6.A10) become

$$J_1 = -nJ_{b^-} + nJ^e_{a\pm}X_a/RT \tag{6.A11}$$

$$J_2 = J^e_{a\pm}X_a/RT + J_{c^+} \tag{6.A12}$$

130   Bioenergetics and linear nonequilibrium thermodynamics

Since cycles b and c are operating virtually unidirectionally, their corresponding cycle forces no longer appear in the flux equations. We may conveniently rewrite Eqs. (6.A11) and (6.A12) in the following form:

$$J_1 = n^2 J_{a\pm}^e X_1/RT + n J_{a\pm}^e X_2/RT + n(-J_b^-) \quad (6.A13)$$

$$J_2 = n J_{a\pm}^e X_1/RT + J_{a\pm}^e X_2/RT + J_{c^+} \quad (6.A14)$$

Equations (6.A13) and (6.A14) describe the steady-state behavior of the system in the neighborhood of a far-from-equilibrium state in which $X_a = 0$: for highly coupled systems this state will be close to static head. It will be observed that these equations have the form of the Rottenberg flux equations, Eqs. (6.58), and hence they appear to give a physical interpretation to the Rottenberg K's. The range of applicability of Eqs. (6.A13) and (6.A14) would obviously be widest for the case of a unique MIP coincident with the state in which $X_a = 0$. Provided the degree of coupling is high, the third term on the right side of Eq. (6.A13) may be neglected as discussed earlier in connection with Eqs. (6.57) and (6.61). For all experiments in which X_2 remains constant, therefore, (6.A13) and (6.A14) may usefully be expressed as

$$J_1 = n^2 J_{a\pm}^e X_1/RT + n J_{a\pm}^e X_2/RT \quad (6.A15)$$

$$J_2 = n J_{a\pm}^e X_1/RT + J_{a\pm}^e \left(1 + \frac{RT J_{c^+}}{X_2 J_{a\pm}^e}\right) X_2/RT \quad (6.A16)$$

and the apparent degree of coupling under these conditions is given by

$$q_{\text{app}} = \frac{1}{\sqrt{1 + \dfrac{RT J_{c^+}}{X_2 J_{a\pm}^e}}} \quad (X_2 \text{ constant}) \quad (6.A17)$$

As a final observation we note that in studies of (Na, K) ATPase there appears to be some evidence that in the vicinity of the so-called reversal potential, that is, the electrical potential difference at which the hydrolysis of ATP just ceases, the ion fluxes are more or less voltage-independent (see, for example, Marmor, 1971). In terms of our present considerations, if the system is highly coupled and $J_2 \simeq 0$ near static head, we have from Eqs. (6.A11) and (6.A12)

$$J_1 = -n(J_{b^-} + J_{c^+}) \qquad (6.A18)$$

To a first approximation, J_1 is then indeed a voltage-independent unidirectional flux. This argument can be extended to models of the type shown in Fig. 6.4.

7 Energetics of active transport: theory

Because of the inadequacies of the standard formulations for analyzing the energetics of active transport processes, in this chapter we shall take a different approach, based on the general considerations discussed earlier. For convenience we assume a simple model of the transport system which is mathematically tractable; the applicability of the principles to more realistic models will be evident. We rely on the concept of proper trajectories developed in Chapter 6, believing that experimental perturbations can perhaps often be chosen so as to be on such pathways. Experimental evidence bearing on this point will be presented in the next chapter.

7.1 Background

Many people have employed NET to analyze active transport. Jardetzky and Snell (1960) presented a general theoretical analysis of transport and various metabolic processes utilizing the notation of NET but requiring neither linearity nor the validity of the Onsager reciprocal relations. Kedem (1961) gave a formal description of active transport based on linear NET, assuming the validity of the Onsager relations, and showed how the formalism permits the correlation of different types of measurements in two-flow systems. Hoshiko and Lindley (1967) emphasized the importance of a clear operational definition of active transport and extended the methods of Kedem and Katchalsky (1958) to the active transport of single salt and bi-ionic systems. Procedures were outlined to evaluate the requisite phenomenological coefficients (10 or 15 in number, depending on the circumstances). Heinz (1975) has analyzed sodium-linked amino acid transport in an NET formulation based on that of Rapoport (1970). In order to distinguish between coupled and uncoupled processes a quasi-chemical notation was introduced which treats all uncoupled events in terms of an intrinsically stoichiometric chemical reaction complicated by leakage. This point of view has recently been amplified in a monograph by Heinz (1979).

We have preferred to restrict ourselves to systems with active transport of only one ion, uncoupled to the flows of other species (Essig and Caplan, 1968). The theoretical basis for our work is an extension of Kedem's. In contrast to Rapoport and Heinz, we treat the coupling between transport and metabolism quite generally, admitting the possibility of intrinsic incomplete coupling.

7.2 Model for the active transport of a single ion

In analyzing the energetics of active transport, it is necessary to start with a simple model. In frog skins and toad bladders of appropriate species there appears to be only one significant active transport process, that of sodium, and thus only one significant output for our thermodynamic system. Despite the great complexity of biological tissues, we assume that we can isolate one metabolic process which "drives" active transport. This model is represented in Fig. 7.1. Here one input process, the metabolism of substrate, is linked to one output process, the transport of sodium. In this representation, closely analogous to the general representation of Fig. 4.1, the consumption of M and N to produce P and Q provides the free energy which brings about the active transport of sodium across the membrane.

Similarly, we take a simplified view of the histology of the system (Fig. 7.2). The process of active transport takes place in the rectangu-

134 Bioenergetics and linear nonequilibrium thermodynamics

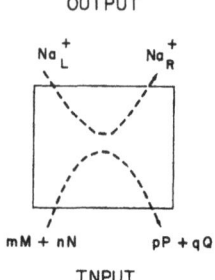

Fig. 7.1. Diagrammatic representation of a biological energy converter which can utilize metabolic energy to do electroosmotic work. The clockwise direction of the lower arrow indicates that the conversion of M and N to P and Q is spontaneous, whereas the counterclockwise direction of the upper arrow indicates that transport of sodium may take place against its electrochemical potential gradient, as a consequence of its coupling to metabolism.

lar box. Although it is not necessary for our analysis, in accord with many experimental observations we represent the outer or apical region as a simple passive barrier across which sodium moves down its electrochemical potential gradient. (For the present we exclude consideration of systems with apical Na–Cl cotransport.) At the inner or basolateral surface is the mechanism responsible for active sodium transport, the so-called sodium pump. Since the active trans-

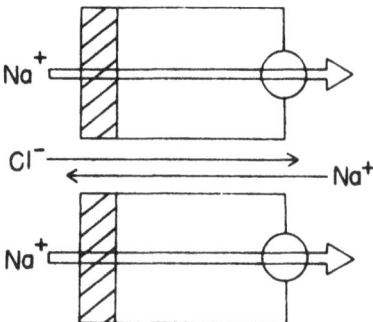

Fig. 7.2. Model of the observable composite transport system, comprising a passive barrier to sodium entry at the apical surface, a sodium pump at the basolateral surface, and a parallel passive channel accessible to all ions in the bathing solutions.

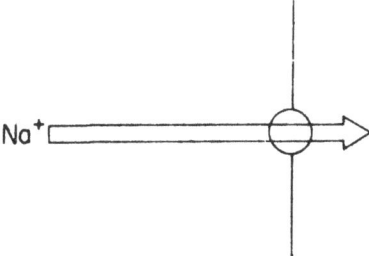

Fig. 7.3. Representation of a sodium (cation) pump.

port system transports only sodium ions, whereas the tissue as a whole reabsorbs sodium chloride, there has to be a pathway across which chloride can move. This is represented as a simple passive channel in parallel with the active transport pathway, which is presumably accessible to all of the ions in the bathing solutions.

Since we have considered active transport to be a two-flow process, there are two pertinent flow equations to be considered, one for cation transport and one for metabolism. (Although we are thinking primarily of sodium transport, the formulation is in principle applicable to the flow of any cation and, for that matter, is readily modified for the case of active transport of any solute.) We could proceed directly to writing phenomenological equations for the system of Fig. 7.2, treating the composite tissue as our black box. However, for some purposes it is useful to relate the behavior of the tissue to that of its elements.

7.2.1 Pump

We consider first the operation of a simple "pump" which carries out the active transport of a single cation, uncoupled to flows of other species (Fig. 7.3). We assume that the metabolic energy necessary for this process is derived from a single driving reaction. For the active transport mechanism, representing the rate of active cation transport by J_+^a and the rate of metabolism by J_r, we have

$$J_+^a = L_+^a X_+ + L_{+r}^a A \tag{7.1}$$

$$J_r = L^a_{+r}X_+ + L^a_r A \qquad (7.2)$$

Here X_+ is the negative electrochemical potential difference of the cation, and A is the affinity of a metabolic reaction which is driving transport. Under the conditions to be considered, the affinity is equivalent to the negative Gibbs free energy change $-\Delta G$ of the driving reaction, as yet undefined in biochemical terms. The L^a's are phenomenological coefficients, assumed constant over the range of forces to be employed. J^a_+ is of course a function of the negative electrochemical potential difference X_+, but to the extent that it is coupled to metabolism, it must also be a function of the affinity A. J_r is obviously a function of A, but to the extent that it is linked to transport, it must also be a function of X_+. By analogy with a variety of transport processes in nonliving systems (Miller, 1960; Blumenthal, Caplan, and Kedem, 1967), the validity of the Onsager reciprocal relation is assumed, that is, the cross coefficients in the two equations are set equal. (Implicit in this assumption is the consideration that we restrict X_+ and A to proper pathways.)

It is frequently convenient to consider the resistance formulation corresponding to Eqs. (7.1) and (7.2), which is given by

$$X_+ = R^a_+ J^a_+ + R^a_{+r} J_r \qquad (7.3)$$

$$A = R^a_{+r} J^a_+ + R^a_r J_r \qquad (7.4)$$

where the phenomenological R coefficients are related to the L coefficients by a simple matrix transformation.

Since there is only a single driving reaction, the rates of consumption and production of metabolites are related stoichiometrically. We may then take any of these processes to represent metabolism, provided that the affinity is expressed appropriately (see the appendix to this chapter). Thus we may, for example, consider J_r to be the rate of consumption of O_2. Then for $J_r A$ to represent the rate of supply of free energy, the affinity A must be expressed as the negative change in free energy of the metabolic reaction per mole of O_2 consumed. For active transport the input $J_r A$ must be positive, whereas the output $-J^a_+ X_+$ may be of either sign. There is a degree of arbitrariness in assigning a polarity to the reactions. We shall adopt the convention that J^a_+, J_r, and A are all > 0 for $X_+ = 0$. Then L^a_{+r} is

> 0, R^a_{+r} is < 0, and coupling is positive in the naturally occurring system. From the positive-definite character of the dissipation function, it follows that the straight phenomenological coefficients (L^a_+ and L^a_r, or R^a_+ and R^a_r) must be > 0.

Certain fundamental points about Eqs. (7.1)–(7.4) deserve emphasis:
1. It is clear that there is no *a priori* assumption of stoichiometry here; the very fact that there are two equations implies that in general we might not expect stoichiometry. As we have seen, stoichiometry will obtain only for a special relationship between the phenomenological coefficients.
2. The equations apply only to steady states.
3. In order for the equations to be helpful experimentally, the phenomenological coefficients must be constant over a range of the forces and flows that is sufficiently large to permit their accurate measurement. This is assured if the forces are restricted to proper pathways. With kinetic rather than thermodynamic linearity (Chapter 6), modification of the treatment is necessary, as discussed in detail for the case of oxidative phosphorylation in Chapter 13.

For certain experimental situations and for theoretical analysis, two combinations of the phenomenological coefficients discussed in Chapter 4 prove useful. For the simple pump these are:

$$Z^a = \sqrt{\frac{L^a_+}{L^a_r}} = \sqrt{\frac{R^a_r}{R^a_+}} \qquad (7.5)$$

and

$$q^a = \frac{L^a_{+r}}{\sqrt{L^a_+ L^a_r}} = -\frac{R^a_{+r}}{\sqrt{R^a_+ R^a_r}} \qquad (7.6)$$

where, of course, as for the general case of Chapter 4,

$$-1 \leq q^a \leq 1 \qquad (7.7)$$

Manipulation of Eqs. (7.3) and (7.4) leads to useful expressions for the flows as functions of the forces and the resistance coefficients:[1]

$$J_+^a = \frac{R_r^a X_+ - R_{+r}^a A}{R_+^a R_r^a - (R_{+r}^a)^2} \tag{7.8}$$

or in terms of the alternative coefficients,

$$J_+^a = \frac{X_+ + (q^a/Z^a)A}{R_+^a[1 - (q^a)^2]} \tag{7.9}$$

Similarly,

$$J_r = \frac{(q^a/Z^a)X_+ + (1/Z^a)^2 A}{R_+^a[1 - (q^a)^2]} \tag{7.10}$$

Although the above model is useful for orientation, biological membranes which transport a variety of substances must of course be more complex. We consider therefore two rather more realistic models.

7.2.2 Composite series membrane

The first modified model, shown in Fig. 7.4, differs from the basic model by the introduction of a barrier in series with the pump. For this composite membrane the formal description changes only in the modification of the straight phenomenological coefficient of the test species; hence Eqs. (7.3) and (7.4) continue to apply if R_+^a is now taken to represent the total series resistance of the active transport pathway. This is seen by considering that at the barrier, $X_+^{\text{barrier}} = R_+^{\text{barrier}} J_+^a$, and at the pump, $X_+ - X_+^{\text{barrier}} = R_+^{\text{pump}} J_+^a + R_{+r}^a J_r$, where we have written R_+^{pump} for R_+^a. Adding,

$$X_+ = (R_+^{\text{barrier}} + R_+^{\text{pump}})J_+^a + R_{+r}^a J_r \tag{7.11}$$

Equation (7.4) is of course unchanged. The degree of coupling is now given by

$$q^a = -\frac{R_{+r}^a}{\sqrt{(R_+^{\text{barrier}} + R_+^{\text{pump}})R_r^a}} \tag{7.12}$$

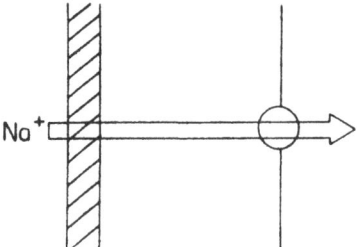

Fig. 7.4. Representation of a sodium (cation) pump and a passive series barrier.

and is thus less than that of the simple pump (unless the system is completely coupled).[2]

7.2.3 Composite membrane with series and parallel elements

Since we consider active transport to be restricted to a single ion, whereas the tissue transports salt, there must be a passive channel accessible to ions, for otherwise the accumulation of net charge would establish a static head, bringing transport to a halt. Combining a parallel leak pathway with the composite series membrane just discussed gives us the model of the observable composite transport system shown in Fig. 7.2. In analyzing the observable system, we shall use the parameters of the generalized NET formulations of Chapter 4, restricting the use of superscripts a and p to characterization of the active and passive elements, respectively.

Combining the appropriate phenomenological equations of the elements leads readily to those of the composite system. Formally it is immaterial whether leak is intracellular or intercellular. If the passive flow J_+^p is uncoupled to the flow of other species,

$$X_+ = R_+^p J_+^p \tag{7.13}$$

Since $J_+ = J_+^a + J_+^p$, the membrane is again characterized by linear equations.[3]

It is useful to express the phenomenological coefficients of the composite membrane in terms of those of its elements. Combining Eqs. (7.9) and (7.13):

$$J_+ = \left(\frac{1}{R^a_+[1 - (q^a)^2]} + \frac{1}{R^p_+}\right)X_+ + \left(\frac{q^a/Z^a}{R^a_+[1 - (q^a)^2]}\right)A \quad (7.14)$$

For given forces, J_r is unaffected by the presence of a leak, and Eq. (7.10) continues to apply.

Although we may be unable to characterize all the permeability barriers, the above treatment suggests that with appropriate precautions even complex systems may be treated by linear overall relations. Presuming that this is the case for the system of interest here, we may employ the generalized parameters and write:

$$J_+ = \frac{X_+ + (q/Z)A}{R_+(1 - q^2)} \quad (7.15)$$

and

$$J_r = \frac{(q/Z)X_+ + (1/Z^2)A}{R_+(1 - q^2)} \quad (7.16)$$

Two reference stationary states of fundamental importance in the general treatment of coupled processes are "level flow" ($X_i = 0$) and "static head" ($J_i = 0$) (Kedem and Caplan, 1965). Transport across epithelial tissues often approximates level flow. This appears to be the case, for example, in the proximal tubule of mammalian kidney, where some 70% of the filtered sodium is reabsorbed, apparently by an active process. The associated water movement is sufficiently rapid to maintain near-isotonicity of the tubular fluid, and the electrical potential difference is small. In contrast, symmetrical cells such as red blood cells and muscle cells generally establish static head for a variety of solute species, and in some circumstances polar tissues may approximate this state as well.

The generalized equations which describe the two stationary states are, for cations,

$$(J_+)_{X_+=0} = \frac{(q/Z)A}{R_+(1 - q^2)} \quad \text{(level flow)} \quad (7.17)$$

and

$$X_+^0 = (X_+)_{J_+ = 0} = -\frac{qA}{Z} \quad \text{(static head)} \tag{7.18}$$

It is important to emphasize that considering the nature of our model, Eqs. (7.15) and (7.16) could have been written immediately, without separate consideration of the influence on transport of the pump, the series barrier, and the leak pathway. Indeed, Eqs. (7.15) and (7.16) are evidently applicable to any linear two-flow system in which active transport of a single species is the consequence of a single metabolic process. To characterize and compare such systems, it is useful to carry out an analysis in terms of the general parameters. However, given the fact that the model of Fig. 7.2 is widely accepted as a reasonable representation of several important systems (for example, sodium transport in the toad bladder), it is appropriate also to examine the dependence of various characteristics on the parameters of the elements. In particular, it is important to examine systematically the influence of leak pathways, since there is often a tendency to consider leakage as representing a departure from ideal behavior, accounting for departures from stoichiometry and decrease in efficiency.

The degree of coupling for active cation transport is indeed decreased by a leak, being given by

$$q^2 = \frac{(q^a)^2}{1 + (R_+^q/R_+^p)[1 - (q^a)^2]} \tag{7.19}$$

We see that as is necessary for consistency, with decrease of leak $(R_+^p \to \infty) q \to q^a$. In the presence of leak, q becomes less than q^a. For effective operation q^a should be as large as possible, and it might be thought that similarly the overall degree of coupling should be as large as possible. However, we have seen that with nonselective leak pathways the effective production of a large flow requires that R_+^p be fairly small, and therefore q^2 will be appreciably less than $(q^a)^2$. Thus for a system whose function is salt transport, the optimal degree of coupling of the overall *cation* transport mechanism may not be the largest value. Clearly, the optimal degree of coupling is determined by the function of the system. Therefore, rather than assume complete coupling, we should attempt to learn the degree of coupling and the implications of this value.

The magnitude of leak has no effect on level flow, as is intuitively evident and is seen by setting $X_+ = 0$ in Eq. (7.14). With X_+ fixed at a given value less than zero, J_+ is of course diminished by back flux through the leak pathway. If, however, as *in vivo*, X_+ is not fixed experimentally, the situation may be quite different. In this case the leaks will permit not only back flux but also partial dissipation of the adverse electrical potential difference resulting from active transport. Consider for simplicity a membrane separating identical solutions of NaCl, and assume that in addition to active transport of sodium, there is passive transport of both Na^+ and Cl^-. If the permeability to water is very high, so that transport of NaCl is associated with transport of sufficient water to maintain near-isotonicity, $X_w \simeq 0$, and

$$X_+ \simeq -F\Delta\psi \simeq -X_- \qquad (7.20)$$

where F is the faraday and $\Delta\psi$ the difference in electrical potential across the membrane. Representing the phenomenological resistance coefficient of chloride by R_-^p, the permselectivity of the passive pathway is characterized by R_-^p/R_+^p, which we shall call Z_{+-}^2. (In the absence of coupled water flow, Z_{+-}^2 is the ratio of the transport numbers, τ_+/τ_-.)

The chloride flux is given by

$$J_-^p = \frac{X_-}{R_-^p} \qquad (7.21)$$

In the steady state, when the net current is zero, Eqs. (7.14), (7.20), and (7.21) give

$$(J_+)_{I=0} = \frac{(q^a/Z^a)A}{R^a(1 + Z_{+-}^2)[1 - (q^a)^2] + R_-^p} \qquad (\Delta c = 0) \qquad (7.22)$$

It is seen that when X_+ is not fixed experimentally, for a given permselectivity of the passive pathway, isotonic saline reabsorption is directly related to the magnitude of the leak. (This can be shown to be true even if the flows of ions and solvent are coupled.) It must be appreciated that this conclusion is valid only if the factors changing the extent of leak have no effect on permselectivity or the phenomenological coefficients of the active transport system. As has been pointed out by Boulpaep, the depressed Na reabsorption noted with increased paracellular conductance following saline loading in *Necturus* is probably associated with a change in permselectivity (Boulpaep, 1972).

The magnitude of static head is of course influenced by the magnitude of the leak, as can be shown from Eq. (7.14), which gives

$$X_+^0 = -\frac{(q^a/Z^a)A}{1 + (R_+^a/R_+^p)[1 - (q^a)^2]} \tag{7.23}$$

As mentioned earlier, the flow of the cation is inversely related to R_+^a. On the other hand, Eq. (7.23) shows that if R_+^p is large in comparison with R_+^a, X_+^0 is insensitive to the latter, being equal to $(R_{+r}^a/R_+^a)A$. The insensitivity of X_+^0 to antidiuretic hormone, an agent which stimulates sodium transport in the toad bladder, has been cited in support of the view that this substance may act by decreasing the resistance of a series permeability barrier, without having a significant direct effect on the sodium pump (Civan, Kedem, and Leaf, 1966).

7.3 Experimental characterization of the transport system

From our point of view, one would ideally wish to characterize a transport system by determining its phenomenological coefficients and the forces in all pertinent modes of operation. In practice, however, in most situations of interest such complete descriptions are lacking. We shall examine first the data which can be readily acquired by the use of commonplace electrical techniques and their interpretation from the classical and NET viewpoints. We shall then consider the means which might be employed to achieve a more adequate characterization. In doing so, however, it is necessary to accept certain working hypotheses in order to make the analysis of the transport system tractable.

The transport system is characterized by phenomenological coefficients which define its behavior for all values of two independent variables. In practice, however, since only X_+ can be readily controlled, the response of the system is unpredictable without knowledge of the value of a second independent variable. If J_r were maintained constant, as in a constant-current electrical system, the system could be readily characterized (Hoshiko and Lindley, 1967). Another possibility, which seems to us *a priori* more likely, is that the affinity of some region of the metabolic chain may be constant. This could be the case, for example, for the entire (global) metabolic reaction if the pools of substrate and product were large. Alternatively, the affinity of some more local region of the metabolic chain might be maintained constant by the action of a regulator. In this section we shall consider that the affinity of some region of the metabolic chain is in fact constant and shall take this region to represent the driving reaction of the transport system (see the appendix to this chapter). Experimental results bearing on this issue will be presented in Chapter 8.

7.3.1 The short-circuit current and the open-circuit potential

It has become routine to study epithelial membranes such as frog skin and toad bladder by exposing them to identical physiological saline solutions at each surface and measuring the short-circuit current ($I_{\Delta\psi=0}$), the open-circuit potential ($\Delta\psi_{I=0}$), and the electrical resistance, $-dI/d(\Delta\psi)$. For systems in which active transport is limited to a single ion (in this case Na$^+$), these quantities are readily interpreted in terms of the formulations presented above. From Eqs. (7.14) and (7.15) it is seen that the short-circuit current is given by

$$I_0 \equiv F(J_+)_{\Delta\psi=0} = \frac{F(q/Z)A}{R_+(1-q^2)}$$

$$= \frac{F(q^a/Z^a)A}{R_+^a[1-(q^a)^2]} \quad (\Delta c = 0) \quad (7.24)$$

When $\Delta c = 0$, I_0/F is simply level flow. From Eqs. (7.14), (7.20), and (7.21), the open-circuit potential is given by

$$(\Delta\psi)_{I=0} = \frac{(q/Z)A}{F[1 + (R_+/R_-^p)(1-q^2)]} \quad (\Delta c = 0) \quad (7.25)$$

Thus, whereas the short-circuit current is a characteristic only of the sodium active transport system, the open-circuit potential reflects the magnitude of ion leakage as well. Combining Eqs. (7.18) and

Active transport: theory 145

(7.25) gives the relationship between the open-circuit potential and the magnitude of static head:

$$(\Delta\psi)_{\substack{I=0\\\Delta c=0}} = -\frac{X_+^0}{F[1 + (R_+/R_-^p)(1 - q^2)]} \quad (7.26)$$

From Eqs. (7.17) and (7.18) we obtain the relationship between the short-circuit current and the magnitude of static head:

$$(I)_{\substack{\Delta\psi=0\\\Delta c=0}} = (I)_{X_+=0} = -\frac{FX_+^0}{R_+(1 - q^2)} \quad (7.27)$$

7.3.2 Evaluation of the electrical resistance

The electrical resistance is readily evaluated by means of the simple techniques employed above. However, its interpretation in terms of fundamental parameters offers a clear example of the significance of the response of the system to the manipulations used to characterize it. If J_r could be maintained constant, one could measure a pure phenomenological resistance coefficient. Neglecting activity coefficients and assuming the solutions to contain only a single uni-univalent salt, X_+ is given by $-X_+ = RT\Delta \ln c + F\Delta\psi$.[4] From Eqs. (7.15) and (7.16) we obtain

$$-\frac{1}{F}\left(\frac{\partial J_+}{\partial(\Delta\psi)}\right)_{\Delta \ln c, J_r} = \frac{1}{R_+} \quad (7.28)$$

If, however, as seems to us *a priori* much more likely, J_r is not constant, the variation of J_+ with $\Delta\psi$ would reflect not only R_+ but also the degree of coupling. For example, if the affinity remains constant, Eq. (7.15) gives

$$-\frac{1}{F}\left(\frac{\partial J_+}{\partial(\Delta\psi)}\right)_{\Delta \ln c, A} = \frac{1}{R_+(1 - q^2)} \quad (7.29)$$

For the anion, which is not actively transported,

$$\frac{1}{F}\left(\frac{\partial J_-}{\partial(\Delta\psi)}\right)_{\Delta \ln c} = \frac{1}{R_-^p} \quad (7.30)$$

146 Bioenergetics and linear nonequilibrium thermodynamics

We now stipulate that perturbations of $\Delta\psi$ must be sufficiently long that transient K and Cl fluxes across cell membranes will have disappeared, and so Na flow will be conservative. Then, combining Eqs. (7.29) and (7.30) we have for the electrical resistance, which we designate $\mathcal{R}$,

$$\left(\frac{1}{\mathcal{R}}\right)_{\Delta \ln c, A} = -\left(\frac{\partial I}{\partial(\Delta\psi)}\right)_{\Delta \ln c, A}$$

$$= \left(\frac{1}{R_+(1-q^2)} + \frac{1}{R_-^p}\right) F^2 \quad (7.31)$$

Thus in this case the electrical resistance depends both on "passive" parameters and on the degree of coupling. If neither J_r nor A remains constant on perturbation of $\Delta\psi$, the dependence becomes still more complex.

If there is linearity between I and $\Delta\psi$,

$$-\frac{\partial I}{\partial(\Delta\psi)} = \frac{(I)_{\Delta\psi=0}}{(\Delta\psi)_{I=0}} \quad (7.32)$$

or

$$(\Delta\psi)_{I=0} = (I)_{\Delta\psi=0}\,\mathcal{R} \quad (7.33)$$

Equation (7.32) provides an experimental test for linearity. If the derivative and the ratio differ, the ratio is not a well-defined electrical resistance. If they agree, this suggests that either A or J_r is constant or that they have a linear dependence on $\Delta\psi$.

7.3.3 Affinity A and its evaluation

In general it is not possible to vary the affinity of the driving reaction in controlled fashion experimentally. However, a given hormone or drug might well exert its effect by this means. From Eqs. (7.15), (7.16), and (7.18) it is seen that J_+, J_r, and $-X_+^0$ are all increased by increasing A.[5] The effects on the efficiency and efficacy parameters are, however, more complex; these will be discussed below.

If the affinity of some region of the metabolic chain is constant, it can be evaluated. For example, with identical solutions on each side of the membrane, Eqs. (7.15), (7.16), and (7.18) give

$$A = -\frac{Z}{q} X_+^0 = \left(\frac{\partial J_+}{\partial J_r}\right)_A F(\Delta\psi)_{J_+ = 0} \qquad (7.34)$$

If $(\partial J_+/\partial J_r)_A$ is determined in the neighborhood of static head, the correct value of A should be obtained even if the phenomenological coefficients are strong functions of state.

Alternatively, assuming that with variation of $\Delta\psi$ at constant concentration X_+ lies on a proper pathway, Eqs. (7.15) and (7.16) provide an independent means of evaluating the affinity:

$$A = -\frac{(I)_{\Delta\psi=0}}{[\partial J_r/\partial(\Delta\psi)]_A} \qquad (7.35)$$

Agreement between the values of A determined by these two independent methods would confirm constancy of the affinity and consistency of the techniques employed. (However, it would not demonstrate the validity of the Onsager relation.) It should be noted that both methods are applicable even with coupling of ion and water flows in the passive pathway (Essig and Caplan, 1968).

Admittedly, for the present the affinity evaluated by the above means is a rather vague quantity, being defined only in abstract thermodynamic terms. Nevertheless, it is of physiological interest, since it must reflect the substrate–product concentration ratio of some critical reaction in the metabolic pool which supports active transport, and it must incorporate the influence of poorly characterized local pH, standard free energies, and activity coefficients. This is in contradistinction to mean cell concentration ratios of various substrates and products, including nucleotides such as ATP and ADP, and creatine phosphate and creatine (Handler, Preston, and Orloff, 1969). Although attempts have been made to study the driving forces for transport by such measurements, mean concentration ratios may well depend importantly on tissue functions other than transepithelial transport. Attempts to evaluate cytoplasmic ATP/(ADP × P_i) (Veech, Raijam, and Krebs, 1970) also involve theoretical and experimental difficulties. It must be appreciated, however, that values of A evaluated from measurements of O_2 consumption can be interpreted in terms of the free energy of ATP hydrolysis only given knowledge of the P/O ratio, which in principle can vary significantly under different operating conditions.

7.3.4 Evaluation of phenomenological coefficients

Given a linear system for which the affinity is constant on variation of X_+ along a proper pathway, it is quite practical to evaluate the phenomenological coefficients of the conductance formulation. Again, this is most readily done by exposing the tissue to identical solutions at each surface and varying $\Delta\psi$. Equations (7.1) and (7.2) then give

$$L_+^a = \frac{\partial J_+^a}{\partial X_+} = \frac{-\partial J_+^a}{\partial (F\Delta\psi)} = \frac{-\Delta J_+^a}{\Delta(F\Delta\psi)} \quad (\Delta c = 0, A \text{ constant}) \quad (7.36)$$

and

$$L_{+r}^a = \frac{\partial J_r}{\partial X_+} = \frac{-\partial J_r}{\partial (F\Delta\psi)} = \frac{-\Delta J_r}{\Delta(F\Delta\psi)} \quad (\Delta c = 0, A \text{ constant}) \quad (7.37)$$

Having determined A and L_{+r}^a as above, L_r is then given by

$$L_r^a = \frac{J_{r0}^{sb}}{A} \quad (\Delta c = 0) \quad (7.38)$$

Here J_{r0}^{sb} represents the rate of suprabasal metabolism at $\Delta c, \Delta\psi = 0$. (Possible experimental means for determining J_+^a, J_r, and J_r^{sb} will be discussed below.)

The phenomenological resistance coefficients cannot be as easily determined as the L coefficients for a system with constant A, since this would require independent control of the two flows. Given the values of L_+^a, L_{+r}^a, and L_r^a, however, the corresponding R coefficients can be calculated by matrix transformation.

7.3.5 Evaluation of the degree of coupling

If one knows the values of the phenomenological coefficients of either the conductance or resistance formulation, it is also possible to calculate the degree of coupling of active transport by applying Eq. (7.6). Occasionally, however, it may be convenient to determine q without first determining the phenomenological coefficients and A. Given constancy of A, this is possible by either of two methods (Kedem and Caplan, 1965):

$$q_1 = \sqrt{1 - \frac{(J_r^{sb})_{J_+^a=0}}{J_{r0}^{sb}}} \qquad (7.39)$$

where $J_+^a = 0$ represents a state in which active cation transport is abolished by the use of an appropriate electrochemical potential difference of the cation. Alternatively,

$$q_2 = \sqrt{\frac{J_{+0}^a/J_{r0}^{sb}}{(\partial J_+^a/\partial J_r^{sb})_A}} \qquad (7.40)$$

7.4 Stoichiometry and the degree of coupling

An implicit aspect of classical bioenergetic formulations is the expectation, in analogy to tightly coupled chemical reactions, of a fixed stoichiometric ratio of rates of transport and metabolism. Such a fixed ratio is considered to obtain both in different tissues and in a given tissue studied under varying conditions. First, in the case of different tissues studied in the short-circuited state with identical solutions at each surface, it is clear that there is no thermodynamic limitation on the flow ratio, since under these conditions no electroosmotic work is performed. Although there must of course be mechanistic considerations influencing the magnitude of (J_{+0}/J_{r0}), in principle this quantity might well differ from one tissue to another.

When transport and metabolism are varied by perturbation of the forces, still another consideration becomes relevant. If an active transport system were completely coupled, that is, if $(q^a)^2 = 1$, J_+^a/J_r would be identically equal to Z^a, as can be seen from Eqs. (7.9) and (7.10). Under these circumstances it is appropriate to speak of a stoichiometric ratio. However, J_+^a/J_r has a unique value only if the active transport system is completely coupled. There is no *a priori* reason to assume that this is the case, and it is quite possible that there may be significant benefits associated with nonstoichiometry (Stucki, 1980).

Nevertheless, estimates of the "stoichiometric ratio" have commonly been made on the basis of observed linear relationships between flows. From Eqs. (7.9) and (7.10) it is seen, however, that if A is constant, linearity follows directly from the linearity of the phenomenological equations, irrespective of the degree of coupling. Thus, although $(\partial J^a_+/\partial J_r)_A$ is identically equal to Z^a/q^a, (J^a_+/J_r) is constant only if q^a is unity. The behavior of (J^a_+/J_r) in the more general case is described by Eq. (4.17).

Although for effective functioning the coupling in the active transport pathway must be rather tight, and may in some cases even be complete, clearly the observable system is incompletely coupled, if only because of the leak pathway. This point is of interest relative to the common notions concerning stoichiometry. Apparently stoichiometry is often assumed because of the natural analogy with *in vitro* chemical reactions, where for effectiveness one seeks to eliminate side reactions. It might seem, therefore, that in the case of transport processes it would be desirable to eliminate the equivalent of side reactions, namely leaks. Yet we see that elimination of the leak pathway, although it would impart stoichiometry to the system (if the active transport mechanism itself were completely coupled), would at the same time bring net salt transport to a halt. Therefore it is clear that we cannot consider stoichiometry the be-all and end-all of an ideal transport system. Rather we must take into account the system's mode of operation and what it is "trying to do." If a system is "designed" to maintain a static head, as is the case with symmetrical cells, leak is undesirable. If, however, its function is to maintain a large salt flow against only a small chemical potential gradient, as with epithelial tissues such as the proximal tubule of the kidney, a degree of leakiness is necessary, to the extent that it is consistent with other requirements. If the function of a system is to maintain neither a static head nor a large, near-level flow, but something in between, that is, an appreciable flow against appreciable gradients, still some other pattern for the leak pathways would be desirable, and the optimal situation would change with a change in the system's mode of operation, say in the presence of hormones, drugs, or nervous stimuli stimulating transport. This aspect of active transport mechanisms has not received the attention it would seem to deserve.

7.5 Effectiveness of energy conversion

The formulation developed above permits a self-consistent evaluation of the effectiveness of energy conversion of the active transport system, whatever its state. No unique criterion is adequate for this purpose. Rather the conditions of operation of the system dictate the appropriate criterion to be employed.

When the rate of performance of electroosmotic work is appreciable, we are interested in the efficiency, given here by

$$\eta = -\frac{J_+X_+}{J_r A} = -\frac{(ZX_+ + qA)X_+}{[qX_+ + (1/Z)A]A} \qquad (7.41)$$

Variations of resistance coefficients which increase J_+ also increase η (for fixed $X_+ < 0$ and fixed A); η increases with q^2, but for any finite J_+ it is always less than unity. The nature of the dependence of η on $Z(X_+/A)$ for various values of q may be appreciated by referring back to Fig. 4.3.

Symmetrical cells in the steady state will be at static head, $X_+ = X_+^0$, where the function of the transport system is apparently the maintenance of an electrochemical potential difference. Here it is useful to consider the force developed per given rate of expenditure of metabolic energy, the efficacy of force:

$$\epsilon_{X_+} = -\frac{X_+}{J_r A} \qquad (7.42)$$

The efficacy of force is a monotonically increasing function of $-X_+$, reaching its maximal value, $-R_{+r}/A$, at static head. It is of interest that $\epsilon_{X_+^0}$ is independent of the magnitude of both straight resistance coefficients of the observable system. As with X_+^0, $\epsilon_{X_+^0}$ is decreased by an increase in the magnitude of the leak (Fig. 7.5).

It should be noted that although ion transport is often analyzed in terms of an electrochemical cell without side reactions, for which no expenditure of energy is required to maintain static head, this is not true for an incompletely coupled system. In this case, metabolism must occur even in the absence of transport, as shown by Eqs. (7.15) and (7.16), giving $(J_r)_{J_+=0} = A/R_r$.

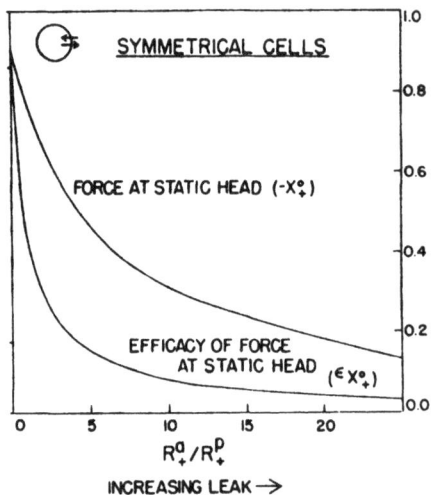

Fig. 7.5. Example of the effect of leak on the force at static head and the efficacy of force at static head for an arbitrarily chosen set of values of the parameters: $q^a = 0.9$, $Z^a = 1.0$, $R^a_+ = 1.0$ kcal · cm² · sec · mole⁻². $A = 1.0$ kcal · mole⁻¹, $R^p_+/R^p_- = 1$.

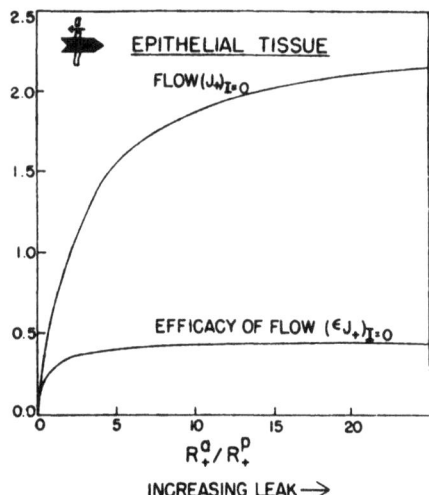

Fig. 7.6. Example of the effect of leak on isotonic saline reabsorption and efficacy of flow at open circuit. The values of the parameters are the same as in Fig. 7.5. See comments in text.

Epithelial tissues often operate near level flow, where the function of the transport system appears to be transport of large quantities of material. In this case a parameter of interest is the rate of transport per given rate of expenditure of metabolic energy,

$$\epsilon_{J_+} = \frac{J_+}{J_r A} = \frac{ZX_+ + qA}{[qX_+ + (1/Z)A]A} \tag{7.43}$$

As with η, for given forces in the driving region ϵ_{J_+} is increased by changes of resistance coefficients which increase the rate of transport. Unlike η, for fixed A, ϵ_{J_+} is a monotonically increasing function of X_+ (unless $q^2 = 1$), reaching qZ/A at level flow.

It was shown above that if X_+ is not fixed experimentally, isotonic saline reabsorption is directly related to the magnitude of the leak, presuming constant permselectivity of the leak pathway and constancy of X_w. The same is true of the efficacy of flow under these conditions (Fig. 7.6).

$$(\epsilon_{J_+})_{I=0} = \frac{-R^a_{+r}}{[R^a_+(1 + Z^2_{+-}) + R^p_-]A} \quad (\Delta c = 0) \tag{7.44}$$

Again, a similar result holds if ion and solvent flows are coupled.

With the aid of the two efficacy functions we are able to evaluate systematically the effectiveness of energy conversion even when the function of a system is not performance of electroosmotic work. However, unlike the efficiency, these quantities are not normalized or dimensionless. Nevertheless, they may be of some use in comparing the adequacy of energy utilization in different systems. Their application is less ambiguous in comparing the function of a given system in different states. Like X^0_+, the functions ϵ_X and ϵ_J are fundamental parameters which can be useful irrespective of mechanisms, linearity, or the response of the metabolic free energy to the change of operating conditions.

7.6 The equivalent circuit model

Many workers have found it natural and useful to analyze transepithelial active Na transport in terms of an analogous electrical

154   Bioenergetics and linear nonequilibrium thermodynamics

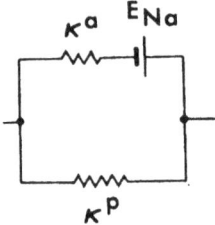

Fig. 7.7. A simple equivalent circuit model for active sodium transport. In accordance with the model of Fig. 7.2, for purposes of analysis we assume that entry of sodium across the apical barrier is passive and that the electromotive force of sodium transport E_{Na} is associated with the function of a sodium pump at the basolateral membrane. Any electrochemical potential difference which may appear across the apical membrane does not constitute an independent contribution to E_{Na}, since it is a consequence of the operation of the pump.

system, a linear equivalent circuit (Fig. 7.7). It is important to consider how the parameters of this model relate to those of the NET formulation and to compare their utility in the interpretation of energetics. For this purpose it is necessary, of course, that both sets of parameters be evaluated during steady states.

In the equivalent circuit model the active pathway consists of a conductance element, to which in the steady state we can assign the value κ_{Na}^a, in series with an "electromotive force of sodium transport" E_{Na}. Salt transport requires the existence of a parallel passive element of conductance κ^p. The electrical current is then given by

$$I = \kappa_{Na}^a E_{Na} - \kappa\Delta\psi = \kappa_{Na}^a(E_{Na} - \Delta\psi) - \kappa^p\Delta\psi \qquad (7.45)$$

and at short circuit

$$I_0 = \kappa_{Na}^a E_{Na} \qquad (7.46)$$

Thus the evaluation of I_0 and $\kappa_{Na}^a \equiv \kappa - \kappa^p$ provides the value of E_{Na}. Setting $I = 0$ in Eq. (7.45) gives the relationship between the open-circuit potential and E_{Na}.

$$(\Delta\psi)_{I=0} = \frac{\kappa_{Na}^a}{\kappa} E_{Na} \qquad (7.47)$$

In applying the above and related formulations, it has been considered that E_{Na} represents the driving force of active sodium transport, and it has been suggested that if one desires to characterize the effect of a given agent it would be appropriate to distinguish between an effect on E_{Na} and an effect on κ_{Na}^a (Ussing and Zerahn, 1951; Ussing and Windhager, 1964). Thus the equivalent circuit model indicated the importance of both kinetic and energetic factors and led to useful experiments attempting to differentiate between the two. However, examination of Fig. 7.7 shows that describing the active transport mechanism by this means fails to take explicitly into consideration the existence of two flows, one of sodium, the other of metabolism. This being the case, the output parameter E_{Na} must in general fail to provide a precise value for the energetic parameter of prime interest—the free energy of the metabolic input process driving transport. This is readily seen by combining Eqs. (7.1), (7.36), and (7.46), giving

$$I_0 = F(J_+)_{\Delta\psi=0} = FL_{+r}^a A = F^2 L_+^a E_{Na}$$

or

$$E_{Na} = \left(\frac{1}{F}\right)\left(\frac{L_{+r}^a}{L_+^a}\right) A \qquad (7.48)$$

Thus, in contrast to the affinity A, the electromotive force E_{Na} reflects both kinetic and energetic factors. This is true even for a completely coupled system; in this case $L_{+r}^a/L_+^a = 1/Z_+^a = J_r/J_{Na}^a$, the reciprocal of the stoichiometric ratio. (For more complex equivalent circuit models which are often invoked, the dependence of E_{Na} on kinetic factors will of course be still more complex. Such models are conveniently dealt with by the techniques of network thermodynamics [Mikulecky, Huf, and Thomas, 1979].)

7.7 Dependence of J_{Na}^a and J_r on A

Both the NET and equivalent circuit analyses above depend on the assumption that perturbations of $\Delta\psi$ of appropriate duration and magnitude will not affect the phenomenological coefficients, A, or

E_{Na}. In Chapter 8 we shall consider experimental results indicating that this assumption is often appropriate.[6] Where this is so, assuming proper pathways as discussed in Chapter 6, Eqs. (7.1) and (7.2) permit appropriate characterization of a system and evaluation of A, irrespective of the effect which an arbitrary perturbation of A would have on J_{Na}^a and J_r.

At this point it is of interest to inquire into the relationship between the flows and the affinity as the affinity varies, either physiologically or as a result of experimental manipulation. This could occur, for example, following the administration of substrates or metabolic inhibitors, drugs, or hormones. While these maneuvers might well influence the phenomenological coefficients also, it is pertinent first to investigate the possible effects of variation only of A.

In general, it is not to be assumed that flows will be linearly dependent on A except near equilibrium, since for an isolated chemical reaction linearity requires $A \ll RT$, where R is the gas constant and T the absolute temperature. The biological situation, however, is of course much more complex and, as was pointed out by Prigogine years ago, biochemical reactions with large A often consist of a large number of elementary reactions in series (Prigogine, 1961). In such cases the A's of the elemental reactions may be sufficiently small that the elemental reactions show linearity. Since in the steady state all series reactions occur at the same rate, it was considered that J_r may then be linear in the overall affinity, which is the sum of the affinities of the individual reactions. This point of view overlooks the fact that the phenomenological L coefficients of the elemental reactions are parameters of state and so might well change appreciably when an alteration of the overall affinity changes the concentrations of metabolic intermediates.

While this means that series reactions in themselves do not necessarily result in linearity in the overall affinity, it remains possible that in certain instances linearity will obtain. For an example of conditions in which this would be the case, consider the simple elemental reaction

$$A \rightleftarrows 2B \qquad (7.49)$$

assuming that Michaelis-Menten kinetics apply. Kinetic analysis then leads to the result (see, for example, Blumenthal, Caplan, and Kedem, 1967):

$$v = \frac{V[1 - (c_B^2/Kc_A)]}{1 + (K_m/c_A) + (V/V')(c_B^2/Kc_A)} \quad (7.50)$$

Here v is the rate of net reaction, V and V' are, respectively, the maximum (saturation) forward and reverse velocities of the reaction, K_m and K are the appropriate Michaelis-Menten and equilibrium constants, and c_A and c_B represent the concentrations of A and B. (When $c_B \simeq 0$, Eq. 7.50 reduces to the Michaelis-Menten equation.) In sufficiently dilute solutions the affinity can be written

$$A = RT \ln (Kc_A/c_B^2) \quad (7.51)$$

Introducing Eq. (7.51) into Eq. (7.50) gives

$$v = \frac{V(1 - e^{-A/RT})}{1 + (K_m/c_A) + (V/V')e^{-A/RT}} \quad (7.52)$$

From this result it is seen that the elemental reaction will indeed be linear if $A \ll RT$, since on expanding the exponential in Eq. (7.52)

$$v = L_r A \quad (7.53)$$

where

$$L_r = \frac{V}{RT[1 + (K_m/c_A) + (V/V')]} \quad (7.54)$$

In order for the overall reaction to be linear it is necessary (and sufficient) that the L_r's of the elemental reactions be insensitive to variation of the overall affinity (with resultant effects on substrate concentrations). For the reaction considered, this condition for a proper pathway is satisfied if

$$K_m \ll c_A[1 + (V/V')] \quad (7.55)$$

Since biological reactions often comprise large numbers of enzymatic

158   Bioenergetics and linear nonequilibrium thermodynamics

reactions in series, of which many depart only slightly from equilibrium (Hess and Brand, 1965), the above arguments suggest that it may not be unreasonable to expect that certain reactions and active transport processes may be near-linear in A over ranges of biological interest. It is not suggested, of course, that this will generally be the case. Adequate investigation of this point will require more detailed information of reaction systems than is yet available. One system of interest is that of mitochondrial oxidative phosphorylation, for which the force–flow relations have been studied by various workers, whose findings will be discussed in Chapter 13.

7.8 Coupled solute and volume flow

An important aspect of active transport is its close correlation with volume flow in the absence of favorable transepithelial osmotic or hydrostatic pressure gradients. In an attempt to explain these relationships, Curran and his coworkers suggested an ingenious model comprising two membranes in series separated by a closed compartment (Fig. 7.8). Membrane I carries out active transport of solute s from compartment 1 to 2, thereby setting up an osmotic pressure difference generating volume flow. The crucial feature of the formulation is that whereas in general it might be expected that the active transport process would be associated with volume flow into the central region, compartment 2, this model predicts volume flow across the composite membrane from compartment 1 to compartment 3. The basis for this behavior is the difference in "effective osmotic

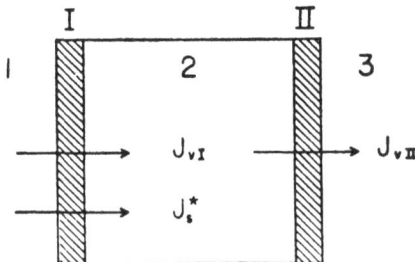

Fig. 7.8. Series membrane system. The central compartment is closed, and the solute is actively transported across membrane I from compartment 1 to compartment 2. (From Katchalsky and Curran, 1965.)

pressure" $\Delta\pi_{\text{eff}} \equiv \sigma RT \Delta c$ induced by solute s across the two series membranes. Thus at membrane I (representing the cellular barriers and the "tight" junction between cells), $\sigma_I \sim 1$, so transport of s from compartment 1 to compartment 2 is associated with water flow between these two regions (from the mucosal bathing solution into the lateral intercellular and/or subepithelial spaces; Fig. 7.9). This water flow results in the buildup of the hydrostatic pressure in compartment 2. At membrane II (representing the basal openings of the intercellular spaces, capillary walls, and/or serosal tissues) $0 < \sigma_{II} \ll 1$, so that the effective osmotic pressure difference tending to cause water flow from compartment 3 to compartment 2 is small and is exceeded by the hydrostatic pressure difference of opposite orientation. Thus there is volume flow from compartment 2 to compartment 3.

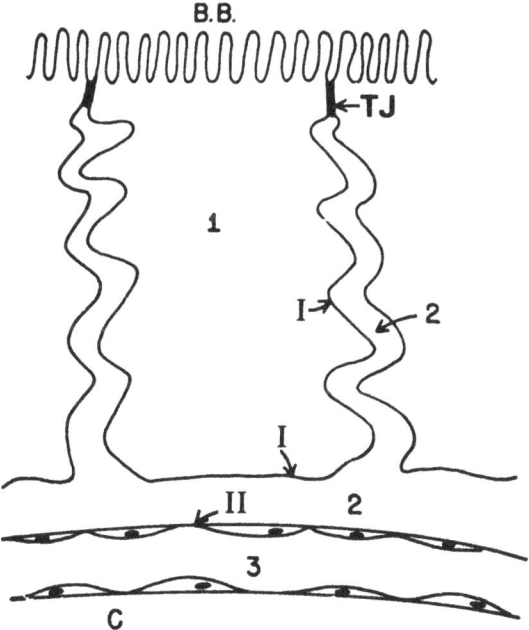

Fig. 7.9. Schematic diagram of the intestinal mucosa indicating possible anatomic counterparts of the double membrane model for water transport. BB is the brush border at the mucosal cell surface, TJ the tight junction, and C a capillary. I and II correspond to the series membranes of Fig. 7.8. (Adapted from Schultz and Curran, 1968.)

160 Bioenergetics and linear nonequilibrium thermodynamics

To express these ideas formally: ignoring the influences of impermeant species and small transepithelial hydrostatic pressure differences, applying Eqs. (3.51) to each membrane and incorporating the rate of active transport J_s^*, leads to Katchalsky and Curran's Eq. (14-73):

$$J_v = \frac{\mathscr{L}(\sigma_I - \sigma_{II})J_s^*}{\omega_I + \omega_{II} + \mathscr{L}c_s(\sigma_I - \sigma_{II})^2} \qquad (7.56)$$

where $\mathscr{L} = L_{pI}L_{pII}/(L_{pI} + L_{pII})$. Thus, as is found experimentally in many epithelia, the rate of volume flow is proportional to the rate of active solute transport. The proportionality coefficient is determined by the properties of the two membranes, in particular the difference in their reflection coefficients.

7.9 Influence of more complex pump mechanisms

For simplicity, the analysis of Sec. 7.2.1 dealt with a pump which carries out active transport of a single cation, uncoupled to flows of other species. Since the mechanism of pump function may well be more complex, it might seem that the present treatment is highly unrealistic. It can be shown, however, that irrespective of specific mechanistic details, the simple linear treatment remains applicable, provided that each elemental function is itself linear. An example is provided by considering a common model for the basolateral Na$^+$ pump, in which Na$^+$ efflux is coupled to K$^+$ influx, with active K$^+$

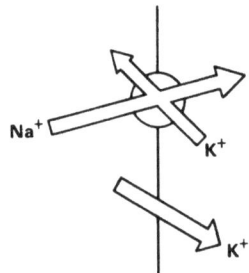

Fig. 7.10. Coupled Na$^+$-K$^+$ pump-leak active transport mechanism at basolateral cell surface.

uptake being compensated by leakage (Fig. 7.10). The fundamental relationships are then for the pump pathway,

$$J_{Na}^a = L_{Na}^a X_{Na} + L_{Na,K}^a X_K + L_{Na,r}^a A \qquad (7.57)$$

$$J_K^a = L_{Na,K}^a X_{Na} + L_K^a X_K + L_{K,r}^a A \qquad (7.58)$$

$$J_r = L_{Na,r}^a X_{Na} + L_{K,r}^a X_K + L_r^a A \qquad (7.59)$$

and for the basolateral leak,

$$J_K^l = L_K^l X_K \qquad (7.60)$$

Considering now the steady state, in which $J_K \equiv J_K^a + J_K^l = 0$, since transepithelial active cation transport $J_+^a \equiv J_{Na}^a$, combining Eqs. (7.57)–(7.60) gives Eqs. (7.1) and (7.2), in which $L_+^a = L_{Na}^a - (L_{Na,K}^a)^2/(L_K^a + L_K^l)$; $L_{+r}^a = L_{Na,r}^a - L_{Na,K}^a L_{K,r}^a/(L_K^a + L_K^l)$; and $L_r = L_r^a - (L_{K,r}^a)^2/(L_K^a + L_K^l)$.

7.10 Summary

1. The energetics of active transport processes are usefully analyzed in terms of a two-flow system in which flow of a single cation is driven by a single metabolic reaction.

2. It is postulated that the electrochemical potential difference X_+ and the affinity of the metabolic driving reaction A can be restricted to "proper" pathways such that the flows J_+ and J_r are linear functions of X_+ and A.

3. The formulation is applicable either to the cation pump or to the composite system comprising series and parallel elements.

4. No assumption is made of completeness of coupling (stoichiometry). Coupling is evaluated by the dimensionless parameter q: $-1 < q < 1$.

5. Important reference stationary states are level flow ($X_+ = 0$) and static head ($J_+ = 0$).

6. The criterion for effectiveness of a transport system depends on its function. Performance of electroosmotic work is evaluated by the efficiency $\eta = -J_+ X_+/J_r A$. Near level flow and static head η is

small; suitable criteria are then the efficacy of flow $\epsilon_{J_+} = J_+/J_r A$ and the efficacy of force $\epsilon_{X_+} = -X_+/J_r A$, respectively.

7. Perturbation of X_+ along a proper pathway while A remains constant permits experimental evaluation of the phenomenological (conductance) L or (resistance) R coefficients, the degree of coupling q, and the affinity A. In contrast to estimates of free energy based on mean tissue concentration ratios, the value of A obtained by this (nondestructive) means reflects the substrate–product activity ratio of a metabolic reaction driving transport, incorporating the influence of local pH, standard free energies, and activity coefficients.

Appendix

An example of the type of system with which we are concerned is shown in Fig. 7.11. This is a scheme for utilization of metabolic energy in the transport of Na^+ from the left side to the right side of a membrane. At some point in the main sequence of reactions there is a subsidiary sequence of consecutive steps directly coupled to some part of the pump mechanism. (It is immaterial which steps are involved.) Our formalism may then be applied either to this local region or to the overall system. The local process driving the transport of Na^+ is the reaction $mM + nN \rightleftharpoons pP + qQ$, for which the affinity[7] A is given by $(m\mu_M + n\mu_N) - (p\mu_P + q\mu_Q)$. M and N may represent ATP and H_2O, while P and Q represent ADP and P_i. From the point of view of the local system, it does not matter how the concentrations of the reactants and products are maintained. (For example, we may have a compartment in which the reactants and products are each close to equilibrium with large reservoirs.) Stoichiometric conversion of M and N to P and Q requires that there be no "branches" in this region of the chain; this is our fundamental assumption. As indicated, however, there may be a number of alternative closed pathways (loops). In the stationary state, consumption of M and N would still be related stoichiometrically to production of P and Q. Some of these loops may not be coupled to the pump, and it is clear that although they would not influence the validity of the formalism, they would decrease the degree of coupling between metabolism and transport. It should be appreciated that even locally the rate of dissipation of free energy must be expressed as the input minus the out-

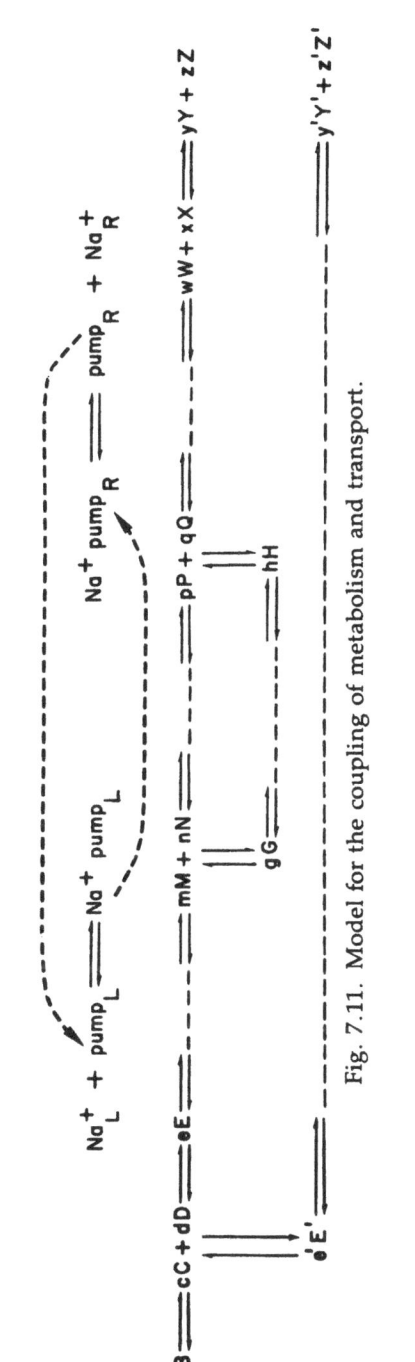

Fig. 7.11. Model for the coupling of metabolism and transport.

put. The input term is $J_r A$, where the affinity A, which has been defined above, constitutes the driving force of the process. In this view the system may be regarded as a black box, with $mM + nN \to pP + qQ$ as the input, and the transport of Na^+ from left to right as the output (Fig. 7.1).

It might seem that it would be useful to regard the interconversion of the species $pump_L$, $pump_R$, $Na^+ pump_L$, and $Na^+ pump_R$ as contributing to the input. That this is not the case can be seen from the following argument. The dissipation function for the black box of Fig. 7.1 is

$$\Phi = J_{Na} X_{Na} + J_r A \qquad (7.A1)$$

Alternatively we may consider reactions taking place within the black box:

(α) $\qquad\qquad Na_L^+ + pump_L \to Na^+ pump_L$

(β) $\qquad\qquad Na^+ pump_L \to Na^+ pump_R$

(γ) $\qquad\qquad Na^+ pump_R \to pump_R + Na_R^+$

(δ) $\qquad\qquad pump_R \to pump_L$

(ϵ) $\qquad\qquad mM + nN \to pP + qQ$

Each of these reactions may of course be the sum of several subreactions. The combination of (ϵ) and one of the other reactions may possibly represent the actual mechanism of coupling. In any case the dissipation function is given by

$$\Phi = J_\alpha A_\alpha + J_\beta A_\beta + J_\gamma A_\gamma + J_\delta A_\delta + J_r A \qquad (7.A2)$$

where the affinities, expressed in terms of the electrochemical potentials, are

$$A_\alpha = \tilde{\mu}_{Na_L^+} + \tilde{\mu}_{pump_L} - \tilde{\mu}_{Na^+ pump_L}$$

$$A_\beta = \tilde{\mu}_{Na^+ pump_L} - \tilde{\mu}_{Na^+ pump_R}$$

$$A_\gamma = \tilde{\mu}_{Na^+pump_R} - \tilde{\mu}_{pump_R} - \tilde{\mu}_{Na_R^+}$$

$$A_\delta = \tilde{\mu}_{pump_R} - \tilde{\mu}_{pump_L}$$

In the stationary state,

$$J_\alpha = J_\beta = J_\gamma = J_\delta = J_{Na} \qquad (7.A3)$$

and therefore

$$\Phi = J_{Na}(A_\alpha + A_\beta + A_\gamma + A_\delta) + J_r A \qquad (7.A4)$$

It is readily seen that Eqs. (7.A1) and (7.A4) are identical.

We have considered $J_r A$ the input and $-J_{Na}X_{Na}$ the output. Alternatively, since for example the terms $J_\alpha A_\alpha$, $J_\gamma A_\gamma$, and $J_\delta A_\delta$ may be positive, that is, represent spontaneous processes, it might be possible to consider these, as well as $J_r A$, as inputs. The output would then be $-J_\beta A_\beta$. This approach seems to us unnatural and unproductive, since the pump is involved in a cyclic process. As pointed out by Katchalsky and Spangler (1968), the circulation of a species makes no formal contribution to the dissipation function, although it influences the phenomenological coefficients.

For some purposes it is more convenient to take a more global view than that considered above and to analyze the system in terms of the affinity over some larger region of the chain, providing of course that this region encompasses the reaction $mM + nN \rightleftharpoons pP + qQ$ and includes no branches which fail to return. This would be true of the whole reaction sequence if Y' were identical to Y, and Z' were identical to Z, and both were in the same compartment (for example, CO_2 and H_2O). In this case there is a single metabolic reaction, the affinity of which is given by $(a\mu_A + b\mu_B) - (y\mu_Y + z\mu_Z)$. Another possibility whereby the entire reaction sequence could represent a single metabolic input would be if E were produced wholly by breakdown of C, and E' wholly by breakdown of D. In this case the overall reaction is $aA + bB \rightleftharpoons yY + zZ + y'Y' + z'Z'$, and the affinity is given by $(a\mu_A + b\mu_B) - (y\mu_Y + z\mu_Z + y'\mu_{Y'} + z'\mu_{Z'})$. In general, the above possibilities do not obtain, and each branch would represent a different reaction, one being characterized by the affinity $(a\mu_A + b\mu_B) - (y\mu_Y + z\mu_Z)$, and the other by the affinity

166   Bioenergetics and linear nonequilibrium thermodynamics

$(a\mu_A + b\mu_B) - (y'\mu_{Y'} + z'\mu_{Z'})$. There is no *a priori* reason for two such reactions to be related stoichiometrically and thus we have, in the global view, two input processes for this scheme. An analysis from this viewpoint is not carried out here. However, in some systems the expenditure of free energy by one or more subsidiary flows may be small enough to be ignored. A possible example is mammalian kidney, where O_2 consumption in the absence of Na^+ reabsorption may fall to as little as 20% of that associated with high rates of Na^+ reabsorption.

Although we are in principle free to take either a local viewpoint or a more or less global viewpoint, the degree of coupling will depend on our choice. Consider the region $cC + dD \rightleftharpoons yY + zZ$. This comprises the following partial sequences:

(λ) $\qquad cC + dD \rightleftharpoons mM + nN$

(σ) $\qquad mM + nN \rightleftharpoons pP + qQ$

(τ) $\qquad pP + qQ \rightleftharpoons yY + zZ$

From the local point of view we may apply Eqs. (7.3) and (7.4), where the reaction (σ) is the metabolic input, that is,

$$X_{Na} = R^a_{Na} J^a_{Na} + R^a_{Na,r} J_r \qquad (7.A5)$$

$$A_{(\sigma)} = R^a_{Na,r} J^a_{Na} + R^a_{r(\sigma)} J_r \qquad (7.A6)$$

From the global point of view we must consider the additional phenomenological relations:

$$A_{(\lambda)} = R^a_{r(\lambda)} J_r \qquad (7.A7)$$

$$A_{(\tau)} = R^a_{r(\tau)} J_r \qquad (7.A8)$$

Adding Eqs. (7.A6), (7.A7), and (7.A8), we have

$$A_{global} = A_{(\lambda)} + A_{(\sigma)} + A_{(\tau)}$$
$$= R^a_{Na,r} J^a_{Na} + (R^a_{r(\lambda)} + R^a_{r(\sigma)} + R^a_{r(\tau)}) J_r \qquad (7.A9)$$

Equations (7.A5) and (7.A9) are the phenomenological equations of the global system. The degree of coupling in the global viewpoint,

$$\frac{-R^a_{\text{Na},r}}{\sqrt{R^a_{\text{Na}}(R^a_{r(\lambda)} + R^a_{r(\sigma)} + R^a_{r(\tau)})}}$$

is clearly less than that in the local viewpoint except in the case of complete coupling (see note 1 to this chapter). Although the local and global treatments lead to different values of q, Z, and A, no inconsistency arises.

In attempting to apply the present formulation experimentally, it seems natural to look at the overall system and to consider the concentrations of exogenous substrates and of final end products. However, this may not always be possible. For example, if there is branching in the metabolic chain, then in general the system cannot be described in terms of a single input, as discussed above. Another possibility is that the affinity over some subsidiary portion of the chain is maintained constant by a self-regulatory process (which would generally imply nonlinearity in the remainder of the chain). If so, this affinity is a characteristic property of the system and would be the parameter determined by application of the linear Eqs. (7.34) and (7.35). Even in systems in which a global view is experimentally practicable, it may be useful to analyze subsystems as well. This approach has been taken with considerable success in the case of glycerin-extracted muscle fibers which have no endogenous source of ATP. For such fibers it is possible in principle to regulate the affinity of the ATP hydrolysis reaction experimentally. A similar technique may one day be applicable to epithelial tissues, but this would require major technical advances.

8 Energetics of active transport: experimental results

In the previous chapter, in reaction to the inadequacies of classical approaches to the energetics of active transport processes, we attempted to provide a comprehensive self-consistent formulation. It was shown that if active transport and the associated metabolism can be characterized by a simple linear formalism, pertinent kinetic and energetic parameters can be evaluated under a variety of circumstances. In addition to providing a basis for a meaningful analysis of energetic considerations per se, such an approach should lead to insights concerning the mechanism of action of factors influencing active transport. In the present chapter we shall examine the experimental evidence concerning the validity and utility of the linear NET formulation for active transport, primarily for the case of the sodium ion in anuran epithelia, and we shall consider various problems in attempting to test and apply the theoretical formalism. Although, for reasons which will be discussed, the process of testing is as yet incomplete, the results to date are sufficiently encouraging to justify continuing efforts to apply linear NET to a variety of systems.

8.1 Validation of linear phenomenological equations

8.1.1 Choice of experimental preparations

In principle any biological tissue permits a valid and meaningful test of the range of validity of the linear NET formulation of active transport, but in practice it is important to consider practical considerations which may strongly affect the feasibility of such a program. For initial studies it is advantageous to employ an epithelial tissue exposed to physiological solutions at each surface. Since both solutions are readily accessible, it is possible to monitor and regulate certain of the pertinent forces, the transepithelial electrochemical potential differences of any transported species of interest. With the use of appropriate volumes and suitable measuring techniques, the solution composition may be maintained nearly constant for extended periods, while carrying out precise measurements of rates of transport and metabolism. Unfortunately, such preparations do not permit fixing the affinities of metabolic reaction at desired levels; indeed, given the complexity of tissue and cellular structure, the mere evaluation of pertinent affinities will generally prove impossible with existent biochemical techniques. In contrast to asymmetric epithelial tissues, symmetrical cells such as red blood cells and muscle present important technical difficulties in the measurement of transmembrane electrochemical potential differences. Furthermore, in long-term steady states, net flows must be zero, and measurements of rates of oppositely directed active and passive components are difficult and imprecise; the imposition of conditions resulting in net flow must soon change intracellular parameters.

An important practical consideration in the choice of experimental preparations is that the formulation to be tested must be of the simplest possible character, with a single transport process being driven by a single metabolic reaction. Fortunately for our purposes, Ussing and his colleagues' studies with frog skin and Leaf's studies with Dominican Republic toad urinary bladder have demonstrated that in appropriate species and circumstances, transepithelial active transport is restricted almost entirely to sodium and is largely attributable to oxidative metabolism (Ussing and Zerahn, 1951; Zerahn, 1956; Leaf, Anderson, and Page, 1958). The study of such simple systems, in addition to being of interest in its own right, should pro-

170 Bioenergetics and linear nonequilibrium thermodynamics

vide a useful basis for later analysis of more complex systems involving multiple and/or coupled transport processes.

For these various reasons we have chosen to test the linear NET formulation for active transport by the study of sodium transport and associated oxygen consumption in frog skins and toad urinary bladders. A diagram of the set-up used for these purposes is shown in Fig. 8.1. Since we can vary only the electrochemical potential difference X_{Na} independently, it is not possible to test the formalism completely. As we shall see, however, under certain circumstances it is reasonable to expect near-constancy of the affinity, in which case it is possible to apply the formalism in a meaningful way.

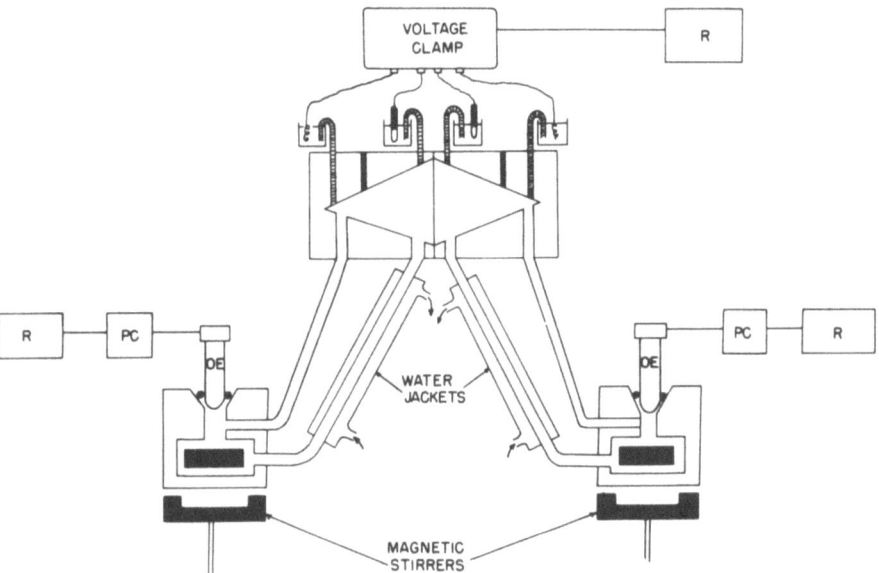

Fig. 8.1. Apparatus for simultaneous measurement of electrical parameters and oxygen consumption in the frog skin or toad urinary bladder. When either of these tissues is exposed to identical sodium–Ringer's solutions at each surface, an electrical potential difference is generated, reflecting active sodium transport from the outer surface of the frog skin (or mucosal surface, M, of the toad bladder) to the inner or serosal surface (S). OE = oxygen electrode; PC = polarographic circuit; R = recorder. (Adapted from Vieira, Caplan, and Essig, 1972a.)

8.1.2 Choice of experimental formulation

The linear phenomenological equations may be written with either the forces or flows as independent variables, that is, as L (conductance) or R (resistance) formulations, respectively. From a theoretical point of view the two formulations are equivalent. From the experimental point of view the choice is a matter of convenience. In view of the ease of manipulating the electrochemical potential difference of sodium X_{Na}, and the presumed constancy of the affinity A on brief perturbation of X_{Na}, it is convenient to employ the L formulation and to determine the effects of variation of X_{Na} on the flows of sodium and metabolism. Thus, for the active sodium transport system our fundamental equations are:

$$J_{Na}^a = L_{Na}X_{Na} + L_{Na,r}A \tag{8.1}$$

$$J_r^{sb} = L_{Na,r}X_{Na} + L_rA \tag{8.2}$$

Since in this chapter we deal primarily with ion flows by way of the active pathway, we shall omit the superscript a of the phenomenological coefficients. J_{Na}^a is taken as positive in the direction from the outer (mucosal) to the inner (serosal) surface of the tissue. J_r^{sb} represents the rate of suprabasal oxygen consumption, assumed to be independent of oxygen consumption associated with metabolic functions other than transport. The basis for assuming equality of the cross coefficients is the requirement that X_{Na} be restricted to a proper pathway.

8.1.3 Effect of experimental protocol on parameters of NET formulation

In attempts to characterize transport and metabolism in terms of the linear treatment discussed above, it is evidently essential that the phenomenological parameters of the system to be evaluated remain near-constant in the course of perturbations of the external variables. Only in this case can J_{Na}^a and J_r be expected to be linear functions of X_{Na}. In general, however, the L's (R's) and A will be functions of state and may therefore quite possibly be influenced by manipulations which appreciably alter tissue configuration and/or composition.

172 Bioenergetics and linear nonequilibrium thermodynamics

The importance of such considerations was shown by the early observations of a "memory effect" following prolonged perturbations of $\Delta\psi$. Thus, if the potential is perturbed so as to greatly enhance Na transport and the associated metabolism for an extended period, on return to the short-circuit state both I_0 and J_{ro} are less than they were initially. These results are shown by the open circles of Fig. 8.2. The solid circles show the converse effect, noted after perturbing $\Delta\psi$ so as to slow transport and metabolism. These alterations of I_0 and J_{ro} from control levels clearly indicate changes in $L_{Na,r}$, L_r, and/or A, and emphasize the significance of the experimental means employed to vary X_{Na}, and thus J_{Na}^a and J_r, in the characterization of active transport.

A priori it seems that manipulations of transmembrane electrical

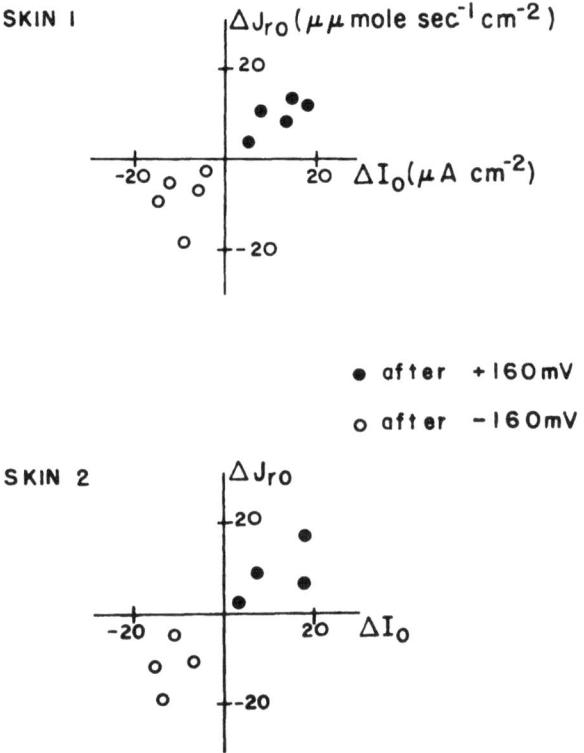

Fig. 8.2. Effect of electrical potential perturbations on subsequent values of short-circuit current I_0 and rate of oxygen consumption J_{ro} in the frog skin. (From Vieira, Caplan, and Essig, 1972b.)

potential difference are less likely to alter system parameters than alterations of bathing fluid Na concentration, which may well affect intracellular composition and the size and organization of both paracellular and cellular compartments. Accordingly, in our studies we have varied X_{Na} by perturbation of $\Delta\psi$ alone. Equations (8.1) and (8.2) then become

$$J_{Na}^a = L_{Na}(-F\Delta\psi) + L_{Na,r}A \qquad (8.3)$$

$$J_r^{sb} = L_{Na,r}(-F\Delta\psi) + L_r A \qquad (8.4)$$

(Here $\Delta\psi = \psi^{in} - \psi^{out}$.)

Even when the bath concentrations are maintained constant, the considerations implicit in the memory effect result in practical difficulties: on the one hand, in order to characterize the NET parameters precisely, the system must be in a quasi-steady state during observations both before and after alteration of $\Delta\psi$, hence the period of perturbation must not be too short. On the other hand, in order that the NET parameters remain near-constant, the period of perturbation must not be too long. An additional complication is instability of the epithelia under study, meaning that a complete sequence of perturbations must be carried out sufficiently expeditiously to avoid significant variation in base-line function.

The conditions we have found most appropriate for analysis in terms of a linear formalism are discussed in detail below.

8.1.4 Dependence of J_{Na}^a on $\Delta\psi$

With identical solutions at each surface, if indeed the rate of active sodium transport is a linear function of the forces promoting transport, one might expect to find steady-state linear current–voltage relationships. In practice, as would be expected, the nature of the $I-\Delta\psi$ relationship is strongly influenced by the protocol employed. Thus if $\Delta\psi$ is varied monotonically at intervals of fractions of a second, the $I-\Delta\psi$ curve consists of discrete linear segments joined at "break points" (Civan, 1970; Helman and Miller, 1971). If $\Delta\psi$ is varied symmetrically relative to 0 mV at 5–10 s intervals, the $I-V$ curve is near-linear over a range of some $-75 \rightarrow 100$ mV (Saito, Lief, and Essig, 1974; Wolff and Essig, 1980). These observations suggest lin-

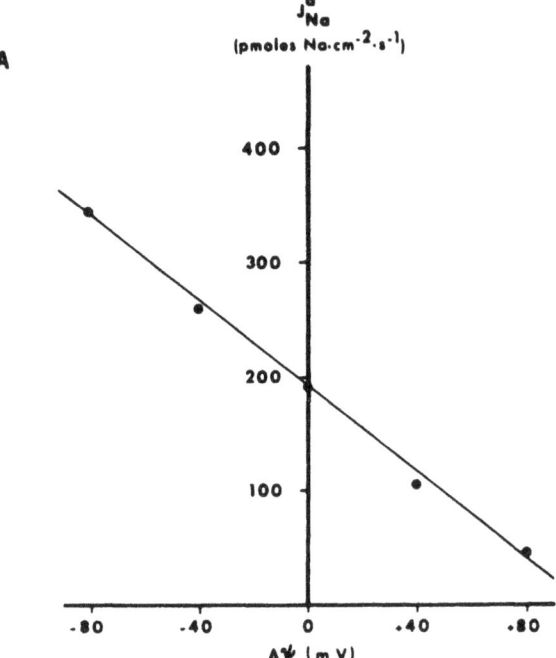

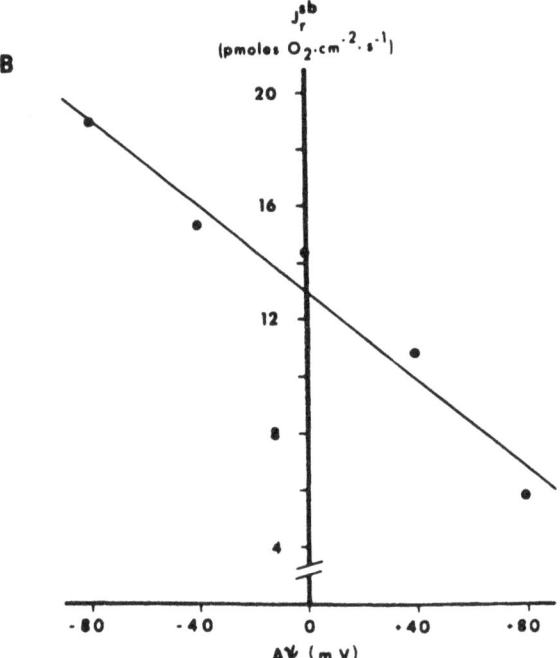

earity of active Na transport but are not completely convincing, since with brief perturbations of $\Delta\psi$ it is not clear to what extent I represents transient K and Cl fluxes rather than Na flux at the basal lateral surface. If, however, the potential is perturbed for extended periods, the $I-\Delta\psi$ curve is clearly nonlinear; with 15-min intervals there is saturation at negative settings of $\Delta\psi$, suggesting effects on cellular parameters (Wolff and Essig, 1980).

Perhaps the most convincing examples of linearity of J_{Na}^a in $\Delta\psi$ occur in studies in the toad urinary bladder and frog skin in which $\Delta\psi$ was perturbed symmetrically at 6-min intervals over the range $0 \to \pm 80$ mV (Lang, Caplan, and Essig, 1977b). Since in Dominican toad urinary bladders mounted in chambers, only sodium is transported actively across the epithelium at an appreciable rate, the dependence of J_{Na}^a on $\Delta\psi$ can be determined explicitly from the steady-state current–voltage relationship, given a means for evaluating the contribution of passive ionic flows. Thus

$$I = I^a + I^p$$
$$= FJ_{Na}^a - \kappa^p \Delta\psi$$

$$J_{Na}^a = \frac{1}{F(1 + \kappa^p \Delta\psi)} \tag{8.5}$$

Since the application of 10^{-5} M amiloride to the mucosal surface of the toad urinary bladder reduces the rate of active sodium transport to near-zero, without evident effect on κ^p (Hong and Essig, 1976; Labarca, Canessa, and Leaf, 1977), this agent may be used to evaluate J_{Na}^a. The demonstration of linear current–voltage relationship 4–6 min after symmetrical perturbation of $\Delta\psi$, in both the presence and absence of amiloride, indicated that J_{Na}^a was indeed a linear function of $\Delta\psi$ under these circumstances (Fig. 8.3), particularly when considered in conjunction with the studies of rates of oxygen consumption to be discussed below (Lang, Caplan, and Essig, 1977b).

The demonstration of linearity of the active transport process indicates that the phenomenological coefficients and the affinity were

---

Fig. 8.3. (A) Relationships between J_{Na}^a and $\Delta\psi$ and (B) relationships between J_r^{ab} and $\Delta\psi$ in the toad urinary bladder. (From Lang, Caplan, and Essig, 1977b.)

unaffected by perturbations of $\Delta\psi$ of the magnitudes and duration employed. This suggests that with perturbation of $\Delta\psi$ at constant Δc_{Na}, X_{Na} may lie on a proper pathway. Alternative possibilities, that the phenomenological coefficient $L_{Na,r}$ or the affinity may be a linear function of $\Delta\psi$, seem unlikely.

8.1.5 Dependence of J_r^{sb} on $\Delta\psi$

Having presented experimental evidence supporting the validity of Eq. (8.3) for the analysis of active transport in conditions of physiological interest, we turn now to the analogous relation for the metabolism associated with active transport, Eq. (8.4). In the tissues under study, active sodium transport is largely attributable to oxidative metabolism. As shown in Fig. 8.1, in the studies to be described J_r was determined by the use of oxygen electrodes; the slope of the plot of oxygen tension against time, evaluated over a period of 2 minutes, provided a measure of the rate of oxygen consumption. In evaluating J_r in this manner it is critically important, as in the study of transport, that the tissue be in a quasi-steady state during the course of measurements, both in the control period and following perturbation of $\Delta\psi$. To facilitate the achievement of steady states, measurements were again made during the interval 4 to 6 minutes after perturbations of $\Delta\psi$; possible long-term changes in tissue function were monitored by repeated determinations of the short-circuit current I_0 and the associated rate of oxygen consumption J_{r0}.

First studies were performed in frog skins. Characteristically, with symmetrical perturbations the relationship between J_r and $\Delta\psi$ was linear over a range of at least ± 70 mV, occasionally ± 100 mV. In order to examine the relationship between J_r and $\Delta\psi$ in greater detail, we must take several determinations of J_r, requiring long periods during which appreciable spontaneous decline of sodium transport and metabolism may occur. Since such instability interferes with the determination of the relationship between J_r and $\Delta\psi$, it was important to search for a more stable preparation. For this purpose we employed prolonged exposure to 10^{-6} M aldosterone in glucose–Ringer's solution. This resulted in stability, with a significantly larger value of I_0 than in an untreated skin from the same animal. Furthermore, J_{r0} was comparable to that in freshly mounted skins and was also relatively stable.

Active transport: experimental results 177

With the use of the more stable preparation, linearity was demonstrable over a large range of $\Delta\psi$. Figure 8.4 shows the result of one such experiment. As is seen, linearity was observed over a range of ±160 mV. (In all such studies we avoided several positive or negative periods in succession, as, for example, 0, −40, −80, −120, −160, −200, followed by the positive values, in order to prevent systematic effects of polarity.) Further observations have confirmed the linearity of the relationship between the rate of oxygen consumption and $\Delta\psi$, when $\Delta\psi$ is perturbed symmetrically and suitable precautions are taken to assure stability in both the frog skin (Saito, Essig, and Caplan, 1973; Owen, Caplan, and Essig, 1975b; Lahav, Essig, and Caplan, 1976), and the toad urinary bladder (Lang, Caplan, and Essig, 1977b).[1]

In relating the rate of metabolism to forces across the membrane, it is of course important that the results of variation of $\Delta\psi$ be specific,

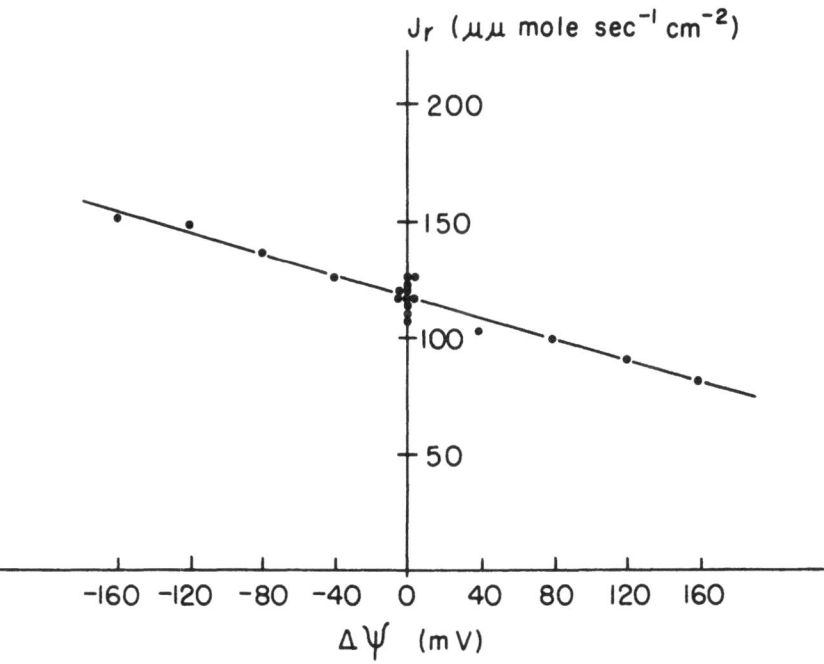

Fig. 8.4. Dependence of the rate of oxygen consumption on the electrical potential difference $\Delta\psi$ in the presence of aldosterone. (From Vieira, Caplan, and Essig, 1972b.)

178 Bioenergetics and linear nonequilibrium thermodynamics

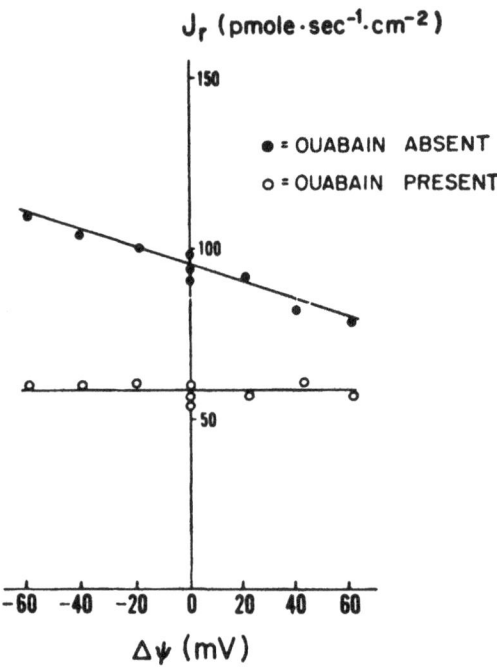

Fig. 8.5. Dependence of the rate of oxygen consumption J_r on the electrical potential difference $\Delta\psi$ in the aldosterone-treated frog skin; influence of ouabain (10^{-3} M) added 30 min prior to perturbation of $\Delta\psi$. (From Saito, Essig, and Caplan, 1973.)

reflecting intrinsic changes in the function of the active transport system. Specificity was shown by the insensitivity of J_r to $\Delta\psi$ after sodium transport had been blocked by ouabain (Vieira, Caplan, and Essig, 1972b; Saito, Essig, and Caplan, 1973) or amiloride (Lahav, Essig, and Caplan, 1976; Labarca, Canessa, and Leaf, 1977). This is seen in Fig. 8.5.

Various lines of evidence suggest that the rate of oxygen consumption after the blockage of active sodium transport by ouabain or amiloride is a good estimate of the basal rate of oxygen uptake unrelated to transepithelial sodium transport (Lau, Lang, and Essig, 1979). Therefore, the results obtained following administration of ouabain indicate that the basal rate of oxygen consumption was unaffected by changes in the electrical potential difference across the skin. Taken together with observations such as those shown in Fig.

8.4, the results suggest linearity of the rate of suprabasal oxygen consumption in $\Delta\psi$. As for the analogous case of linearity of the rate of active sodium transport, the results indicate constancy of the phenomenological coefficients and the affinity during perturbations of $\Delta\psi$ of the magnitudes and duration employed, suggesting again that in this case X_{Na} may lie on a proper pathway.

8.1.6 Dependence of J_{Na}^a and J_r^{sb} on X_{Na}

The above demonstrations of linearity of J_{Na}^a and J_r^{sb} in $\Delta\psi$ are gratifying in that they point to the possibility of meaningful characterization of the parameters of active transport mechanisms by means of Eqs. (8.3) and (8.4). While the use of these equations would provide in itself a basis for the systematic study of a transport system under varied circumstances, we are often interested in tissues whose opposite surfaces are exposed to differing conditions that may change with time. Accordingly, it is pertinent to inquire to what extent active sodium transport may be characterized more generally by means of Eqs. (8.1) and (8.2), describing the dependence of transport and metabolism on the total electrochemical potential difference of sodium, incorporating the contribution of the chemical potential difference of sodium $\Delta\mu_{Na}$, as well as the electrical potential difference, $\Delta\psi$. As indicated above, since the phenomenological coefficients are, in principle, functions of state, it cannot be assumed that rates of transport and metabolism will show the same response to a given thermodynamic force X_{Na}, irrespective of the relative contributions of $\Delta\mu_{Na}$ and $F\Delta\psi$. This is a matter which can be determined only by experiment.

This issue has been studied by Vieira and colleagues, working with abdominal toad skins. For convenience, studies were carried out with the transmembrane electrical potential difference clamped at zero. It was first shown that over the range of external sodium concentrations employed (some 5–110 meq/L, as compared with 110 meq/L in the internal solution) the rate of passive efflux was nearly constant, and the rate of active sodium transport was reasonably well predicted from the short-circuit current. A study was then made of the dependence of the rate of active sodium transport (short-circuit current) on the external sodium concentration while maintaining the internal sodium concentration constant. As is seen in Fig. 8.6,

180 Bioenergetics and linear nonequilibrium thermodynamics

short-circuit current was a linear function of ln (c_{Na}^e/c_{Na}^i), that is, the rate of active sodium transport was a linear function of the chemical potential difference of sodium $\Delta\mu_{Na}$ across the membrane, in consistency with Eq. (8.1). In an associated series of studies (Fig. 8.7), it was concluded that the rate of suprabasal oxygen consumption is also a linear function of $\Delta\mu_{Na}$ (Danisi and Vieira, 1974).

On the basis of these observations in conjunction with the earlier findings, it seems reasonable to conclude that in the anuran epithelia studied the rates of both active sodium transport and associated oxygen consumption are linear functions of the electrochemical potential difference of sodium, whether X_{Na} is perturbed by varying $\Delta\psi$ or by varying the external Na concentration. A note of caution is in order, however, since it has not yet been demonstrated that in a given tissue the response to perturbation of the electrical potential difference is quantitatively the same as that to a thermodynamically

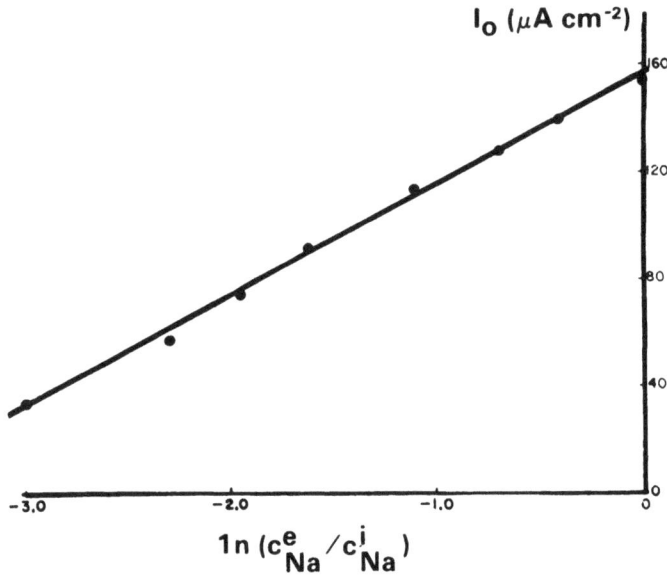

Fig. 8.6. Dependence of the rate of active sodium transport (short-circuit current) on the chemical potential difference of sodium $\Delta\mu_{Na}$ across the toad skin. The external sodium concentration was varied, while the internal sodium concentration was kept constant. The electrical potential difference $\Delta\psi$ was zero. (Adapted from Danisi and Vieira, 1974.)

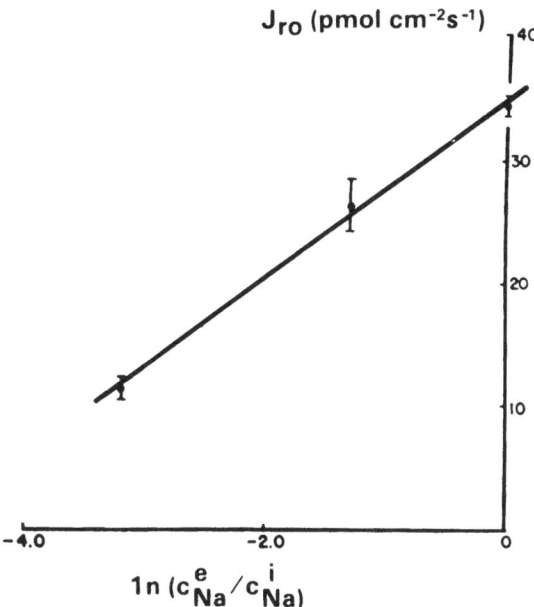

Fig. 8.7. Dependence of the mean rate of suprabasal oxygen consumption on the chemical potential difference of sodium $\Delta\mu_{Na}$ across the toad skin. The external sodium concentration was varied, while the internal sodium concentration was kept constant. The electrical potential difference was zero. (Adapted from Danisi and Vieira, 1974.)

equivalent change in the chemical potential difference. This must await combined studies of J_{Na}^a and J_r^{ab} during perturbation of both $\Delta\psi$ and $\Delta\mu_{Na}$ in the same tissue.

When X_{Na} was altered by change of internal Na concentration, linearity was no longer observed (Varanda and Vieira, 1978); hence Eqs. (8.1) and (8.2) are inapplicable. This nonlinearity might well result from changes in tissue microstructure and composition.

8.2 Evaluation of the NET parameters

8.2.1 Evaluation of the affinity

The results discussed above are consistent with the validity of a linear NET analysis of active transport, provided that X_{Na} is varied

182 Bioenergetics and linear nonequilibrium thermodynamics

along a proper pathway (thermodynamic linearity), while A remains constant. In the case of kinetic linearity, modification of the treatment will be necessary, as is discussed in Chapter 13.

Assuming that perturbation of $\Delta\psi$ provides a proper pathway, we can evaluate the affinity by use of either of the two expressions developed in Chapter 7, Eq. (7.34) or (7.35). (For tissues in which perturbation of the external Na concentration also provides a proper pathway, it is possible to supplement these equations with appropriate expressions in terms of $\Delta\mu_{Na}$.) Experimentally, the most convenient expression will generally be that involving the short-circuit current and the voltage dependence of the rate of oxygen consumption:[2]

$$A = \frac{-I_0}{dJ_r/d(\Delta\psi)} \quad (A \text{ constant}) \quad (8.6)$$

It should be recalled that A represents the free energy change for a characteristic region of the metabolic chain for which A remains constant on perturbation of $\Delta\psi$. We shall occasionally use the symbol A_{O_2} to remind us that in our studies A was expressed in terms of the free energy change per mole of O_2. Values of A_{O_2} in untreated frog skins and toad urinary bladders have ranged from some 20–80 kcal · mole^{-1} O_2. Since it is thought that active transport in these tissues is driven by ATP, we would like to reexpress this affinity in terms of ATP utilization, but we are hampered by lack of knowledge of the stoichiometry of oxidative phosphorylation in intact epithelia. If, however, we assume a P/O ratio of 3, as may be observed in healthy mitochondria metabolizing appropriate substrates, each mole of O_2 consumed will result in synthesis (and in the steady state, utilization) of 6 moles of ATP, giving values of A_P ranging from about 3 to 13 kcal · mole^{-1} ATP. These values are in the neighborhood of those estimated from tissue measurements, for example, some 7.6 kcal · mole^{-1} ATP under standard conditions and Wilson et al.'s 11.4 kcal · mole^{-1} ATP under conditions obtaining in isolated rat liver cells (Wilson et al., 1974). The theoretical advantages of analyzing energetics by the evaluation of A rather than from measurements of mean tissue metabolite concentrations have been considered in Sec. 7.3.3. Clearly, however, a complete understanding of the energetics of active Na transport requires knowledge not only of A_{O_2}, but also of

Active transport: experimental results 183

the P/O ratio obtaining in the conditions under study, thus permitting the calculation of A_P. Such information is as yet unavailable.

8.2.2 Effects of model compounds on A

Definitive clarification of the significance of the affinity calculated by the thermodynamic method above must await experiments correlating thermodynamic studies with a variety of biochemical procedures. Meanwhile, it is of interest to examine effects on the affinity in several circumstances in which transport and metabolism were altered by various means. Two studies employed agents which depress transport without direct effects on metabolism. One such agent is the cardiac glycoside ouabain, an inhibitor of the sodium-potassium-ATPase generally identified with the sodium pump. When ouabain is administered in a concentration adequate to eliminate active sodium transport, the dependence of oxygen consumption on the electrical potential difference is abolished. In order to apply the formulation for evaluation of the affinity, it is necessary to depress active sodium transport substantially, but not completely, so that A in Eq. (8.6) will remain determinate. This was accomplished by using a low concentration of ouabain, 10^{-7} M. Within 2½ hours of exposure to this concentration of ouabain, the short-circuit current in the treated experimental hemiskin had gradually fallen to a level about half that in the paired control hemiskin. This was accompanied by a significant depression of the sensitivity of oxygen consumption to perturbation of $\Delta\psi$, as is shown in Fig. 8.8, representing a typical result in nine experiments. Incomplete inhibition of sodium transport for a relatively short period was not associated with a significant effect on the affinity (Fig. 8.9).

A second inhibitor of transport, the diuretic amiloride, interferes with passive Na entry at the outer surface. In studies with frog skins, amiloride was applied to the outer surface in concentrations sufficient to depress I_0 to about a third of the initial level for 4 hours (Fig. 8.10). Initially, the apparent affinity was the same in the control and experimental tissues. One hour after the administration of amiloride, A was not affected demonstrably (Fig. 8.11). Four hours after the administration of amiloride, however, A in the treated tissues was significantly greater than initially and significantly greater than simultaneously in the paired control tissues. Elevation of A might explain

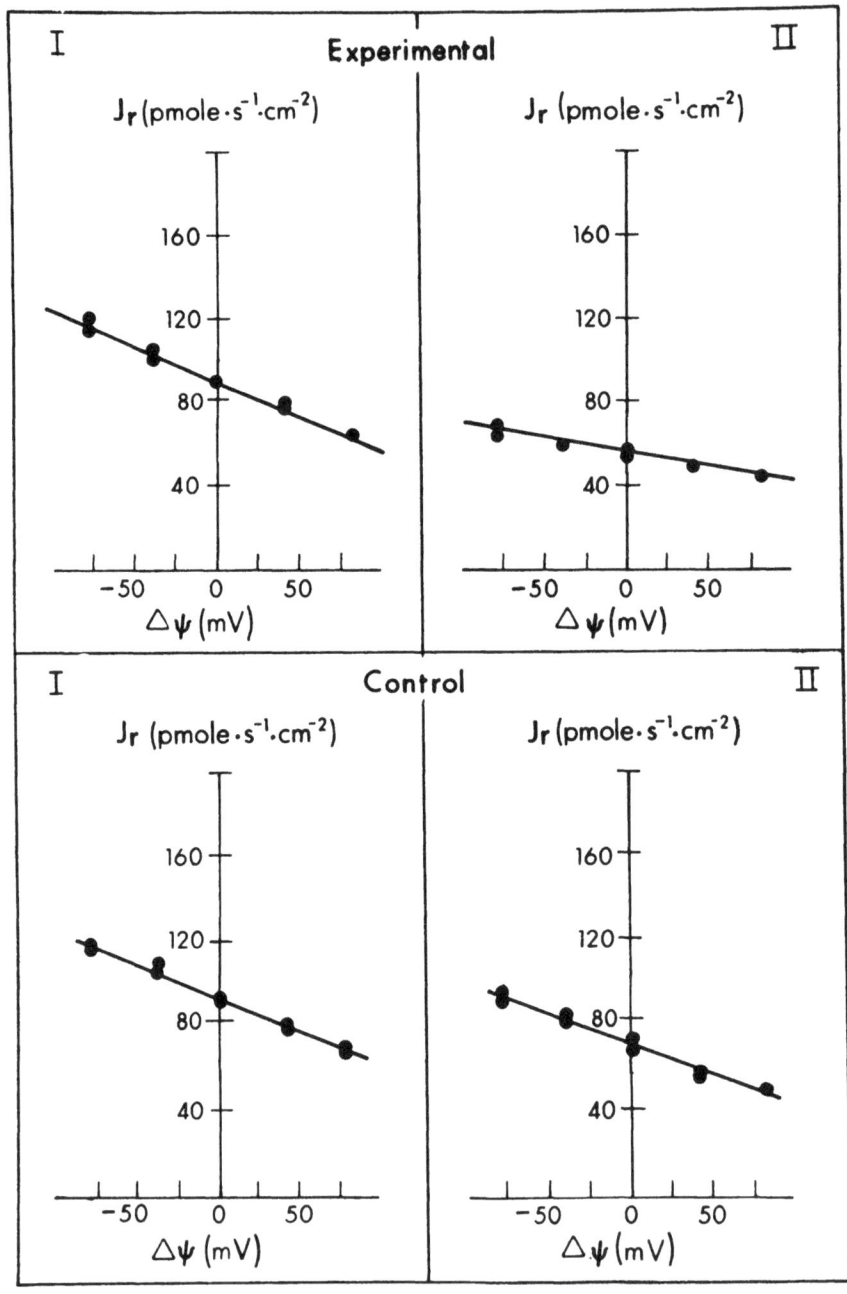

Active transport: experimental results 185

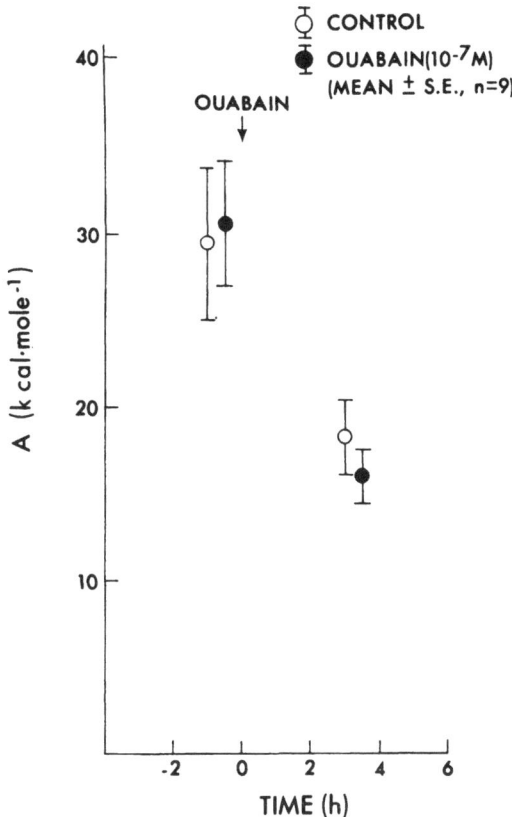

Fig. 8.9. Effect of 10^{-7} M ouabain on the affinity in the frog skin. (Adapted from Owen, Caplan, and Essig, 1975b.)

the fact that on removal of the amiloride following some 4 hours of exposure, I_0 in the treated tissues rose to a level significantly higher than in the initial period and higher than observed simultaneously in the paired control tissues (Fig. 8.10). But of course the overshoot might very possibly reflect kinetic factors as well.

Fig. 8.8. Effect of 10^{-7} M ouabain on dependence of the rate of oxygen consumption J_r on the electrical potential difference $\Delta\psi$. Initially the slopes $dJ_r/d(\Delta\psi)$ in paired hemiskins (Control I and Experimental I) differed insignificantly. Following the administration of ouabain, $|dJ_r/d(\Delta\psi)|$ was significantly less in the treated tissue (Experimental II) than in the untreated tissue (Control II). (From Owen, Caplan, and Essig, 1975b.)

186 Bioenergetics and linear nonequilibrium thermodynamics

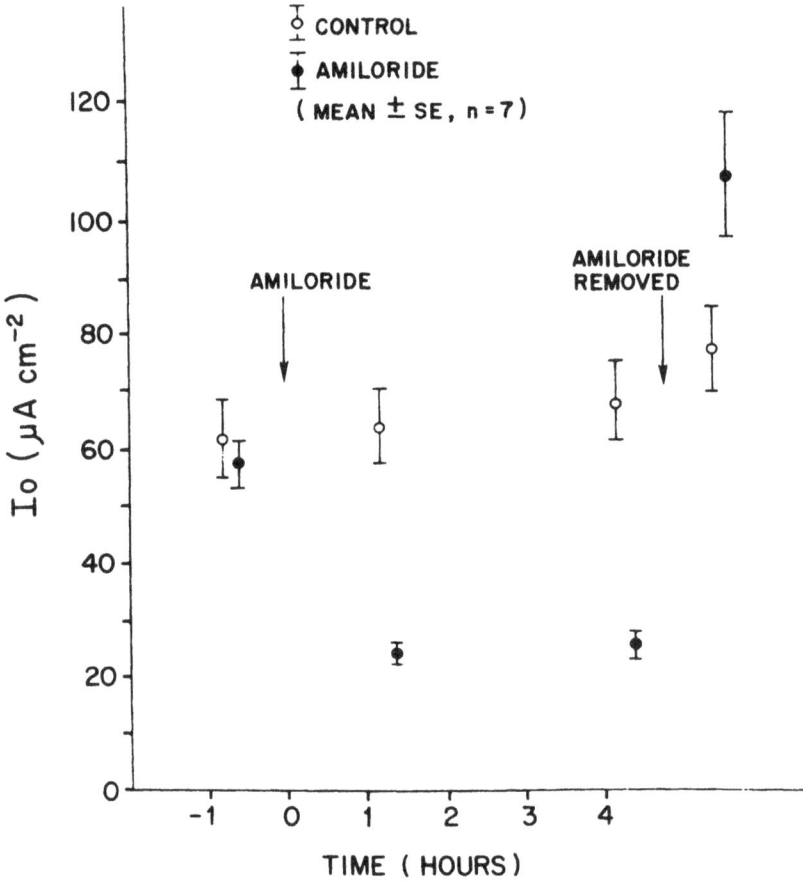

Fig. 8.10. Effect of 10^{-7} M–10^{-5} M amiloride on short-circuit current in the frog skin. (From Saito, Essig, and Caplan, 1973.)

Also of interest are substances which act directly to depress metabolism. One such is the sugar 2-deoxy-D-glucose (2DG). This substance interferes with cellular energy metabolism in three principal ways: competition with glucose for cell uptake, competition with glucose for phosphorylation by hexokinase, and blockage by 2-deoxy-D-glucose 6-phosphate of the isomerization of glucose 6-phosphate to fructose 6-phosphate. Since 2-deoxy-D-glucose 6-phosphate is not metabolized, these effects should promote depletion of ATP and depression of active transport. In studies of frog skins exposed to 1 mM glucose, a concentration of 16 mM 2DG depressed active sodium transport, as measured by the short-circuit current, to an average of 58% of the control level. This was associated with a significant decrease in the affinity, in this case to 53% of control level (Fig. 8.12). This depression of A is readily explicable in terms of the known effects of 2DG on metabolism.

It is instructive to compare the above response with that following another metabolic inhibitor, rotenone. This substance inhibits oxidative metabolism by blocking electron transport between NAD and cytochrome b. The effect of rotenone was studied by exposing frog skins to concentrations of 5–10 μM, in both the presence and absence of aldosterone. In contrast to the case with 2DG, depression of transport and suprabasal O_2 consumption to half or less of control level was associated with only a minor (and statistically insignifi-

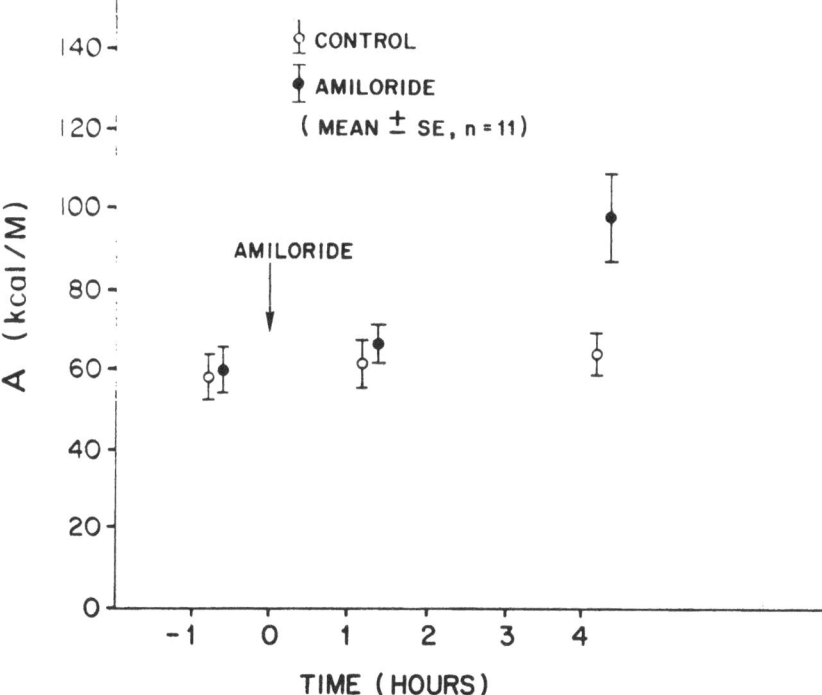

Fig. 8.11. Effect of amiloride on the affinity in the frog skin. (From Saito, Essig, and Caplan, 1973.)

188 Bioenergetics and linear nonequilibrium thermodynamics

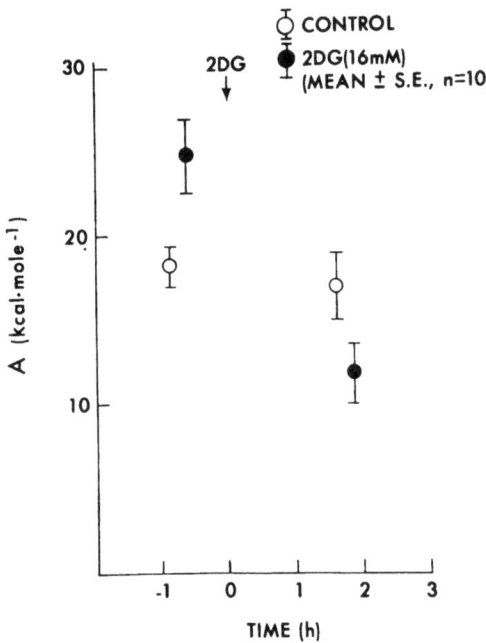

Fig. 8.12. Effect of 16 mM 2-deoxy-D-glucose (2DG) on the affinity in the frog skin. (Adapted from Owen, Caplan, and Essig, 1975b.)

cant) effect on A, as calculated according to Eq. (8.6) (Lau, Lang, and Essig, unpublished results). While a definitive understanding of this discrepancy must await further studies, it should be noted that the different effects of the two inhibitors are consistent with present knowledge of the interaction between the oxidative and glycolytic pathways for ATP formation. Thus partial depression by rotenone of oxidative metabolism, and hence of the affinity A_P, would be expected to cause marked stimulation of glycolysis (Nishiki, Erecińska, and Wilson, 1979). The resultant increase in A_P would partially restore active Na transport (and thus I_0), but should have little effect on $dJ_r/d(\Delta\psi)$, as evaluated from the potential-dependence of O_2 consumption. Accordingly, the inappropriate application of Eq. (8.6) under circumstances where glycolytic metabolism is contributing substantially to I_0 would lead to an overestimate of A. In contrast, in the presence of inhibitory concentrations of 2DG, which depress ATP concentration by diverse means, no compensatory mechanisms are available to restore A_P toward normal. (As will be seen in Sec. 8.2.4, overestimation of A will lead to underestimation of the phenomenological coefficient L_r.)

8.2.3 Effects of hormones on A

Estimates of the affinity have also been used in an attempt to elucidate the effects of aldosterone and antidiuretic hormone (ADH), physiological substances which stimulate active Na transport by mechanisms which are incompletely understood. Three general possibilities are shown in Fig. 8.13. Mechanism 1 is facilitation of Na entry across the outer passive permeability barrier, mechanism 2 is stimulation of phosphorylation of ADP, with increase in the affinity A_P, and mechanism 3 is a direct stimulation of the Na pump.

As is shown in Fig. 8.14, 14–18 hours of exposure of frog skins to 5×10^{-7} M aldosterone, with stimulation of J_{Na0} and J_{r0}, was associated with elevation of the affinity A. Figure 8.15 shows a similar response following exposure to 100 mU/ml ADH. (The possibility that elevation of A was in response to a primary effect on Na permeability and Na transport appears to have been ruled out, at least for the case of ADH, since this agent caused an increase in A even in the presence of amiloride, with rates of transport below control level.)

In drawing the inference of a primary action on energetic factors, it is important to inquire whether in the presence of the test agent the application of Eq. (8.6) for the evaluation of A remains legitimate.

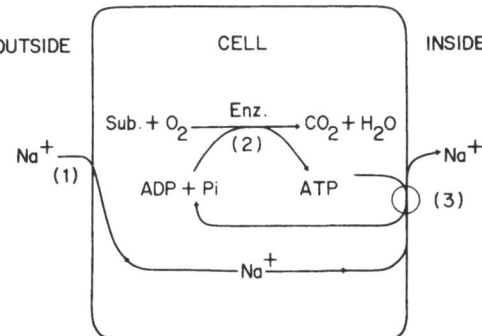

Fig. 8.13. Model of the active transport system; possible mechanisms of regulation of transport. (Adapted from Fanestil et al., 1968.)

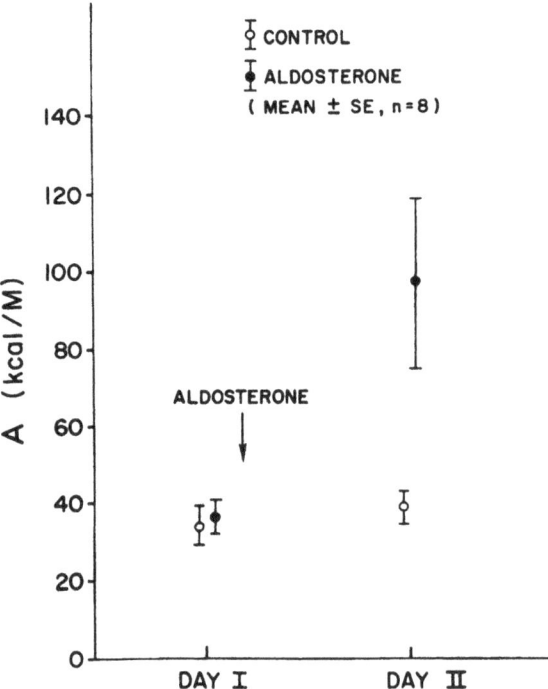

Fig. 8.14. Effect of 5×10^{-7} M aldosterone on the affinity in the frog skin. (From Saito, Essig, and Caplan, 1973.)

In particular, as pointed out for the case of rotenone, it is important to assess to what extent glycolysis may invalidate the use of I_0 to evaluate the rate of active Na transport attributable to oxidative metabolism. There is good reason to believe that in the control state aerobic glycolysis is too insignificant to lead to an important error in the evaluation of A (Vieira, Caplan, and Essig, 1972a). Furthermore, if glycolytic metabolism is constant, it must lead to an underestimate of the extent to which substances enhance A in stimulating transport. A special problem arises with aldosterone and ADH, however, since there is good evidence (in studies of the toad urinary bladder) that both hormones stimulate glycolysis (Handler, Preston, and Rogulski, 1968; Handler, Preston, and Orloff, 1969). Since glycolysis per se results in ATP formation, it might seem that both agents could stimulate active Na transport significantly even without an effect on oxidative metabolism. The resultant enhancement in I_0, without a com-

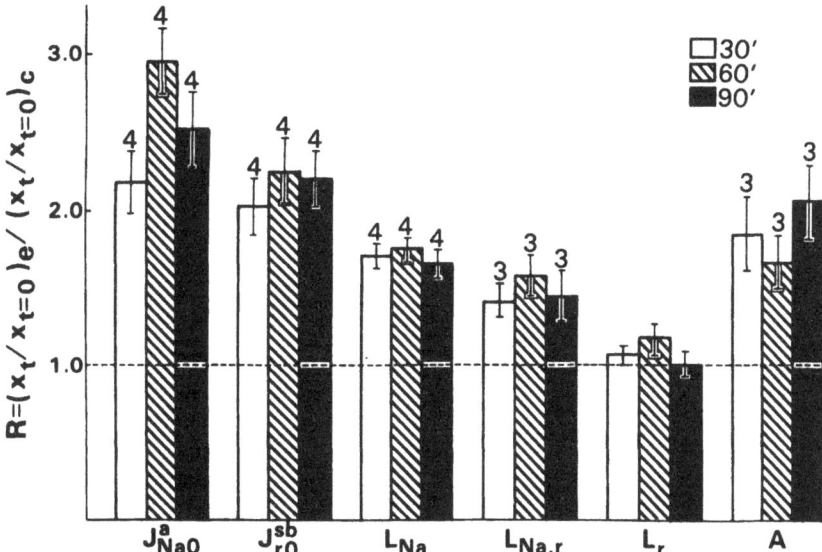

Fig. 8.15. Effect of 100 mU/ml ADH on J_{NaO}, J_{rO}, and NET parameters in frog skin 30, 60, and 90 min after administration. The calculation of the L coefficients is described in Sec. 8.2.4. In these studies of paired tissues, an attempt was made to correct for spontaneous variation with time by relating observations in experimental (e) and control (c) tissues, giving the doubly normalized quantity $R = (x_t/x_{t=0})_e/(x_t/x_{t=0})_c$. The numerals at the top of the columns indicate the significance of the difference of R values from 1.00; 3: $p(\Delta) < 0.01$; 4: $p(\Delta) < 0.001$. The absence of a numeral indicates nonsignificance. (From Lau, Lang, and Essig, 1981.)

mensurate effect on the potential-dependence of O_2 consumption, $dJ_r/d(\Delta\psi)$, would then lead to an erroneous impression that there was an increase in A (see Eq. 8.6). However, such an error must be small, since the metabolism of a molecule of glucose provides only 2 molecules of ATP as a result of glycolysis, as compared with some 32 molecules of ATP on complete oxidation. Accordingly, since both hormones stimulated suprabasal O_2 consumption appreciably (Figs. 8.14 and 8.15), the contribution of ATP formed from glycolysis must have been unimportant.

Also implicit in inferring an increase in the affinity of ATP hydrolysis on the basis of the above findings is the assumption that aldosterone and ADH do not cause an increase in the P/O ratio. Given the evidence that oxidative phosphorylation is not completely coupled (see Chapter 13), variation of J_P/J_O with change of physiological conditions is a possibility. This important issue remains to be explored.

Finally, we emphasize again that the above-noted changes in affinity and phenomenological coefficients under diverse conditions cannot be considered to provide unambiguous insights into mechanisms without far more physiological and biochemical information than is yet available.

8.2.4 Evaluation of the phenomenological (L) coefficients

Given the applicability of Eqs. (8.3) and (8.4) with constancy of A, evaluation of the L coefficients is straightforward, as discussed in Sec. 7.3.4.

$$L_{Na} = -\left(\frac{1}{F}\right)\left(\frac{dJ_{Na}^a}{d(\Delta\psi)}\right) \quad (\Delta c = 0, A \text{ constant}) \quad (8.7)$$

$$L_{Na,r} = -\left(\frac{1}{F}\right)\left(\frac{dJ_r}{d(\Delta\psi)}\right) \quad (\Delta c = 0, A \text{ constant}) \quad (8.8)$$

$$L_r = \frac{J_{r0}^{sb}}{A} \quad (\Delta c = 0) \quad (8.9)$$

It is seen that $(F^2)L_{Na}$ is a conductance, and $L_{Na,r}$ reflects the linkage between transport and metabolism; L_r may be considered a generalized rate constant for the metabolic driving reaction. Although precise understanding of the nature of the phenomenological coefficients must await the development of a detailed model of the active transport system, it would appear that quantification of the L (or R) coefficients as well as A should distinguish between influences on kinetic and energetic factors. A listing of representative values of L coefficients in the toad urinary bladder is given in Table 8.1.

A summary of the qualitative effects of various agents in the frog skin is given in Table 8.2. As is seen, in several instances both inhibitors and stimulants of transport affect both phenomenological coefficients and A. Possibly in some cases such combined effects reflect mechanisms altering transepithelial transport while maintaining

Table 8.1. NET characterization of active sodium transport in toad urinary bladder.[a]

Parameter	Value
I_0	$25.6 \pm 2.2 \; \mu A \cdot cm^{-2}$
J_r^b	$24.3 \pm 2.9 \; pmole \cdot cm^{-2} \cdot s^{-1}$
J_{r0}^{sb}	$18.1 \pm 2.3 \; pmole \cdot cm^{-2} \cdot s^{-1}$
$(\Delta\psi)_{J_{Na}=0}$	$117 \pm 8 \; mV$
L_{Na}	$103.9 \pm 12.5 \; \mu mole^2 \cdot cm^{-2} \cdot s^{-1} \cdot kcal^{-1}$
$L_{Na,r}$	$5.41 \pm 0.83 \; \mu mole^2 \cdot cm^{-2} \cdot s^{-1} \cdot kcal^{-1}$
L_r	$0.369 \pm 0.073 \; \mu mole^2 \cdot cm^{-2} \cdot s^{-1} \cdot kcal^{-1}$
A_1	$56.0 \pm 5.8 \; kcal \cdot mole^{-1} \; O_2$
A_2	$58.2 \pm 6.5 \; kcal \cdot mole^{-1} \; O_2$
q_1	0.87 ± 0.05
q_2	0.85 ± 0.05
$(J_r^{sb})_{J_{Na}^a=0}$	$4.4 \pm 1.9 \; pmole \cdot cm^{-2} \cdot s^{-1}$
dJ_{Na}^a/dJ_r^{sb}	$21.5 \pm 2.0 \; mole \; Na^+/mole \; O_2$
J_{Na0}^a/J_{r0}^{sb}	$14.8 \pm 0.9 \; mole \; Na^+/mole \; O_2$

Source: Lang, Caplan, and Essig, 1977b.
a. Mean values of parameters in 11 tissues.

near-constancy of intracellular electrolyte concentrations. Thus, for example, following aldosterone administration the enhancement of A and $L_{Na,r}$, tending to lower cellular Na, might be compensated for by an increase in apical conductance (manifested by an increase in L_{Na}), tending to raise cellular Na. Further study will be necessary to determine under what circumstances parallel effects on A and the L's represent discrete processes, as against interactions between kinetic and energetic factors modulating transport. Be this as it may, it is seen that the NET formalism permits a systematic characterization of the function of the active transport system under diverse conditions.

8.2.5 Evaluation of the degree of coupling: stoichiometry

Following the treatment of Sec. 7.3.5, we can see that if the phenomenological coefficients are known, it is possible to calculate the degree of coupling q directly from Eq. (7.6). Alternatively, it may be convenient to evaluate q without determining all the phenomenological

Table 8.2. Effects of agents on NET parameters of frog skin.

Agent	Concentration	Time	J_{Na0}	J_{r0}^{sb}	L_{Na}	$L_{Na,r}$	L_r	A	n	Source
Ouabain	10^{-7} M	2.5 hr	↓[a]	↓	—	↓	—	0[a]	9	Owen, Caplan, and Essig, 1975b
Amiloride	$1-5 \times 10^{-6}$ M	30 min	↓	↓	↓	↓	↓	0	9	Lau et al., unpublished
		60 min	↓	↓	↓	↓	↓	0	9	
		90 min	↓	↓	↓	↓	↓	0	4	
	$10^{-7} - 10^{-5}$ M[b]	1 hr	↓	↓	—	↓	↓	0	11	Saito, Essig, and Caplan, 1973
		4 hr	↓	0	—	0	—	↑[a]	11	
2-deoxyglucose	16×10^{-3} M[c]	1 hr	↓	↓	—	0	—	↑	10	Owen, Caplan, and Essig, 1975a, b
Aldosterone	5×10^{-7} M	14-18 hr	↑	↑	—	0	—	↑	8	Saito, Essig, and Caplan, 1973
Antidiuretic hormone	100 mU/ml	30 min	↑	↑	↑	↑	0	↑	9	Lau, Lang, and Essig, 1981
		60 min	↑	↑	↑	↑	0	↑	9	
		90 min	↑	↑	↑	↑	0	↑	9	
	10 mU/ml	30-120 min	↑[d]	↑[d]	↑[d]	↑[d]	0	↑	6	
Antidiuretic hormone (amiloride)[e]	10 mU/ml	30-90 min	0	0	↑	↑	↓	↑	4-9	Lau, Lang, and Essig, 1981
Cyclic-AMP	10^{-2} M	60-90 min	↑	↑	↑	↑	0	↑	6	Lau, Lang, and Essig, 1981

a. The symbols ↓, 0, and ↑ represent a significant decrease, no effect, and a significant increase, respectively, compared with the control value.
b. The test agent was studied in the presence of 5×10^{-7} M d-aldosterone.
c. The test agent was studied in the presence of 10^{-3} M glucose.
d. Effects were smaller than in tissues exposed to 100 mU/ml ADH.
e. In order to examine the effect of ADH on A in the absence of elevated rates of transport and metabolism, tissues were exposed to $1-5 \times 10^{-6}$ M amiloride prior to and during exposure to ADH.

coefficients, using the method of either Eq. (7.39) or Eq. (7.40), given here by

$$q_1 = \sqrt{1 - \frac{(J_r^{sb})_{J_{\text{Na}}^a=0}}{J_{r0}^{sb}}} \quad (8.10)$$

$$q_2 = \sqrt{\frac{J_{\text{Na}0}^a/J_{r0}^{sb}}{dJ_{\text{Na}}^a/dJ_r^{sb}}} \quad (8.11)$$

Table 8.1 compares values of q_1 and q_2 measured in studies of toad urinary bladders (Lang, Caplan, and Essig, 1977b). As would be expected for a linear system, agreement between the two methods was good. The mean value of 0.86 differed quite significantly from 1.00, indicating incompleteness of coupling. A similar degree of coupling has been demonstrated in the frog skin (Lahav, Essig, and Caplan, 1976).

With an incompletely coupled system, metabolic energy must be expended to maintain an electrochemical potential difference of Na even in the absence of active transport, that is $(J_r^{sb})_{J_{\text{Na}}^a=0} \neq 0$ (Table 8.1). This being so, although both J_{Na}^a and J_r^{sb} vary linearly with $\Delta\psi$, there is no fixed stoichiometric ratio. Figure 8.16 shows how J_{Na}^a/J_r^{sb} varies in a representative tissue as J_{Na}^a is varied by perturbation of $\Delta\psi$.[3] Furthermore, in analyzing graphs of J_{Na}^a versus J_r^{sb}, one must take care to avoid confusion arising from comparing estimates of stoichiometry based on slopes $(dJ_{\text{Na}}^a/dJ_r^{sb} = L_{\text{Na}}/L_{\text{Na},r})$ with estimates obtained at short circuit $(J_{\text{Na}0}^a/J_{r0}^{sb} = L_{\text{Na},r}/L_r)$, since these must differ when $q \neq 1$ (see Table 8.1).

The efficiency of energy conversion is a sensitive function of the degree of coupling; for $q = 0.86$ the maximum efficiency is only 33%. Nevertheless, it is conceivable that there are also biological advantages associated with a slight degree of uncoupling (Stucki, 1980). Present knowledge of metabolic mechanisms is, however, too limited to settle this point. In our studies q was not closely correlated with $J_{\text{Na}0}$, J_{r0}^{sb}, or A.

In principle, incompleteness of coupling could come about in several ways. One possibility is recirculation of transported sodium (Fig. 8.17). To the extent that a passive serosal leak permits reentry of transported sodium into the active transport pool, setting $\Delta\psi$ at the value appropriate to make $J_{\text{Na}}^a = 0$ would not result in inactivity of

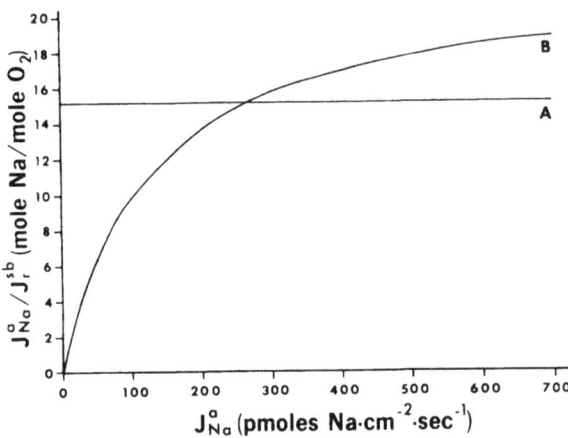

Fig. 8.16. Examples of the relationship between Na^+/O_2 ratio and active sodium transport. The Na^+/O_2 ratio (J^a_{Na}/J^{sb}_r) at different transport rates (J^a_{Na}) is shown. When $\Delta\psi = 0$, Na^+/O_2 is constant (line A). When $\Delta\psi$ is varied, Na^+/O_2 increases with the rate of active transport, indicating incomplete coupling between transport and metabolism (line B). (From Lang, Caplan, and Essig, 1977a.)

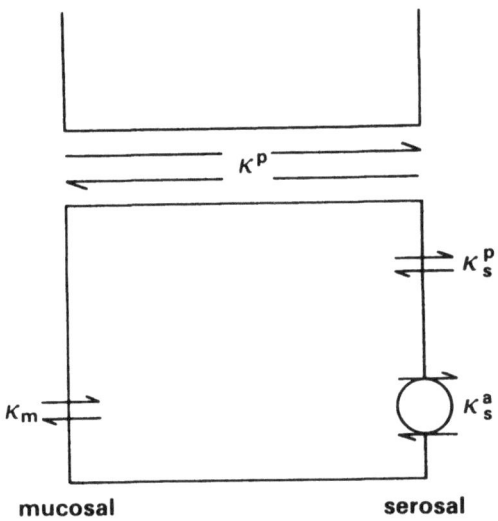

Fig. 8.17. Model of transepithelial sodium transport systems. Sodium may cross the membrane via either an active or a passive pathway, of conductance κ^a and κ^p, respectively. Under usual operating conditions sodium enters the active pathway by way of a passive permeability barrier of conductance κ_m at the mucosal surface and leaves by way of the sodium active transport mechanism of conductance κ^a_s at the serosal surface. There may be some degree of back leakage through a serosal leak of conductance κ^p_s. (From Lang, Caplan, and Essig, 1977a.)

the pump, but rather in active transport exactly compensated by leak. Accordingly, the rate of suprabasal oxygen consumption $(J_r^{sb})_{J_{Na}^a=0}$ would not be zero. This possibility appears to have been ruled out, however, for the toad urinary bladder by studies of the rate of CO_2 production in tissues in which transepithelial active sodium transport was abolished by the administration of amiloride; since removal of serosal Na then had little effect on J_{CO_2} it appears that recycling of serosal Na must be minimal (Canessa, Labarca, and Leaf, 1976). Earlier evidence for recycling of Na through the basolateral membrane of frog skin (Biber and Mullen, 1977) has also been contested (Corcia, Lahav, and Caplan, 1980). There remain the possibilities that partial uncoupling of oxidative metabolism and transport might reflect either incomplete coupling of oxidative phosphorylation or incomplete coupling of the mechanism linking ATP utilization and translocation of Na.

Over and above variation in J_{Na}^a/J_r^{sb} attributable to variation of $\Delta\psi$ in uncoupled tissues, it has been found that the ratio of rates of transport and metabolism differs in different tissues even under short-circuit conditions (Fig. 8.18). Thus, in studies of frog skins, when J_{Na0} and J_{r0} varied spontaneously or as the result of the administration of antidiuretic hormone or ouabain, although the value of dJ_{Na0}/dJ_{r0} for each tissue was characteristic, values in different tissues ranged from 7.1 to 30.9 (Vieira, Caplan, and Essig, 1972a). Similarly, the value of J_{Na0}^a/J_{r0}^{sb} in toad urinary bladders studied with the aid of amiloride was characteristic for each tissue but varied for different tissues from 9.3 to 20.4 (Lang, Caplan, and Essig, 1977a). This variability remains to be explained. Since J_r here was evaluated from the rate of O_2 consumption, the variable ratios cannot be attributed to variability of the respiratory quotient, as has been invoked to explain variability of $J_{Na}^a/J_{CO_2}^{sb}$ (Al-Awqati, Beauwens, and Leaf, 1975; Canessa et al., 1978). One possibility is that differences in the substrates utilized by different tissues may result in differences in the P/O ratio and thus in the Na/O_2 ratio. Also, for the case of frog skin, recirculation of Na^+ has not been entirely ruled out.[4]

8.2.6 Comparison of E_{Na} and A

In the analysis of active transepithelial sodium transport, many have utilized an equivalent circuit model (Ussing and Zerahn, 1951). In

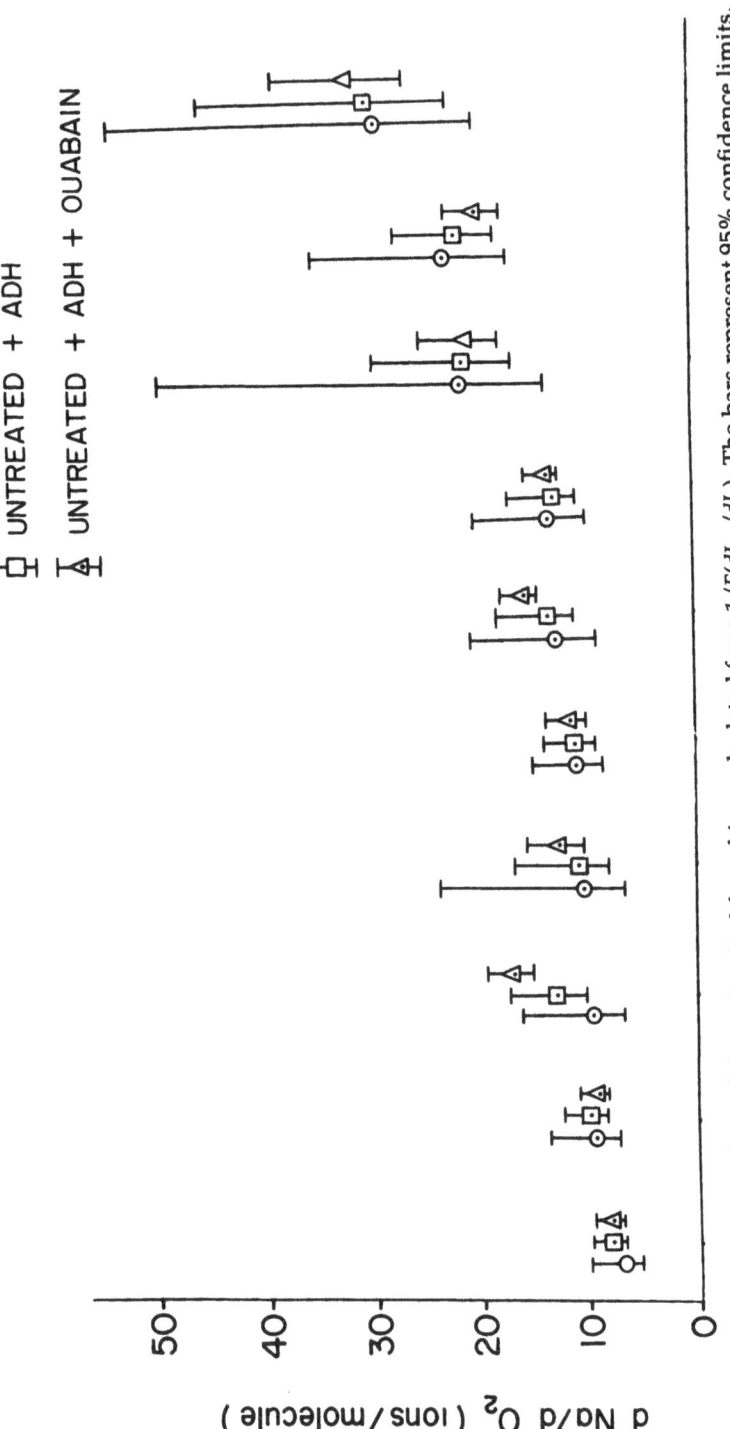

Fig. 8.18. Values of $d\text{Na}/d\text{O}_2$ of short-circuited frog skins, calculated from $1/F(dJ_{r0}/dI_0)$. The bars represent 95% confidence limits. (From Vieira, Caplan, and Essig, 1972a.)

this model the active pathway consists of a conductance element κ_{Na}^a in series with the electromotive force of sodium transport E_{Na}, in parallel with a passive pathway of conductance κ^p (Fig. 7.7). The short-circuit current I_0 is then given by

$$I_0 = \kappa_{Na}^a E_{Na} \qquad (8.12)$$

As discussed above, attempts to use the equivalent circuit parameters to distinguish between kinetic and energetic factors modulating transport appear inappropriate, since E_{Na} is not a purely energetic quantity (see Eq. 7.48). Thus in principle it is possible for a given substance to alter E_{Na} by influencing either phenomenological coefficients or A. For this reason it was of interest to compare the effects of various agents on E_{Na} and A (Hong and Essig, 1976).

The use of Eq. (8.12) to calculate E_{Na} requires making a distinction between current attributable to active Na transport and that resulting from other ion flows. Hong and Essig used the diuretic amiloride for this purpose. Adequate concentrations of this agent depress entry of Na into the active pathway almost completely; the residual conductance then gives the "passive" conductance κ^p, and combination with the original total conductance κ gives the amiloride-sensitive "active" conductance κ^a:

$$\kappa^a = \kappa - \kappa^p \qquad (8.13)$$

In order to evaluate parameters repeatedly and frequently, it was convenient to evaluate κ from $-\Delta I/\Delta(\Delta\psi)$, setting $\Delta\psi$ sequentially at $+20, 0, -20, 0$ mV for 5 sec each at 1-min intervals. Presuming then that the value of κ^a evaluated in this way gives κ_{Na}^a allows the calculation of E_{Na}.

This technique was used to study the effects of antidiuretic hormone, ouabain, amiloride, and 2DG in the toad urinary bladder. The results are shown in Table 8.3. Comparative effects on the apparent value of E_{Na} and A are shown in Table 8.4, which summarizes the effects of three of the above agents, as well as those of aldosterone, analyzed in earlier studies. These observations suggest that E_{Na} cannot be used reliably to monitor the free energy of the metabolic driving reaction. The analysis is not completely convincing, however, since it depends on the use of the amiloride-sensitive conductance κ^a eval-

Table 8.3. Effects of agents on equivalent circuit parameters of toad urinary bladder.[a]

Agent	Concentration	t(min)	$I_{0,t}/I_{0,0}$	κ_t^a/κ_0^a	"$E_{Na,t}$"/"$E_{Na,0}$"	n
Antidiuretic hormone	100 mU/ml	7	2.25 ± 0.11^4	2.53 ± 0.19^4	0.89 ± 0.03^2	17
Control	—	7	1.04 ± 0.01	1.08 ± 0.02	0.97 ± 0.02	17
Ouabain	10^{-4} M	30	0.41 ± 0.06^4	0.54 ± 0.07^4	0.76 ± 0.07^2	8
Control	—	30	0.99 ± 0.02	1.04 ± 0.03	0.95 ± 0.03	8
Amiloride	5×10^{-7} M	5	0.41 ± 0.04^4	0.21 ± 0.05	1.92 ± 0.33^3	6
Control	—	5	1.00 ± 0.01	1.02 ± 0.02	0.98 ± 0.02	6
2-deoxyglucose	7.5×10^{-3} M	60	0.33 ± 0.03^4	0.38 ± 0.03^4	0.89 ± 0.05	7
Control	—	60	1.05 ± 0.05	1.11 ± 0.08	0.94 ± 0.05	7

Source: Hong and Essig, 1976.
a. The data are expressed as mean $x_t/x_{t=0} \pm$ S.E.M. Superscripts indicate significant differences between control and experimental values; 2: $p(\Delta) < 0.025$; 3: $p(\Delta) < 0.01$; 4: $p(\Delta) < 0.001$.

Table 8.4. Comparative effects of agents on "E_{Na}" and A.[a]

Agent	"E_{Na}"				A			
	Concentration	t	"E_{Na}"	Source	Concentration	t	A	Source
Ouabain	10^{-4} M	5–30 min	↓[b]	Hong and Essig, 1976	10^{-7} M	2.5 hr	0[b]	Owen, Caplan, and Essig, 1975b
Amiloride	5×10^{-7} M	1–60 min	↑[b]	Hong and Essig, 1976	10^{-7}–10^{-5} M	1 hr	0	Saito, Essig, and Caplan, 1973
Amiloride	—	—	—	—	10^{-7}–10^{-5} M	4 hr	↑	Saito, Essig, and Caplan, 1973
2-deoxy-glucose	7.5×10^{-3} M	5–60 min	0	Hong and Essig, 1976	1.6×10^{-2} M	1 hr	↓	Owen, Caplan, and Essig, 1975b
Aldosterone	5×10^{-7} M	1–6 hr	0	Saito and Essig, 1973	5×10^{-7} M	14–18 hr	↑	Saito, Essig, and Caplan, 1973

a. "E_{Na}" was determined in toad urinary bladder; A was determined in frog skins.
b. The symbols ↓, 0, ↑ represent a significant decrease, no effect, and a significant increase, respectively, compared with the control value.

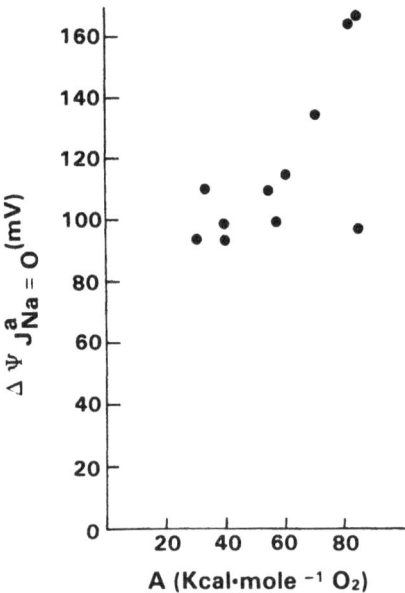

Fig. 8.19. Relationship between $(\Delta\psi)_{J_{Na}^a=0}$ (static head) and A. The correlation coefficient $r = 0.63$. (Adapted from Lang, Caplan, and Essig, 1977b.)

uated from 5-sec perturbations of $\Delta\psi$ and requires that this value of κ^a provide an adequately accurate approximation to the value of the sodium conductance κ_{Na}^a which would obtain under steady-state conditions, when transcellular Na flow is conservative and given by the amiloride-sensitive current I^a. When this is not the case, "E_{Na}" cannot be considered to provide an accurate estimate of E_{Na} (that is, the value of $\Delta\psi$ adequate to make the steady-state rate of active Na transport zero) (Wolff and Essig, 1980). This issue is currently under investigation.

Possibly more meaningful in this regard is the relationship between values of A and values of E_{Na} estimated from the intercept of the 6-min $I^a-\Delta\psi$ plot, $(\Delta\psi)_{J_{Na}^a=0}$ (static head) (Fig. 8.19). As is seen, the two quantities are correlated, but not closely.

8.3 Analysis of active transport in other epithelia

Linearity of current–voltage relationships in several epithelial tissues suggests that the formalisms applied above to active sodium transport in the frog skin, toad skin, and toad urinary bladder may possibly be applicable to other transport systems as well. For the most part such linearity has led to analysis only in terms of the equivalent circuit formulation. Schultz, Frizzell, and Nellans (1977) demonstrated linearity between the rate of active sodium transport and $\Delta\psi$ in the rabbit colon and on this basis developed an equivalent circuit model which was interpreted in terms of the phenomenological resistance coefficients and affinity of a linear NET formulation. However, no experimental determination of the NET parameters was carried out.

8.3.1 Active H^+ transport in the turtle urinary bladder

One active transport system which has been studied systematically from the points of view of both the NET and equivalent circuit formulations is the hydrogen ion transport system of the turtle urinary bladder (Steinmetz, 1974; Al-Awqati et al., 1976; Beauwens and Al-Awqati, 1976; Al-Awqati, Mueller, and Steinmetz, 1977; Al-Awqati, 1977). This tissue carries out active transport of both Na^+ (from mucosa to serosa) and H^+ (from serosa to mucosa). In the presence of adequate CO_2, and with control of the transepithelial electrical potential difference, the two transport processes are independent. (For our analysis it is immaterial whether the molecular event is outward H^+ transport, as indicated here, or inward OH^- or HCO_3^- transport, as has been suggested.)

Since the proton conductance of the parallel passive pathway in the turtle bladder is very small, the rates of net H^+ transport J_H and active H^+ transport J_H^a can be considered essentially equal. Accordingly, the rate of active H^+ transport can be evaluated either by the use of a pH-stat or, in the absence of active Na^+ transport, from the short-circuit current, that is, the current measured at $\Delta\psi = 0$ with the surfaces of the membrane bathed by solutions identical in composition, except possibly for pH; with the low passive proton conductance, passive transepithelial proton flow is negligible. Active Na^+ transport is easily eliminated by the use of appropriate concentrations of ouabain.

With the use of CO_2-free bathing solutions, the rate of oxidative metabolism may be evaluated from J_{CO_2}, the rate of total CO_2 production measured by a conductometric method as previously in the study of active sodium transport. H^+ transport is more vigorous,

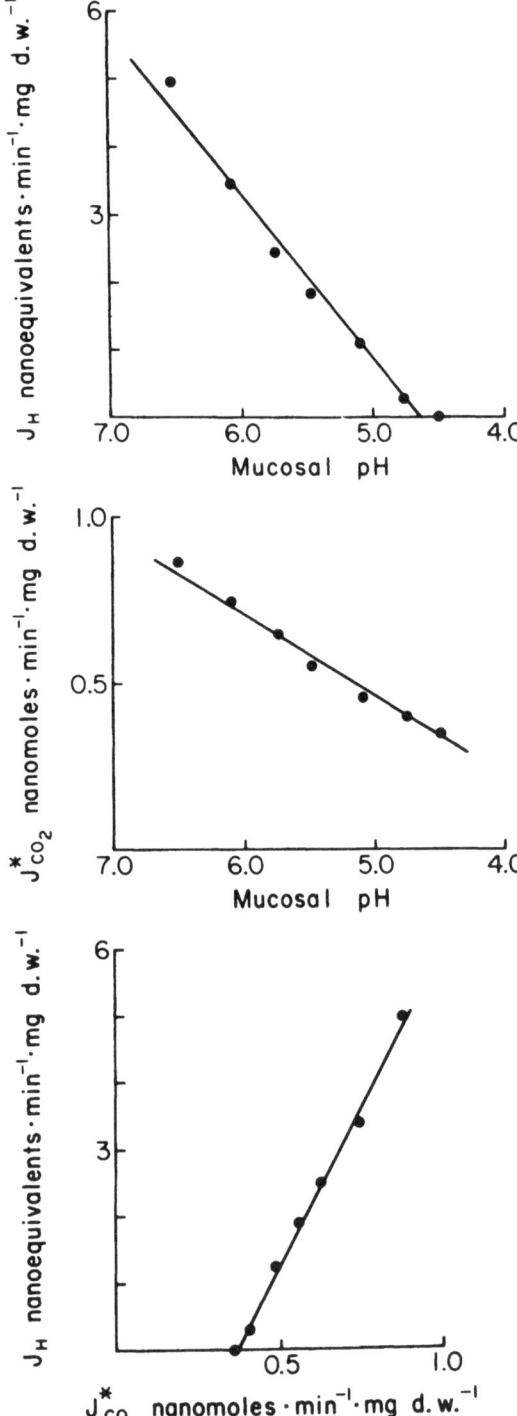

however, in the presence of ambient CO_2; in this case the rate of oxidative metabolism was estimated from measurement of the rate of $^{14}CO_2$ generation from ^{14}C-labeled glucose, $J^*_{CO_2}$.

Transport and metabolism are readily perturbed by changing the ambient pCO_2 (with the electrochemical potential difference X_H clamped, conveniently, at zero) or by altering X_H, either by perturbing $\Delta\psi$ or by altering the mucosal (lumenal) pH.[5] With change of X_H, J^a_H reaches a new steady state within less than a minute; the apparent rate of CO_2 production changes more slowly, reaching a new steady state only after 10–15 min.[6]

In the steady state a highly linear relationship was noted between J_H and both J_{CO_2} and $J^*_{CO_2}$; values of dJ_H/dJ_{CO_2} (or $dJ_H/dJ^*_{CO_2}$) were similar in hemibladders from the same animal. With evidence for complete coupling of H^+ transport and oxidative metabolism in this tissue (see below), it was considered that these slopes evaluate a stoichiometric ratio, irrespective of which of the above means was employed to vary transport and metabolism. Values in different animals varied greatly: In the presence of 1% CO_2, $dJ_H/dJ^*_{CO_2}$ varied from 4.4 ± 1.2 to 38.2 ± 3.5 ($n = 40$). In CO_2-free media, dJ_H/dJ_{CO_2} varied from 1.41 ± 0.26 (S.D.) to 11.23 ± 1.17 ($n = 6$). This degree of variability is greater than can reasonably be attributed to variability of the respiratory quotient. Thus, as for the sodium transport system of anuran epithelia, there appears to be no unique stoichiometric ratio relating rates of transport and oxidative metabolism.

8.3.2 NET analysis of active H^+ transport

Other studies indicate that H^+ transport and the associated metabolism in the turtle bladder are amenable to analysis by means of linear NET, since J_H, J_{CO_2}, and $J^*_{CO_2}$ are linear functions of mucosal pH (Fig. 8.20). Furthermore, in a comparison of the dependence of J_H on mucosal pH and on $\Delta\psi$, the values of $dJ_H/d(\Delta pH)$ and $dJ_H/d(\Delta\psi)$ were statistically the same, as were the intercepts at which the rate of active H^+ transport became zero. Accordingly, in analogy to the case for active Na^+ transport, we may write

$$J^a_H = L_H X_H + L_{Hr} A \tag{8.14}$$

Fig. 8.20. The relation of the rate of H^+ secretion J_H and the rate of $^{14}CO_2$ production $J^*_{CO_2}$ to the mucosal pH and to each other in the turtle urinary bladder. (From Beauwens and Al-Awqati, 1976.)

$$J_r^{sb} = L_{Hr}X_H + L_r A \qquad (8.15)$$

In order to conform to the polarity conventions of Al-Awqati et al., which are the opposite of those adopted for Na⁺ transport above, J_H is considered positive when H⁺ moves from serosa to mucosa, and $X_H \equiv -\Delta\tilde{\mu}_H = -(RT\Delta \ln a_H + F\Delta\psi)$ is taken as

$$X_H = (\log_e 10)RT\Delta pH - F\Delta\psi$$
$$= 2.3\,RT(pH^m - pH^s) - F(\psi^m - \psi^s) \qquad (8.16)$$

Presuming proper pathways, we again assume the validity of the Onsager reciprocal relationship, and we assume that for metabolism, as for transport, it is immaterial whether X_H is altered by variation of mucosal pH or of $\Delta\psi$. In analogy to the case for Na⁺ transport in toad skin, the equations do not apply in the case of alteration of serosal [H⁺].

8.3.3 The degree of coupling of the H⁺ transport system

The above formalism was first employed to evaluate the extent of coupling of the transport system. If coupling is complete, rates of transport and metabolism will be in a fixed stoichiometric ratio under all circumstances, whereas if coupling is incomplete, $dJ_H^a/dJ_{CO_2}^*$ will depend on the method employed to perturb transport. Beauwens and Al-Awqati used two means for this purpose: (1) variation of mucosal pH (X_H variable), and (2) variation of the pCO_2, maintaining mucosal and serosal pH constant and equal ($X_H = 0$). Presuming that variation of X_H is without effect on A, as suggested by the linear relationships observed, Beauwens and Al-Awqati considered that the first technique evaluates

$$\left(\frac{\partial J_H^a}{\partial J_{CO_2}^*}\right)_A = \frac{L_H}{L_{Hr}} \qquad (8.17)$$

whereas the second technique gives

$$\left(\frac{\partial J_H^a}{\partial J_{CO_2}^*}\right)_{X_H=0} = \frac{L_{Hr}}{L_r} \qquad (8.18)$$

Referring to Eq. (4.12), it is seen that

$$\frac{(\partial J_H^a/\partial J_{CO_2}^*)_{\dot{X}_H=0}}{(\partial J_H^a/\partial J_{CO_2}^*)_A} = \frac{L_{Hr}^2}{L_H L_r} = q^2 \tag{8.19}$$

In a series of 14 experiments, values of $(dJ_H^a/dJ_{CO_2}^*)$ with X_H variable and with X_H constant did not differ significantly; the application of Eq. (8.19) gave a mean value of $q = 0.97 \pm 0.05$ (S.E.) (with a 95% lower confidence limit of 0.90). On this basis it was concluded that coupling was essentially complete. Other evidence may be adduced for completeness of coupling. Thus an inhibitor of the proton pump, dicyclohexylcarbodiimide, gave values of dJ_H^a/dJ_{CO_2} indistinguishable from those obtained by alteration of mucosal pH (Al-Awqati, personal communication). Also, if coupling were incomplete, J_{CO_2} at static head should include a component of suprabasal metabolism, which might be expected to be eliminated by an agent depressing pump function; accordingly, the finding that metabolism at static head was unaffected by 1 mM acetazolamide, an inhibitor of carbonic anhydrase which eliminates J_H^a under usual experimental conditions, was taken as another indication that $q \simeq 1$.[7]

8.3.4 The affinity of the H^+ transport system

The apparent applicability of Eqs. (8.14) and (8.15) also permits evaluation of the affinity, much as for the active Na^+ transport systems (Al-Awqati, 1977). Assuming constancy of A on perturbation of X_H,

$$A = \frac{(J_H^a)_{X_H=0}}{(dJ_r/dX_H)} \tag{8.20}$$

In this case it was convenient to vary X_H by perturbation of mucosal pH and to monitor metabolism by measurements of $J_{CO_2}^*$, the rate of CO_2 production from radioactive glucose. For this approach to be legitimate under the conditions of this study requires that oxidative metabolic energy be obtained solely from glucose oxidation. With this assumption, it was calculated that values of A in control turtle bladders (untreated or unaffected by aldosterone) ranged from 28 to 99 kcal · mole^{-1} CO_2, or assuming a respiratory quotient of 1 and a P/O ratio of 3, from about 5 to 17 kcal · mole^{-1} ATP. These values are in the same range as those cited above for untreated frog skins and toad bladder, some 20–80 kcal · mole^{-1} O_2, or 3–13 kcal · mole^{-1} ATP. While this near-agreement is of interest, it must be remembered that for both systems the precise quantification of energetics in terms of ATP metabolism requires accurate knowledge of the P/O ratio under the conditions of study. In addition, of course, when metabolism is monitored by measurements of CO_2 production, it is necessary to know the respiratory quotient (R.Q.). Both the P/O ratio and the R.Q. can be expected in principle to vary with experimental conditions. That these variations might well be appreciable is suggested by the wide range of values of dJ_H^a/dJ_{CO_2} cited above.

The NET analysis of bladders in which H^+ transport was stimulated by aldosterone is relevant to the mechanism of action of this hormone. Following 1–6 hours of exposure to aldosterone, A was lower in stimulated than in unstimulated tissues, in association with enhancement of $dJ_H^a/d(\Delta pH)$. After 20 hours of exposure, however, the response was as noted for the Na^+ transport system of frog skin, in that A was significantly greater in the stimulated tissues than in paired control tissues. Also of interest was the finding that aldosterone gradually influenced the apparent stoichiometric relationship between H^+ transport and metabolism, such that when these were altered by variation of mucosal pH, following 20 hours (but not 1–6 hours) of exposure to the hormone, $dJ_H^a/dJ_{CO_2}^*$ was significantly greater in the stimulated tissues. These various observations led to the tentative inference that aldosterone acts primarily on the transport apparatus, with a resultant decline in A, followed by adaptation, with an increase in A. However, it is difficult to interpret the validity of this interpretation, owing to the lack of determination of A prior to the administration of aldosterone. Furthermore, additional information is required in order to relate values of A calculated from $^*CO_2$ metabolism to values expressed in terms of ATP metabolism.

8.3.5 Equivalent circuit analysis of active H^+ transport

In close analogy with the case for Na^+ transport systems, the function of the H^+ transport system can be analyzed in terms of an equivalent circuit, consisting of a H^+ pump with a series conductance in the active pathway and a parallel passive pathway. Given the equivalent dependence of J_H on $\Delta\psi$ and mucosal pH, the equivalent circuit can

Active transport: experimental results 209

be conveniently analyzed by two alternative means. It is convenient to consider tissues in which active Na transport has been abolished. Then, in the absence of a pH difference across the membrane, expressing active H$^+$ flow in terms of the electrical current $I_H^a \equiv FJ_H^a$, its dependence on $\Delta\psi$ in the steady state is

$$\kappa_H^a = \frac{-dI_H^a}{d(\Delta\psi)} \tag{8.21}$$

giving on integration

$$I_H^a = \kappa_H^a(E_H - \Delta\psi) \quad (\Delta pH = 0) \tag{8.22}$$

Alternatively, in the absence of an electrical potential difference across the membrane,

$$\kappa_H^a = \left(\frac{1}{2.3RT/F}\right)\left(\frac{dI_H^a}{d(\Delta pH)}\right) \tag{8.23}$$

giving

$$I_H^a = \kappa_H^a[E_H + 2.3(RT/F)\Delta pH] \quad (\Delta\psi = 0) \tag{8.24}$$

(Here, as previously, we adopt Al-Awqati et al.'s polarity conventions, according to which I_H is positive when directed from serosa to mucosa, $\Delta\psi = \psi^m - \psi^s$, and $\Delta pH = pH^m - pH^s$.)

Al-Awqati and his colleagues have evaluated the parameters of such an equivalent circuit model. In their formulation they considered the "force of the pump, the apparent protonmotive force (PMF')." Again neglecting the small contribution of passive transepithelial proton flow, the protonmotive force is given by

$$PMF = (-X_H)_{J_H=0} = FE_H \tag{8.25}$$

(We ignore Al-Awqati's distinction between PMF' and the true PMF, which will arise if the cytoplasmic pH differs slightly from that of the serosal bathing medium.) Values determined from electrical gradients and pH gradients were indistinguishable, being 144 ± 9 mV and 161 ± 8 mV, respectively, expressed in electrical units, or 2.44 ± 0.16 and 2.73 ± 0.13, respectively, expressed in pH units.

Expressing PMF in terms of the linear NET formulation of Eq. (8.14),

$$\text{PMF} = \left(\frac{L_{Hr}}{L_H}\right) A \qquad (8.26)$$

in precise correspondence with Eq. (8.13). It appears, therefore, that just as with E_{Na}, the protonmotive force PMF (or E_H) is not a pure energetic parameter, but depends on kinetic factors as well.

This conclusion was tested by model experiments designed to alter transport by effects on metabolism. As is seen in Table 8.5, the inhibition of metabolism by deoxygenation, 2-deoxyglucose, or substrate depletion all resulted in substantial depression of active H^+ transport and κ_H^a, with relatively smaller effects on PMF. Only with serosal 2,4-dinitrophenol were effects on κ_H^a and PMF of comparable magnitude. (The effect of this agent on the PMF is presumably attributable to its enhancement of proton permeability, increasing back leak of H^+ into the cell via plasma membrane pathways parallel to the pump. This interpretation is supported by the finding that 2,4-DNP reduced PMF significantly [by 12%] even when applied to the mucosal surface in low concentrations [2×10^{-6} M] which failed to affect $J_{H_0}^a$.) Consistent with the above findings, the repletion of depleted tissues by the use of glucose resulted in a marked stimulation of H^+ transport in association with comparable enhancement of κ_H^a, but no increase in PMF.

These various findings all support the interpretation indicated by Eq. (8.26), that PMF, like E_{Na}, is a composite parameter, incorporating both kinetic and energetic factors. The findings are of particular interest in that they avoid the uncertainties of Hong and Essig's studies of "E_{Na}" in which the 5-sec amiloride-sensitive conductance was used to estimate κ_{Na}^a. In contrast, the present studies depend on steady-state currents which very likely represent conservative H^+ flow, so that Eqs. (8.23) and (8.24) can be considered to evaluate κ_H^a and E_H (or PMF) within experimental error.

Also of interest is the observation that the stimulation of H^+ transport following prolonged exposure to aldosterone is associated with comparable enhancement of κ_H^a but no effect on PMF. Although present understanding of the action of aldosterone is inadequate to permit an interpretation of these findings in terms of mechanisms, it

Table 8.5. Effects of experimental treatments on equivalent circuit parameters of turtle urinary bladder.[a]

Treatment	Concentration	$(J_{H_0}^a)_e/(J_{H_0}^a)_c$	$(\kappa^a)_e/(\kappa^a)_c$	$(PMF)_e/(PMF)_c$	n	Source
Deoxygenation	—	0.39	0.44	0.90	6	Al-Awqati, Mueller, and Steinmetz, 1976
2,4-dinitrophenol	10^{-4} M	0.55	0.73	0.77	15	Al-Awqati, Mueller, and Steinmetz, 1976
2-deoxyglucose	10^{-2} M	0.56	0.67	0.87	9	Al-Awqati, Mueller, and Steinmetz, 1976
Depletion	—	0.26	0.29	0.88	6	Al-Awqati, Mueller, and Steinmetz, 1976
Glucose	5×10^{-3} M	1.70	1.84	0.94	6	Al-Awqati, Mueller, and Steinmetz, 1976
Aldosterone[b]	5×10^{-7} M	1.58	1.68	0.96	—	Al-Awqati et al., 1976

a. The data are adapted from the indicated sources, being expressed as the ratio of the mean value in the experimental tissue (e) to the simultaneous mean value in the paired control tissue (c).

b. These studies were performed following 20 hours of exposure to aldosterone. Subsequent studies demonstrated that at this time there is an increase in A (Al-Awqati, 1977).

will be recalled that for the H^+ transport system, as for the Na^+ transport system, prolonged exposure to aldosterone resulted in enhancement of the affinity A, presumably reflecting enhancement of the free energy of the metabolic reaction driving transport. It would be anticipated that substrate depletion, repletion, and the administration of 2-deoxyglucose would also influence this free energy. The demonstration that various perturbations of metabolic factors affect κ_H^a but have little effect on PMF again suggests dynamic interaction between permeability and energetic factors, as noted previously with Na^+ transport.

8.3.6 Biochemical correlates of active H^+ transport

Dixon and Al-Awqati (1979b) have demonstrated that when turtle urinary bladders are poisoned so as to inhibit normal ATP synthesis, the application of an adverse proton electrochemical potential gradient of greater than 180 mV (the magnitude required to abolish tightly coupled H^+ transport and metabolism) results in production of ATP. This was interpreted as indicating that the pump is a reversible proton-translocating ATPase. In order to investigate the relationship between transport and the free energy $-\Delta G_{ATP}$ of ATP hydrolysis, concurrent measurements of J_{H0}^a and the mass action ratio [ATP]/[ADP][P_i] were carried out under varied conditions (Dixon and Al-Awqati, 1979a). Under control conditions $-\Delta G_{ATP}$ was about 46–50 kJ/mole (about 11–12 kcal/mole). (This is to be compared with values of $\sim$20–70 kJ/mole calculated from the control values of A_{ATP} in the NET studies of aldosterone action cited above.) The effects of stimulants and inhibitors of transport were qualitatively appropriate, in that 5% CO_2 depressed $-\Delta G_{ATP}$, whereas acetazolamide and mucosal acidification enhanced it; in each case, however, the change of $-\Delta G_{ATP}$ was of the order of 10% or less. (This is to be compared with estimates of A_{ATP} some 50% higher in aldosterone-treated than in control turtle bladders.)

Although the above results do not permit unambiguous interpretations, they serve to demonstrate the fundamental difficulties associated with the various means presently available for analysis of the energetics of active transport processes. On the one hand, the biochemical techniques provide direct evidence of the involvement of ATP. However, the techniques are destructive, raising the possibility of appreciable change in adenine nucleotide and phosphate content during the course of analysis. Furthermore, in a tissue with significant compartmentalization, the mean mass action ratio will be a complex function of the distribution of ATP, ADP, and P_i in different cell types and subcellular organelles, which may obscure large changes in the region of interest. Finally, the calculation of $-\Delta G_{ATP}$ based on assumed "textbook" values of the standard free energy $-\Delta G^0_{ATP}$ fails to take into account the possible significance of changes of $-\Delta G^0$ attributable to experimental effects on cellular H^+, Mg^{+2}, and Ca^{+2} distribution. On the other hand, the affinity A_{O_2} calculated from the NET formalism is an abstract quantity and, even presuming the validity of the two-flow linear formalism, it cannot be related directly to $-\Delta G_{ATP}$ without knowledge of the P/O ratio under the experimental conditions. However, where the formulation is applicable it would provide a value of $-\Delta G$ referable to the functioning transport system rather than the tissue as a whole and would reflect the influence of factors altering the local $-\Delta G^0$, as well as the mass action ratio.

Given the present highly incomplete understanding of the energetics of active transport processes, both approaches would appear to deserve further extensive investigation.

8.4 Summary

1. Experimental studies indicate that active Na^+ transport in the frog skin and toad urinary bladder, and active H^+ transport in the turtle urinary bladder, may be usefully analyzed in terms of a linear two-flow NET formulation, if appropriate means are employed to vary transepithelial forces.

2. In terms of this formulation, modulation of X_{Na} (or X_H) along proper pathways permits evaluation of the phenomenological L (or R) coefficients and the affinity A of a metabolic driving reaction, thereby differentiating kinetic and energetic factors modulating transport. Kinetic rather than thermodynamic linearity would necessitate modification of the treatment.

3. Values of A_{O_2} (the affinity per mole of suprabasal O_2 consumption) are compatible with biochemical estimates; calculation of A_P

(the affinity per mole of ATP utilization) requires knowledge of the tissue P/O ratio.

4. Effects of model compounds on L's and A_{O_2} are appropriate in terms of known mechanisms of action. Aldosterone and antidiuretic hormone appear to stimulate transport by influencing both kinetic and energetic factors.

5. There is no unique stoichiometric ratio relating rates of Na transport and suprabasal O_2 consumption. Coupling of these processes appears incomplete.

6. Linear NET and equivalent circuit formulations are compared. Theoretical considerations and experimental results indicate that in contrast to A, the electromotive force of Na transport E_{Na} (or E_H) depends explicitly on kinetic parameters.

7. Given the present unsatisfactory understanding of the energetics of active transport processes, both biochemical and formal thermodynamic approaches deserve further extensive investigation.

9 Kinetics of isotope flows: background and theory

In the study of transport processes, analysis of the kinetics of isotope flows has been used in two ways: first, to examine the permeability of membranes or organ systems, and second, to examine the forces promoting transport. Isotope flows can be studied of course without the use of nonequilibrium thermodynamics, and we shall consider some examples of this in the following discussion. However, NET helps to clarify our thinking by providing criteria for testing the legitimacy of our assumptions and a precise formalism for deducing their consequences.

We shall begin by considering how isotopes have been used, mentioning certain anomalies which have been noted and classical explanations proposed to explain these. We shall then consider the application of NET to the general problem of the analysis of isotope flows. In Chapter 10 we look at some specific models. In Chapter 11 we shall examine experimental evidence relevant to the validity of the thermodynamic formulation. Here and there we shall point out what we consider to be inadequacies or oversights in the classical treatment.

216 Bioenergetics and linear nonequilibrium thermodynamics

9.1 Historical survey

Very soon after isotopes (and particularly radioactive isotopes) became available they began to be used widely in the evaluation of the permeability of biological membranes. The reasons for this are evident. Even in the simple case of a substance whose flow is a consequence only of its own concentration gradient, evaluation of its permeability requires measurement of its concentration on both sides of the membrane, as well as its rate of flow. Often such measurements may be in substantial error, if, for example, the concentrations are low or the chemical techniques are inaccurate; furthermore, under physiological circumstances the gradients of concentration may be very small, and flows may be either very small or absent altogether. Any attempt to improve the quality of the measurement by increasing the magnitude of the concentration gradient, and thereby the flow, may produce a very unphysiological state of affairs. By contrast, radioactive isotopes can be added in truly minute quantities, with concentrations often several orders of magnitude smaller than that of the parent substance. Despite the insignificant effect on the total concentration of the chemical species, the gradient of the concentration of the isotope, as well as its rate of flow, may be readily measurable. This permits the calculation of a permeability coefficient.[1]

The quantity desired is the permeability which in principle could be determined from a measurement of the *net* flow induced by a concentration difference of the parent species,

$$\omega = \frac{-J}{RT\,\Delta c} \tag{9.1}$$

where J is the flow from bath I to bath II, and $\Delta c = c^{II} - c^{I}$. (For simplicity we shall omit subscripts whenever this will not be confusing.) This definition is often applied without consideration of the possible influence of coupling of J to the flows of other species. However, this leads to incompatible values for ω determined under different experimental circumstances. In the special case of absence of volume flow and electric current, Eq. (9.1) is equivalent to Eq. (3.51) for J_s, which may be rewritten as:

$$\omega = -\left(\frac{J_s}{\Delta\pi_s}\right)_{J_v, I=0} \qquad (9.2)$$

(In Chapters 9 to 11, to be consistent with our earlier publications on the kinetics of isotope flows, we shall adopt certain polarity conventions that sometimes differ from those used in earlier chapters.)

In the use of tracer isotopes to avoid the difficulties of permeability measurement by chemical techniques, the tracer permeability is taken as

$$\omega^* = \frac{-J^*}{RT\,\Delta c^*} \qquad (9.3)$$

or more precisely, taking the possibility of coupled flows into account,

$$\omega^* = -\left(\frac{J_s^*}{\Delta\pi_s^*}\right)_{J_v, J_s, I=0} \qquad (9.4)$$

If we are dealing with isotopes which show no isotope effects, that is, isotopes whose atomic weights differ so slightly that their kinetic and thermodynamic characteristics are virtually identical, it might seem evident that ω^* from Eq. (9.3) or Eq. (9.4) is always equal to ω, since the tracer molecules act exactly like the abundant molecules. However, this conclusion is incorrect. It is true that if we ignore minute isotope effects an ion of ^{24}Na behaves exactly like an ion of ^{23}Na, but this does not mean that the tracer *species* ^{24}Na behaves exactly like the abundant *species* ^{23}Na. These are different species, with very different concentrations and with very different concentration gradients, and there is therefore no reason to assume that ω^* is necessarily equal to ω. Why this is so will become clearer later.

Early workers did not always appreciate the distinction between the two permeabilities. Indeed, very soon after the measurement of isotope fluxes was introduced, an apparent discrepancy based on these considerations gained wide attention. The permeability of frog skin to water appeared to have two different values, depending on whether the measurement was of osmotic flow (net flow induced by an osmotic pressure gradient) or of the flow of a tracer form of water

218   Bioenergetics and linear nonequilibrium thermodynamics

in the absence of net water flow. This apparent anomaly was clarified by Koefoed-Johnsen and Ussing (1953), who pointed out that a diffusing molecule (in this case a tracer water molecule) "may possess, superimposed upon its rate of diffusion, that rate at which the solvent flows." That is to say, in the case of water flow in frog skins, the difference in the two types of permeability coefficients was related to the phenomenon of coupling of flows.

In this case, the permeability to isotope flow ω^* is less than the permeability to net flow ω. The opposite discrepancy, in which the tracer permeability ω^* is greater than the permeability for net flow ω, has also been noted, for example, in the case of passive sugar transport across the red cell membrane. As will be shown, both of these "discrepancies" are natural consequences of the coupling of flows considered in the framework of NET.

Another broad area to be considered concerns the use of isotopes in attempts to evaluate the *forces* inducing flows across membranes. In this regard, it was pointed out both by Ussing and by Teorell that despite our ignorance concerning the structural characteristics and permeability of biological membranes, it might be possible to find out whether a particular substance moves as a consequence only of its own electrochemical potential gradient (Teorell, 1949; Ussing, 1949, 1952, 1960). This determination involved the measurement of the "flux ratio," the ratio of oppositely directed "unidirectional" fluxes across the membrane. A simple expression was considered to apply, which in our terminology may be written as:

$$RT \ln f = - \Delta\bar{\mu} = X \tag{9.5}$$

where $\ln f$ is the natural logarithm of the flux ratio, and the other terms have their usual significance. In the absence of electrical effects, Eq. (9.5) becomes particularly simple,

$$f = \frac{c^{I}}{c^{II}} \tag{9.6}$$

where c^I and c^{II} are the concentrations of the test species on opposite sides (the outside and inside) of the membrane. (The sign convention used here is illustrated in Fig. 9.1; the flux ratio is defined for our purposes as influx/efflux.)

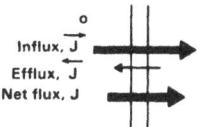

Fig. 9.1. Influx, efflux, and net flux between the outside (o) and inside (i) solutions bathing a membrane.

If Eq. (9.5) or (9.6) applied to a given experimental situation, it seemed reasonable to infer that the test substance moved subject only to its electrochemical potential gradient. If the equations did not apply, this conclusion was considered not to hold, and for such cases the possibility was suggested that other forces influence net flow. Furthermore, it was suggested that the degree to which the flux ratio is "abnormal," that is, the degree to which it deviates from the simple expressions shown, should give an idea of the magnitude of these extra forces, whatever their nature. The use of measurements of the flux ratio in attempts to clarify the nature of water flow led to the concept of "bulk flow," according to which water molecules move across biological membranes not individually but in clusters. In a related series of studies, the demonstration of abnormal flux ratios of acetamide and thiourea in the presence of osmotic water flow led to the inference that these analogues of urea traverse aqueous pathways in permeating epithelial membranes, with "solvent drag" enhancing their flow.

A further use of the flux ratio for studying the forces promoting transport was the attempt of Ussing and his colleagues to characterize the mechanisms responsible for active salt transport across epithelial membranes. Frog skins transport salt against extremely large concentration gradients ($c^{II}/c^{I} \simeq 10^4 - 10^5$). Since in this situation Na$^+$ moves against both a concentration gradient and electrical forces, that is, against an electrochemical potential gradient, its transport would be considered active. Nevertheless, because NaCl might possibly move as an ion pair it was thought important to show that the fluxes of Na$^+$ and Cl$^-$ were independent of each other. To do this, measurements were made of the flux ratios of Na$^+$ and Cl$^-$ associated with the well-defined electrochemical potential differences $\Delta\bar{\mu}_{Na}$ and $\Delta\bar{\mu}_{Cl}$ (Ussing, 1949; Koefoed-Johnsen, Levi, and Ussing, 1952; Ussing, 1960). It was found that the flux ratio of Na was larger than would be predicted from its electrochemical potential difference (that is, $RT \ln f_{Na} > X_{Na}$), as would be consistent with active transport. The flux ratio of Cl^-, on the other hand, was predictable from its electrochemical potential difference ($RT \ln f_{Cl} = X_{Cl}$) (Fig. 9.2). This observation is consistent with passive movement of Cl^- down a favorable electrochemical potential gradient resulting from the active transport of sodium. On this basis it was suggested that the use of the flux ratio would permit the distinction between active and passive transport, and that this should be possible whether transport occurs with or against an electrochemical potential gradient. Furthermore, the extent of deviation of the flux ratio from that consistent with passive diffusion was considered to quantify the metabolic force influencing net flow ("E_{Na}'').

The above approaches have historically been of great value in investigations of the nature of biological transport processes. Nevertheless, it has long been realized that reliance on the flux ratio as a means of evaluating energetic factors can in principle be very misleading. Perhaps the earliest clear example of one limitation in the use of the flux ratio for these quantitative purposes was provided by Hodgkin and Keynes's studies of potassium fluxes in poisoned squid axons, where it was shown that the flux ratio was markedly abnormal, with $RT \ln f$ differing markedly from the electrochemical potential difference (Fig. 9.3), despite the apparent absence of both solvent drag and active transport (Hodgkin and Keynes, 1955). In this

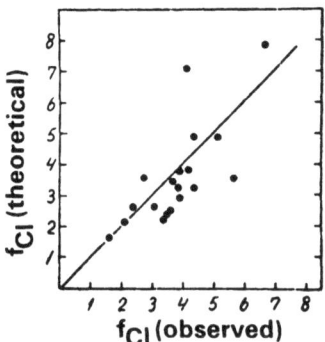

Fig. 9.2 Theoretical versus observed flux ratios for chloride in the frog skin. The theoretical flux ratios were calculated from the unmodified Ussing relation, as in Eq. (9.5). (Adapted from Koefoed-Johnsen, Levi, and Ussing, 1952.)

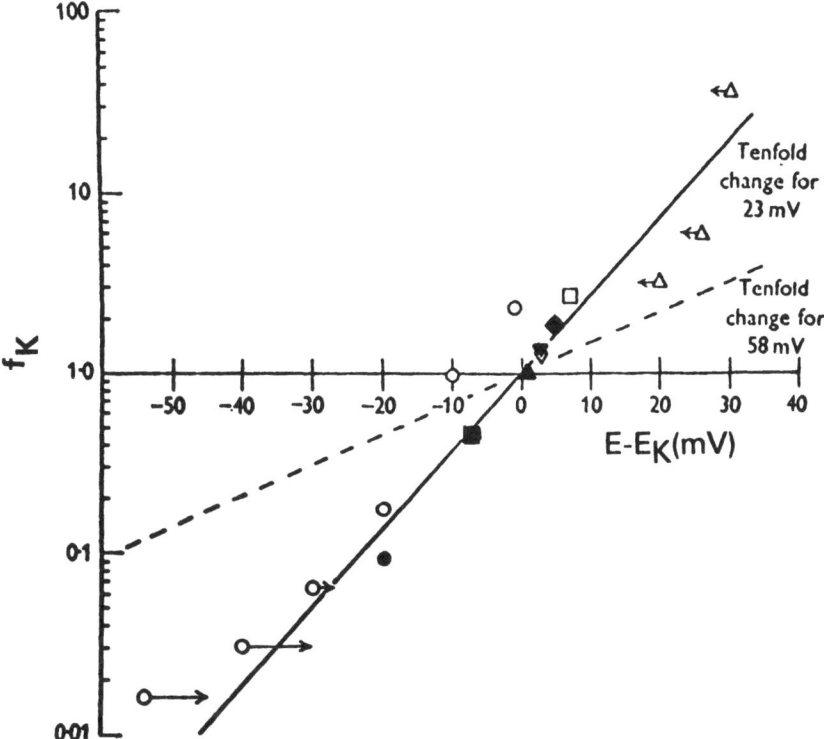

Fig. 9.3 Effect of electrical potential difference on the flux ratio of potassium in fibers of giant *Sepia* axons poisoned with dinitrophenol. E is the potential difference across the membrane (external potential minus internal potential) and $E_K = (RT/F) \ln (c_{K,\text{int}}/c_{K,\text{ext}})$, so that $E - E_K$ equals the negative electrochemical potential difference for potassium, X_K. (From Hodgkin and Keynes, 1955.)

case, generally attributed to "single-file diffusion," the flux ratio overestimates the forces effecting transport. On the other hand, in mechanisms associated with "exchange diffusion," the flux ratio underestimates the forces effecting transport (Levi and Ussing, 1948). Since it is clear that under these circumstances the measurement of the flux ratio does not evaluate the forces which bring about net transport, we must be skeptical as to whether we can use the flux ratio for this purpose in general, in the absence of highly detailed knowledge of the mechanism of permeation.

The failure of the flux ratio to evaluate reliably the forces promoting transport, as well as the difficulties encountered in evaluating membrane permeability, point to the need for a systematic analysis of the kinetics of isotope flows. This should provide a basis for the study of the permeability and energetic factors influencing transport at both a theoretical and an experimental level.

9.2 Thermodynamic analysis

9.2.1 Fundamental equations

The thermodynamic analysis we are going to consider is based on the work of Kedem and Essig (1965). In conformity with our main emphasis, the development is in terms of linear NET. Readers interested in a more fundamental approach are referred to the general treatment of Sauer (1978). As pointed out above, we aim to explain two anomalies: the discrepancy between permeability coefficients determined by isotopic and by nonisotopic techniques and the "abnormality" of the flux ratio in the absence of solvent drag and active transport. The formalism indicates that both of these so-called anomalies can be explained in terms of coupling between flows of different isotopic forms of the test species. This phenomenon has been called isotope interaction.

Since the processes we are going to study depend for the most part on the coupling of conservative flows, it is natural to use a resistance formulation; later we will relate the resistance coefficients to permeability coefficients. We consider first the case of transport by way of *identical pathways*, ignoring for the present the effect of leaks.[2] For sufficiently slow processes it is generally assumed that the electrochemical potential is everywhere definable and that forces and flows are linearly related at every point in a membrane. (The assumption of *local* linearity is a mild assumption; it is not to say that the flows are necessarily linearly related to the *integral* force $\Delta \bar{\mu}$ across the membrane.) In the resistance formulation we start by expressing driving forces as functions of the steady-state flows in the membrane. Considering the x-axis to be perpendicular to the membrane, we write the local flow equations as follows:

$$-\frac{d\bar{\mu}}{dx} = r_{00}J + \sum_{j=4}^{n} r_{0j}J_j \qquad (9.7)$$

and

$$-\frac{d\bar{\mu}_1}{dx} = r_{11}J_1 + r_{12}J_2 + r_{13}J_3 + \sum_{j=4}^{n} r_{1j}J_j \qquad (9.8)$$

$$-\frac{d\bar{\mu}_2}{dx} = r_{21}J_1 + r_{22}J_2 + r_{23}J_3 + \sum_{j=4}^{n} r_{2j}J_j \qquad (9.9)$$

$$-\frac{d\bar{\mu}_3}{dx} = r_{31}J_1 + r_{32}J_2 + r_{33}J_3 + \sum_{j=4}^{n} r_{3j}J_j \qquad (9.10)$$

Here the subscript 0, omitted for convenience in referring to the net flow J, indicates the total test substance, which may be either an ionic or an uncharged species. The subscript 1 refers to the abundant isotope, and 2 and 3 refer to tracer isotopes of the test substance. The r_{ij}'s indicate local resistance coefficients. The flows J_j (taken to be positive in the x-direction) include not only those of all species passing through the membrane, as for example water, but also the movement of mobile membrane components. The sum may also contain terms $r_{ir}J_r$, representing the direct contribution of metabolic reaction to oriented solute flux.

Equations (9.7)–(9.10) are related, as the isotopes are assumed to be indistinguishable in all thermodynamic and kinetic properties. Then for the total test substance,

$$\frac{d\bar{\mu}}{dx} = RT\frac{d\ln c}{dx} + RT\frac{d\ln \gamma}{dx} + zF\frac{d\psi}{dx} + \bar{v}\frac{dp}{dx} \qquad (9.11)$$

and for one of the tracer species,

$$\frac{d\bar{\mu}_2}{dx} = RT\frac{d\ln c_2}{dx} + RT\frac{d\ln \gamma_2}{dx} + \frac{z_2 F d\psi}{dx} + \bar{v}_2\frac{dp}{dx} \qquad (9.12)$$

Since we have assumed the identity of the thermodynamic properties of the different isotopes, we can state that $\gamma = \gamma_2$, $z = z_2$, and $\bar{v} = \bar{v}_2$. Hence, denoting the specific activity c_2/c by ρ_2,

$$\frac{d\bar{\mu}_2}{dx} - \frac{d\bar{\mu}}{dx} = RT\frac{d\ln\rho_2}{dx} \tag{9.13}$$

A similar relation applies to species 1 and 3. Because we have also assumed that the isotopes have identical kinetic properties, we are entitled to claim that no isotope separation can be brought about by the coupled processes. If we consider the case that the specific activities are constant throughout the membrane, the ratio of the flows of isotopes must then be identical with the ratio of their concentrations, whatever the other flows. Equation (9.14) shows that under these conditions the driving forces are equal.

$$\frac{d\bar{\mu}_1}{dx} = \frac{d\bar{\mu}_2}{dx} = \frac{d\bar{\mu}_3}{dx} = \frac{d\bar{\mu}}{dx} \quad \left(\frac{d\rho_i}{dx} = 0\right) \tag{9.14}$$

Turning our attention away from isotopes for the moment, we obtain the relationship between total flow and the difference of electrochemical potential across the membrane by integration of Eq. (9.7):

$$-(\bar{\mu}^{II} - \bar{\mu}^{I}) = X = JR + \int_0^{\Delta x} \sum_4^n r_{0j}J_j dx \tag{9.15}$$

where the integral resistance, which specifies the resistance to net flow, is defined by

$$R = \int_0^{\Delta x} r_{00} dx \tag{9.16}$$

($x = 0$ at the outer surface of the membrane; $x = \Delta x$ at the inner surface of the membrane). In integrating Eq. (9.7) in this manner, we assume that J is constant throughout the membrane in the steady state; however, it is not necessary that the J_j's be conservative.

Both surfaces of the membrane are at equilibrium with adjacent solutions, and it is assumed that $\bar{\mu}$ is continuous throughout the membrane. It is permissible, however, that there be a finite number

Isotope flows: background and theory 225

of discontinuities in $d\bar{\mu}/dx$. This would undoubtedly be the case in a biological membrane, which would be expected to be characterized by discontinuities of local resistances and the presence of metabolic flows. Our considerations are limited to systems in which the test species is present on both sides of the membrane, so that $\Delta\bar{\mu}$ is always finite. From Eq. (9.15),

$$R = \left(\frac{X}{J}\right)_{J_4=J_5=\cdots=J_r=0} \tag{9.17}$$

It is useful at this point to introduce a natural extension of the definition of the permeability coefficient given in Chapter 3. For dilute electrolyte solutions as usually found in biological systems, it is often the case that $J_v \simeq J_w \bar{v}_w$. We may then write

$$\omega = \frac{1}{cR} = \left(\frac{-J}{RT\,\Delta c}\right)_{J_4=J_5=\cdots=J_r=0} \tag{9.18}$$

where c is an average concentration, as discussed in Chapter 3.[3]

Unfortunately, conditions such as those specified in Eq. (9.18) are often not attainable in biological systems. Nor can we always evaluate R or ω from Eq. (9.15). Such a determination requires knowledge of both the driving force and the contribution of coupled flows to the forces promoting transport, but in biological studies this information is frequently unavailable. The hope was, therefore, that it would be possible to obtain the necessary information with the aid of isotope techniques. It is useful to begin an examination of this issue by considering the situation which would exist if isotope interaction were absent.

9.2.2 Absence of isotope interaction

In this case the flow of a species is influenced by its concentration gradient, electrical forces, and possibly by coupling to flows of solvent or even other solute species or metabolism, but the flow of a given isotopic form of the species is uninfluenced by the flow of any other isotopic form of the same species. Therefore Eqs. (9.8)–(9.10) are simplified, due to the consideration that

$$r_{ik} = 0 \quad (i \neq k;\ i,\ k = 1,\ 2,\ 3) \tag{9.19}$$

226 Bioenergetics and linear nonequilibrium thermodynamics

Then, while for net flow J we have as previously (Eq. 9.7),

$$J = -\frac{1}{r_{00}}\left(\frac{d\tilde{\mu}}{dx} + \sum_{j=4}^{n} r_{0j}J_j\right) \qquad (9.20)$$

the abundant and tracer isotope flows J_1 and J_2 are given simply by

$$J_1 = -\frac{1}{r_{11}}\left(\frac{d\tilde{\mu}_1}{dx} + \sum_{j=4}^{n} r_{1j}J_j\right) \qquad (9.21)$$

$$J_2 = -\frac{1}{r_{22}}\left(\frac{d\tilde{\mu}_2}{dx} + \sum_{j=4}^{n} r_{2j}J_j\right) \qquad (9.22)$$

with an analogous expression for the tracer species 3. Note that the straight coefficients, r_{00}, r_{11}, and so on, must be positive according to the second law of thermodynamics, but the cross coefficients may be of either sign.

It was pointed out above that in the absence of a gradient of specific activity, the isotope flows must always be in the ratio of their concentrations. Thus, from Eqs. (9.20) and (9.22), taking into account (9.14),

$$\frac{J_2}{J} = \frac{r_{00}}{r_{22}} \frac{\dfrac{d\tilde{\mu}}{dx} + \sum_{4}^{n} r_{2j}J_j}{\dfrac{d\tilde{\mu}}{dx} + \sum_{4}^{n} r_{0j}J_j} = \rho_2 \qquad \left(\frac{d\rho_2}{dx} = 0\right) \qquad (9.23)$$

and similarly for J_3/J.

As equations of the type of (9.23) must hold for all values of J_j, according to the assumption of kinetic indistinguishability, it follows that

$$r_{1j} = r_{2j} = r_{3j} = r_{0j} \qquad (9.24)$$

and

$$r_{11}\rho_1 = r_{22}\rho_2 = r_{33}\rho_3 = r_{00} \qquad (9.25)$$

We now invoke a fundamental concept of NET, namely, that although small changes in local forces change the flows, they do not change the local parameters of state. (To put it another way, while the local resistance coefficients at x depend on the local concentrations and therefore might change with alteration of the specific activity, they are unaffected by changes in the gradient of specific activity.) Hence Eqs. (9.24) and (9.25) remain true irrespective of whether a gradient in ρ_2 exists or not.

If we now introduce Eqs. (9.13), (9.24), and (9.25) into (9.22), the flow equation for the tracer species becomes

$$J_2 = -\frac{\rho_2}{r_{00}}\left(\frac{d\bar{\mu}}{dx} + \sum_4^n r_{0j}J_j + RT\frac{d\ln\rho_2}{dx}\right) \quad (9.26)$$

This should be compared with the expression for net flow, Eq. (9.20). It then becomes apparent that whenever ρ varies with x, the tracer flow will not be equal to ρJ, but will differ from this value because of isotope exchange. By combining Eqs. (9.20) and (9.26) we obtain

$$J_2 - \rho_2 J = -\frac{RT}{r_{00}}\frac{d\rho_2}{dx} \quad (9.27)$$

This equation shows that while the tracer flows and net flow depend on all the $r_{0j}J_j$, the flow of tracer relative to $\rho_2 J$ depends only on the gradient of ρ_2 and the resistance r_{00}. In the absence of a gradient of specific activity, J_2 and J are seen to be in the ratio ρ_2, as they should be on the basis of our previous assumptions. In the presence of a gradient of specific activity, this relationship no longer applies; the tracer flow now "runs ahead" of the net flow. However, again the relationship between the two flows is intuitively plausible, in that the extent to which the tracer flow runs ahead is proportional to the gradient of specific activity. A point to note in passing is that the proportionality factor relating the specific activity gradient to the exchange flow is here determined by r_{00}, the local resistance coefficient for *net* flow; the significance of this point will become clear later.

Now although Eq. (9.27) is instructive when looked at in this way, in order to be useful experimentally it must be integrated across the membrane or tissue being studied so that we can relate observable flows to other experimentally measurable parameters. Notice that this integration is permissible just as long as we have conservative flows and continuity of the specific activity along our identical parallel pathways. Discontinuities of the local resistance coefficients or the gradient of the specific activity might well occur in moving along the x-axis of various membranes of interest, but these are of no significance here.

Eq. (9.27) is of such a simple form that it is readily integrated. It is useful to consider separately two states: one in which there is no net flow and one in which there is net flow. In the absence of significant net flow we have a simple isotope exchange experiment. In this case the integration of Eq. (9.27) gives, on introducing the resistance to net flow defined in Eq. (9.16),

$$J_2 R = -RT(\rho_2^{II} - \rho_2^{I}) = -RT\,\Delta\rho_2 \quad (J = 0) \quad (9.28)$$

We have here assumed that the specific activity at each surface is equal to that in the contiguous solution: $\Delta\rho$ then refers to the difference of specific activity in the baths. Equation (9.28) shows that under the conditions considered here, that is, for a system with identical pathways in the absence of coupling of isotope flows, a simple isotope exchange experiment would determine the resistance to net flow (and thereby, with Eq. 9.18, the permeability to net flow) even in the absence of knowledge of the driving force of the total test species and with the possibility of coupling to flows of other species or even active transport. Thus for example if $\Delta c = 0$,

$$J_2 c R = -RT\,\Delta c_2 \quad (J = 0) \quad (9.29)$$

and

$$\omega = \frac{1}{cR} = \left(\frac{-J_2}{RT\,\Delta c_2}\right)_{J=0} \quad (9.30)$$

Providing net flow of the test species is zero, it is immaterial what other flows occur. But it is not even necessary to establish this condition. In the presence of net flow, the integration of Eq. (9.29) gives, with its analogue for tracer species 3 and with Eq. (9.16),

Isotope flows: background and theory   229

$$\frac{JR}{RT} = \ln \frac{J_2 - \rho_2^{II} J}{J_2 - \rho_2^{I} J} = \ln \frac{J_3 - \rho_3^{II} J}{J_3 - \rho_3^{I} J} \qquad (9.31)$$

where we have called the specific activity ρ^I at $x = 0$ (outside) and ρ^{II} at $x = \Delta x$ (inside). If a single tracer, say species 2, is added to only one compartment, say the outside, and the inside compartment volume is sufficiently large, we can write

$$\rho_2^{II} \simeq 0 \qquad (9.32)$$

Equation (9.31) then immediately simplifies to

$$\frac{JR}{RT} = \ln \frac{\dfrac{J_2}{\rho_2^I}}{\dfrac{J_2}{\rho_2^I} - J} \qquad (9.33)$$

Thus, in the absence of isotope interaction, the determination of one tracer flux and the net flow determines the resistance to net flow, R. Consequently R can be obtained without knowing either the driving force or the nature of the coupled flows. The permeability to net flow ω may then again be estimated as $1/cR$.

If we measure two unidirectional fluxes, and if each tracer is added to only one compartment such that

$$\rho_2^{II} \simeq \rho_3^{I} \simeq 0 \qquad (9.34)$$

Equation (9.31) reduces to

$$\frac{\dfrac{J_2}{\rho_2^I}}{\dfrac{J_2}{\rho_2^I} - J} = \frac{\dfrac{J_3}{\rho_3^{II}} - J}{\dfrac{J_3}{\rho_3^{II}}} \qquad (9.35)$$

or more simply

$$J = \frac{J_2}{\rho_2^I} + \frac{J_3}{\rho_3^{II}} \equiv \vec{J} - \overleftarrow{J} \qquad (9.36)$$

where in the accepted terminology defined by Ussing and Teorell, $\bar{J} \equiv J_2/\rho_2^I$ is the influx, and $\bar{\bar{J}} \equiv -J_3/\rho_3^{II}$ is the outflux or efflux. Therefore Eq. (9.36) is the widely used relation: net flux = influx − efflux. Putting Eq. (9.36) into Eq. (9.33) gives, with Eq. (9.30),

$$JR = \frac{J}{c\omega} = RT \ln f \equiv RT \ln \left(\frac{\bar{J}}{\bar{\bar{J}}}\right) \qquad (9.37)$$

The above equations show that for the type of system we are considering here, the measurement of two unidirectional fluxes (or alternatively, one unidirectional flux and the net flux) evaluates the resistance (and therefore the permeability) for net flow, despite our ignorance of the forces promoting transport, whatever their nature. Furthermore, the same measurements permit quantification of the net force promoting transport, as we can see by combining Eqs. (9.15) and (9.37):

$$RT \ln f = X - \int_0^{\Delta x} \sum_4^n r_{0j} J_j dx \qquad (9.38)$$

This is analogous to the flux ratio equations of Ussing and Teorell.

9.2.3 Presence of isotope interaction

The treatment up to this point has assumed that although the flow of the tracer is influenced by its electrochemical potential gradient and possibly by the coupled flows of other species and/or metabolism, it is not directly influenced by changes in the net flow of the test species. However, we pointed out earlier that Ussing and his colleagues were able to account for the anomalies in the water permeability of frog skin by considering that the flow of tracer water *was* in fact influenced by the flow of solvent water. We shall consider the possibility of coupling, not only of flows of different isotopic forms of water, but also of flows of different isotopic forms of any solute species. Although this might seem unlikely in dilute aqueous solutions, it might very possibly be the case in highly specialized membranes and carrier systems, so we shall consider the consequences of such isotope interaction quite generally.

As before, Eq. (9.7) applies to the total flow of test species. But for the isotopic components of the flow, Eqs. (9.8)–(9.10) are no longer subject to the condition that the r_{ik}'s be zero as stated in Eq. (9.19). It is important to note that when we say the form of the equation for total flow is unchanged in the presence of isotope interaction, we are making no claim about the possible effects of isotope interaction on the magnitude of the resistance coefficients r_{00} and r_{0j}. In order to make such a statement, we would have to have a very detailed physical picture of a system.

Examining Eq. (9.8), we see that considerations of kinetic indistinguishability require that for given values of $d\bar{\mu}_1/dx$ and J_j, the flow J_1 must depend not on the individual values of J_2 and J_3, but only on their sum, $J_2 + J_3$. Hence $r_{12} = r_{13}$, and this quantity must be independent of the ratio of concentrations (c_2/c_3). Similarly, $r_{21} = r_{23}$, $r_{31} = r_{32}$. We also know that since Eqs. (9.8)–(9.10) relate conjugate forces and flows appearing in the dissipation function of the process, the Onsager reciprocal relation applies: $r_{ik} = r_{ki}$. These equalities taken together lead to the conclusion that all r_{ik}'s are identical ($i \neq k$; $i, k = 1, 2, 3$). Remembering that $J = J_1 + J_2 + J_3$, it follows that $r_{21}J_1 + r_{23}J_3 = r_{ik}(J - J_2)$, and Eq. (9.9) can be rewritten:

$$-\frac{d\bar{\mu}_2}{dx} = (r_{22} - r_{ik})J_2 + r_{ik}J + \sum_{j=4}^{n} r_{2j}J_j \qquad (9.39)$$

As previously, for constant specific activity ρ_i throughout the membrane, $d\bar{\mu}/dx = d\bar{\mu}_i/dx$, and since no isotope separation occurs, $J_i = \rho_i J$. From Eqs. (9.7) and (9.39),

$$r_{00}J + \sum_{j=4}^{n} r_{0j}J_j = (r_{22} - r_{ik})J_2 + r_{ik}J + \sum_{j=4}^{n} r_{2j}J_j \qquad (d\rho_2/dx = 0)$$

and thus

$$\frac{J_2}{J} = \frac{r_{00} - r_{ik}}{r_{22} - r_{ik}} + \frac{\sum_{j=4}^{n}(r_{0j} - r_{2j})}{(r_{22} - r_{ik})}\frac{J_j}{J} = \rho_2 \qquad (9.40)$$

(It is easily shown that with species 2 present only in tracer amounts, the denominators of Eq. 9.40 can never equal zero.) Since these equations must hold for all values of the independent variables, it follows that Eq. (9.24) still holds as before, and that

$$\frac{r_{00} - r_{ik}}{r_{22} - r_{ik}} = \rho_2 \tag{9.41}$$

and similarly for the other isotopic species. As in the earlier case, these relations are valid whether or not specific activity is constant.

The sign of r_{ik} may be confusing. Where there is mutual drag between the isotope flows (positive coupling), r_{ik} is negative; where flow of species i diminishes flow of species k (negative coupling), r_{ik} is positive. As before, when ρ varies with x, the tracer flow will exceed ρJ by the value of isotope exchange. In the present case, however, the resistance to exchange flow will be modified by isotope interaction. Subtracting Eq. (9.7) from Eq. (9.39) we find

$$-\frac{d\bar{\mu}_2}{dx} + \frac{d\bar{\mu}}{dx} = (r_{22} - r_{ik})J_2 - (r_{00} - r_{ik})J + \sum_{j=4}^{n}(r_{2j} - r_{0j})J_j$$

but since Eqs. (9.13) and (9.24) give, respectively,

$$-\frac{d\bar{\mu}_2}{dx} + \frac{d\bar{\mu}}{dx} = -RT\frac{d \ln \rho_2}{dx} \quad \text{and} \quad r_{2j} = r_{0j}$$

we can write

$$(r_{22} - r_{ik})J_2 - (r_{00} - r_{ik})J = -RT\frac{d \ln \rho_2}{dx} = -\frac{RT}{\rho_2}\frac{d\rho_2}{dx}$$

Dividing through by $(r_{22} - r_{ik})$ and introducing Eq. (9.41) leads to the relation

$$J_2 - \rho_2 J = \frac{-RT}{(r_{00} - r_{ik})}\frac{d\rho_2}{dx} \tag{9.42}$$

Here we have an equation analogous to Eq. (9.27). However, there is an important difference: the extent to which a gradient of specific activity results in the tracer's running ahead now depends not only on r_{00}, the local resistance to net flow, but also on r_{ik}, the resistance coefficient embodying the interaction between the different isotopic forms of the test species.

Noting this difference, we can complete our formal treatment just as in the absence of isotope interaction. Again it is often useful to evaluate a resistance by a tracer measurement in the absence of net flow. Integrating Eq. (9.42) for this case gives

$$J_2 \int_0^{\Delta x} (r_{00} - r_{ik}) dx \equiv J_2 R^* = -RT(\rho_2^{II} - \rho_2^{I})$$

$$= -RT\Delta\rho_2 \qquad (J = 0) \qquad (9.43)$$

which should be compared with Eq. (9.28). Clearly the resistance coefficient here is not the integral resistance coefficient for net flow R: in the presence of isotope interaction, a simple isotope exchange experiment does not evaluate the resistance (or permeability) for net flow. We have called the coefficient obtained here the exchange resistance coefficient R^*. In analogy with Eq. (9.16),

$$R^* = \int_0^{\Delta x} (r_{00} - r_{ik}) dx \qquad (9.44)$$

Notice that if $r_{ik} < 0$, which indicates positive coupling, $R^* > R$. On the other hand, if coupling is negative, $r_{ik} > 0$ and $R^* < R$. From Eq. (9.43) we can immediately write down an equation analogous to Eq. (9.30):

$$\omega^* = \frac{1}{cR^*} = \left(\frac{-J_2}{RT\Delta c_2}\right)_{J=0} \qquad (9.45)$$

Here ω^* is the exchange permeability.

Again, we can also consider the local Eq. (9.42) in the presence of net flow and integrate it as previously, obtaining for tracer species 2,

$$\frac{JR^*}{RT} = \ln \frac{J_2 - \rho_2^{II} J}{J_2 - \rho_2^{I} J} \qquad (9.46)$$

and eventually

$$JR^* = \frac{J}{c\omega^*} = RT \ln f \qquad (9.47)$$

in complete analogy with Eq. (9.37). Note that just as in the absence of net flow, a tracer experiment in the presence of net flow does not evaluate the resistance coefficient for net flow R, but the exchange resistance coefficient R^*. This may be either larger or smaller than R.

If we combine Eq. (9.47) with Eq. (9.15), we obtain[4]

$$RT \ln f = \frac{R^*}{R}\left(X - \int_0^{\Delta x} \sum_4^n r_{0j}J_j dx\right)$$

$$= \frac{\omega}{\omega^*}\left(X - \int_0^{\Delta x} \sum_4^n r_{0j}J_j dx\right) \qquad (9.48)$$

It should be noted that this equation, as well as the corresponding expression which applies in the absence of isotope interaction, Eq. (9.38), is valid irrespective of the dependence of the permeability coefficients on the bath concentrations or the electrical potential difference. ω and ω^*, being functions of local phenomenological coefficients, are of course parameters of state, and nothing in our formalism encourages us to think that they may not be sensitive functions of state. Nevertheless, for whatever ω and ω^* apply to the system in the state studied, these equations would be expected to apply.[5]

As is seen, in the presence of isotope interaction the flux ratio cannot evaluate the forces promoting transport. However, Eq. (9.48) is useful in that the factors promoting deviation from the "normal" flux ratio, $\exp(X/RT)$, are clearly seen: (1) coupling with flows of other species, (2) coupling with the flow of metabolism (active transport), and (3) isotope interaction. An additional practical consideration is the influence of parallel leak pathways. The significance of this factor will be examined in Sec. 9.3.

9.2.4 The flux ratio and the energetics of active transport

Because of the importance of analyzing the energetics of active transport, it is often assumed that the influence of isotope interaction and leak are minimal, permitting the use of the flux ratio for this purpose. Thus, following the classical approach of Ussing and Zerahn (1951), the active transport system is considered in terms of an equivalent electrochemical cell, so that in the absence of coupled transepithelial flows the force contributed by the Na⁺ pump is given by FE_{Na}, where E_{Na} represents the "electromotive force of sodium transport." Assuming then that E_{Na} affects the flux ratio of Na⁺ in the same way as an applied electromotive force would affect the flux ratio of a passive ion, it is considered that

$$RT \ln f = X + FE_{Na} \tag{9.49}$$

and that at short circuit (identical bath solutions and $\Delta\psi = 0$)

$$RT \ln f_0 = FE_{Na} \tag{9.50}$$

Equation (9.49) is intuitively plausible and leads to the appropriate conclusion that FE_{Na} is equivalent to static head $(-X^0)$, the force at which J and $\ln f$ are zero. Nevertheless, it is easily seen that it is likely to be inconsistent with Eq. (9.48), even in the absence of isotope interaction. Thus, in the absence of coupled transepithelial flows, Eq. (9.48) may be written as

$$RT \ln f = \frac{\omega}{\omega^*}(X - R_{0r}J_r) \tag{9.51}$$

where R_{0r} represents $\int_0^{\Delta x} r_{0r} dx$. Therefore $FE_{Na} \equiv -X^0 = -R_{0r}(J_r)_{J=0}$ whereas $RT \ln f_0 = -(\omega/\omega^*)R_{0r}(J_r)_{X=0}$. Accordingly, even in the absence of isotope interaction $(\omega/\omega^* = 1)$, Eqs. (9.48) and (9.50) give the same value of E_{Na} only if $(J_r)_{J=0} = (J_r)_{X=0}$. Since this is unlikely to be the case, in general Eq. (9.50) will be incorrect. Indeed, for the linear systems considered here, since by Eq. (7.39), $(J_r)_{J=0} = (1 - q^2)(J_r)_{X=0}$, we see that

$$FE_{Na} = \frac{\omega^*}{\omega}(1 - q^2) RT \ln f_0 \tag{9.52}$$

Given values of E_{Na} commonly in the range of 100–120 mV, Eq. (9.52) suggests that for highly coupled systems the flux ratio of the active pathway might be quite large. Some experimental examples will be considered in Chapter 11.

By this point the reader may be wondering to what extent our emphasis on isotope interaction is realistic. It can be seen that since

NET incorporates explicitly all possible couplings of flows, the consideration of isotope interaction arises automatically. This is not to say, however, that it need necessarily be significant. Some theoretical bases for isotope interaction will be considered in Sec. 9.3 and in Chapter 10, and experimental examples will be discussed in Chapter 11. First, however, it is useful to consider the consequences of the fact that membranes of biological interest are often not homogeneous.

9.3 Function of composite membranes

9.3.1 Parallel arrays

The above treatment is applicable both to an array of parallel identical channels and to composite series membranes, provided there is continuity of the electrochemical potential of the test species. However, with heterogeneous membranes composed of parallel channels in which different factors influence flows, the treatment must be modified. In attempts to analyze the nature of transport processes by measurements of isotope flows, this consideration is not always taken adequately into account. A common example cited immediately above is the use of the flux ratio to evaluate E_{Na}. If this approach is taken, it is reasoned that at least for "tight" epithelia, the resistance of nonspecific leak pathways is sufficiently high that in the absence of an electrochemical potential difference for the test species its unidirectional flows reflect essentially the character of its predominant transport process. This view overlooks the fact that the flux ratio may be highly influenced by the presence of small leak pathways, even when net leak flow is zero (Fig. 9.4).

Another example derives from the use of unidirectional isotope fluxes to evaluate the energetic requirements of putative pump–leak systems in symmetrical cells. Here it is generally reasoned that the rate of isotope uptake quantifies the net rate of leakage into the cell, which in the absence of special transport mechanisms must be equal to the net rate of outward transport by the pump. On this assumption, knowing the electrochemical potential difference of the test species, one is able to calculate the rate of performance of electrochemical work. This approach overlooks the possibility that a sig-

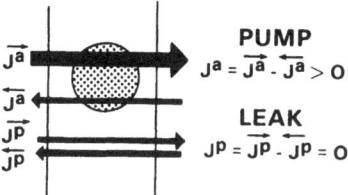

Fig. 9.4. Model of epithelial sodium transport system comprising parallel active transport and leak pathways. In the absence of an electrochemical potential difference for sodium (level flow), net transport occurs only in the active pathway, but passive flows $\vec{J}^p$ and $\overleftarrow{J}^p$ influence the flux ratio.

nificant part of isotope flux into the cell may occur via the active transport mechanism itself, and to the extent that it does, the measurement of isotope uptake will overestimate the need for active transport and the performance of electroosmotic work (Fig. 9.5). (Experimental evidence concerning this point will be presented in Chapter 11.) These various examples can be analyzed precisely by means of our general formalism. Before we consider the specific cases, however, it is useful to develop some general relationships.

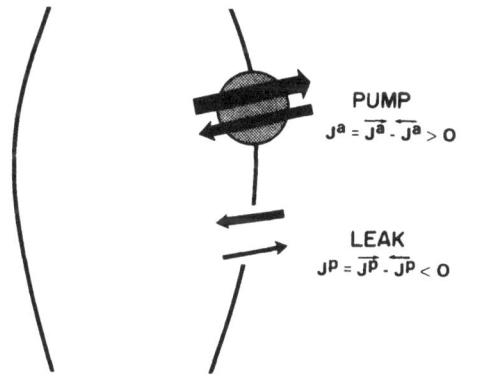

Fig. 9.5. Model of sodium transport system of symmetrical cell comprising parallel active transport and leak pathways. It is suggested that flux in the active pathway is bidirectional (Chen and Walser, 1975; Wolff and Essig, 1977; Dawson and Al-Awqati, 1978). In the absence of net flow of sodium (static head), the net rate of leak into the cell J^p is balanced by the net rate of active transport J^a out of the cell. However, measurements of $J = \vec{J} = \vec{J}^a + \vec{J}^p$ might much exceed $\vec{J}^a - \overleftarrow{J}^a = J^a$ and therefore lead to significant overestimates of the rate of performance of electroosmotic work. (From Essig, 1968.)

238 Bioenergetics and linear nonequilibrium thermodynamics

Combining Eq. (9.36) and the definition of the flux ratio, $f \equiv \bar{J}/\tilde{J}$, gives (for either the elemental pathways or the composite system):

$$\bar{J} = \frac{fJ}{f-1} \tag{9.53}$$

$$\tilde{J} = \frac{J}{f-1} \tag{9.54}$$

Introducing Eq. (9.47),

$$\bar{J} = \left(\frac{f}{f-1}\right) \frac{RT \ln f}{R^*} \tag{9.55}$$

$$\tilde{J} = \left(\frac{1}{f-1}\right) \frac{RT \ln f}{R^*} \tag{9.56}$$

For a composite array of independent passive pathways, each exposed to identical forces, since $J^* \equiv \Sigma J_i^*$, Eq. (9.45) gives $\omega^* = \Sigma \omega_i^*$, and

$$\omega^*/\omega = \Sigma \left(\frac{\omega_i^*}{\omega_i}\right)\left(\frac{\omega_i}{\omega}\right) = \Sigma \left(\frac{\omega^*}{\omega}\right)_i \left(\frac{\omega_i}{\omega}\right) \tag{9.57}$$

so that each element contributes to the overall value of (ω^*/ω) in proportion to its fractional contribution to membrane conductance.

More generally, even when the forces operating on the discrete arrays differ, as with active transport or coupling of flows between different chemical species, a simple relationship between the flux ratios obtains: since $J = \Sigma J_i$, Eq. (9.47) shows that

$$f = \Pi f_i^{(\omega_i^*/\omega^*)} \tag{9.58}$$

(We assume here that c is about the same for all elemental pathways.) In the absence of isotope interaction in any pathway, $\omega^*/\omega = 1$ (see Eq. 9.57), so

$$f = \Pi f_i^{(\omega_i/\omega)} \qquad [(\omega^*/\omega)_i \equiv 1] \tag{9.59}$$

Isotope flows: background and theory 239

It is also useful to express the flux ratio in terms of the elemental unidirectional fluxes. For a system of two discrete parallel arrays α and β,

$$f = \frac{J^\alpha + J^\beta}{\overleftarrow{J}^\alpha + \overleftarrow{J}^\beta} = \frac{\dfrac{f^\alpha J^\alpha}{f^\alpha - 1} + \dfrac{f^\beta J^\beta}{f^\beta - 1}}{\dfrac{J^\alpha}{f^\alpha - 1} + \dfrac{J^\beta}{f^\beta - 1}} \qquad (9.60)$$

Evidently, even in the absence of active transport, if there is isotope interaction in either array, the observed flux ratio may differ appreciably from those in the elemental pathways. Indeed, in the presence of circulating volume flow, the composite system may show isotope interaction despite absence of isotope interaction in either elemental array. This phenomenon will be discussed in Chapter 10.

For our epithelial active transport system the flux ratio will be

$$f = \frac{J^a + J^p}{\overleftarrow{J}^a + \overleftarrow{J}^p} = \frac{\left(\dfrac{f^a}{f^a - 1}\right)\left(\dfrac{\ln f^a}{R^{a*}}\right) + \left(\dfrac{f^p}{f^p - 1}\right)\left(\dfrac{\ln f^p}{R^{p*}}\right)}{\left(\dfrac{1}{f^a - 1}\right)\left(\dfrac{\ln f^a}{R^{a*}}\right) + \left(\dfrac{1}{f^p - 1}\right)\left(\dfrac{\ln f^p}{R^{p*}}\right)} \qquad (9.61)$$

where the superscripts a and p refer to the parallel active and passive pathways respectively. The flux ratio is often measured at level flow ($X = 0$). Assuming, for simplicity, no coupling of passive flows, as $X \to 0$, $f^p \to 1$ (Eq. 9.48), and $\ln f^p \to (f^p - 1)$, so from Eqs. (9.55) and (9.56), $J^p \to \overleftarrow{J}^p \to RT/R^{p*} = RT/R^p$. Writing n for the ratio of exchange resistances of the active and passive pathways ($n \equiv R^{a*}/R^{p*} = R^{a*}/R^p$) and introducing Eq. (9.61),

$$f_{X=0} \equiv f_0 = \frac{n + \left(\dfrac{f_0^a}{f_0^a - 1}\right) \ln f_0^a}{1 + \left(\dfrac{1}{f_0^a - 1}\right) \ln f_0^a} \qquad (9.62)$$

This equation shows the relationship between the observed flux ratio and the flux ratio in the active pathway when the membrane is exposed to two identical bathing solutions and short-circuited

240 Bioenergetics and linear nonequilibrium thermodynamics

($\Delta\psi = 0$). Representative relationships for various values of n are shown in Fig. 9.6. Only in the absence of leak ($n = 0.00$) is the observed flux ratio f the same as f^a of the active pathway. It will be seen that even with rather tight membranes in which $R^p \gg R^{a*}$, flux ratios of the magnitude commonly reported in the literature (~50–100) might differ greatly from f^a. Since it is impossible to be precise about the magnitude of such errors without combined measurements of R^{a*} and R^{p*}, which are as yet unavailable, it would be prudent to be cautious in interpreting the significance of experimental values of f.

For the case of the symmetrical cell, in the steady state net flow is

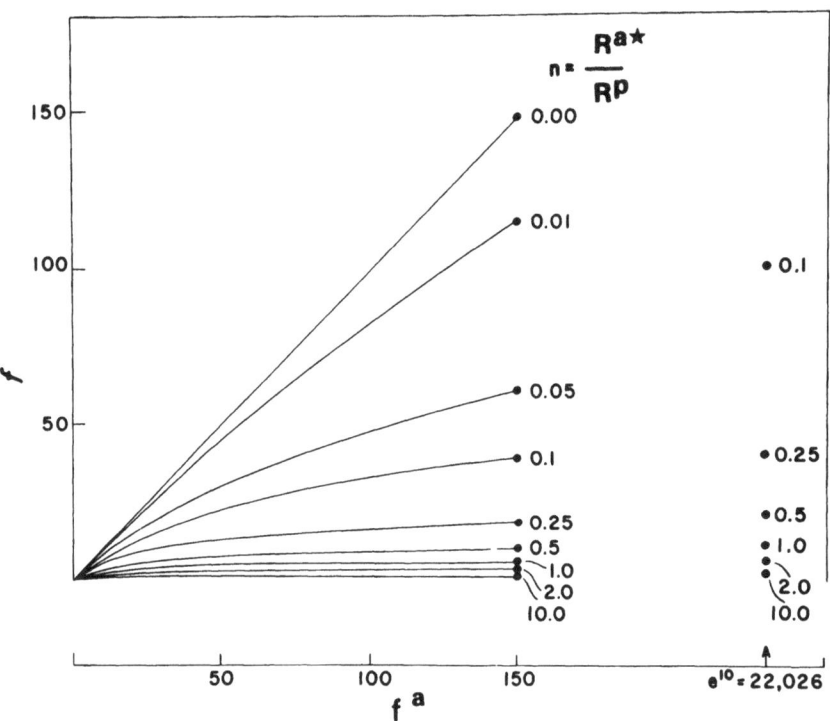

Fig. 9.6. Effect of leak pathway on the relationship between the observed flux ratio f and the flux ratio in the active pathway f^a in the absence of an electrochemical potential difference (short circuit, level flow). Only in the absence of leak ($n = 0.00$) is the observed flux ratio the same as in the active transport pathway. (From Kedem and Essig, 1965.)

at static head, so that $J \equiv J^a + J^p = 0$. Equation (9.54) then gives

$$\tilde{J} = \tilde{J} = \frac{J^a}{f^a - 1} + \frac{J^p}{f^p - 1}$$

$$= J^a \left(\frac{1}{f^a - 1} - \frac{1}{f^p - 1} \right) \quad (J\cdot = 0) \qquad (9.63)$$

Introducing Eqs. (9.15) and (9.47), again assuming no coupling of flows in the passive pathway,

$$\tilde{J} = \tilde{J} = J^a \left(\frac{1}{\exp\left(-\frac{R^{a*}}{R^p}\frac{X}{RT}\right) - 1} - \frac{1}{\exp\left(\frac{X}{RT}\right) - 1} \right) \quad (J = 0) \quad (9.64)$$

Figure 9.7 shows the relationship between $\tilde{J}$ (= $\tilde{J}$) and J^a. As is seen, when either R^{a*}/R^p or $-X$ is low, the unidirectional flux may very much exceed the rate of net active transport. This is readily understood. When R^{a*}/R^p is small, a large fraction of flow into the cell occurs through the active transport pathway, and so the inflow exceeds the net rate of passive leakage, which in the steady state must equal the rate of net active transport. On the other hand, when $-X$ is small, the flux ratio in the passive pathway approaches 1, so again the inflow may much exceed the net rates of leak and active transport.

As has been mentioned, in analyses of the feasibility of membrane pump–leak mechanisms, measurements of unidirectional flux have often been taken to evaluate the rate of active transport. This has led to calculated values of J^a which when combined with commonly cited values of X predict very high energetic requirements, sometimes far in excess of that available. Ling has attributed this discrepancy to overestimation of $-X$. According to his association-induction hypothesis, sodium exists in protoplasm in a very different state from that in aqueous solution. Hence, despite its low concentration and the electronegativity within the cell, it is in equilibrium with external sodium, so there is no active transport across the plasma membrane (Ling, 1962). Ussing has long ago pointed out, however, that "the rate at which Na$^+$ is found to leave muscle fibers—as measured with the tracer method—is not necessarily equal to the rate of [net] active extrusion" and that "part of the ap-

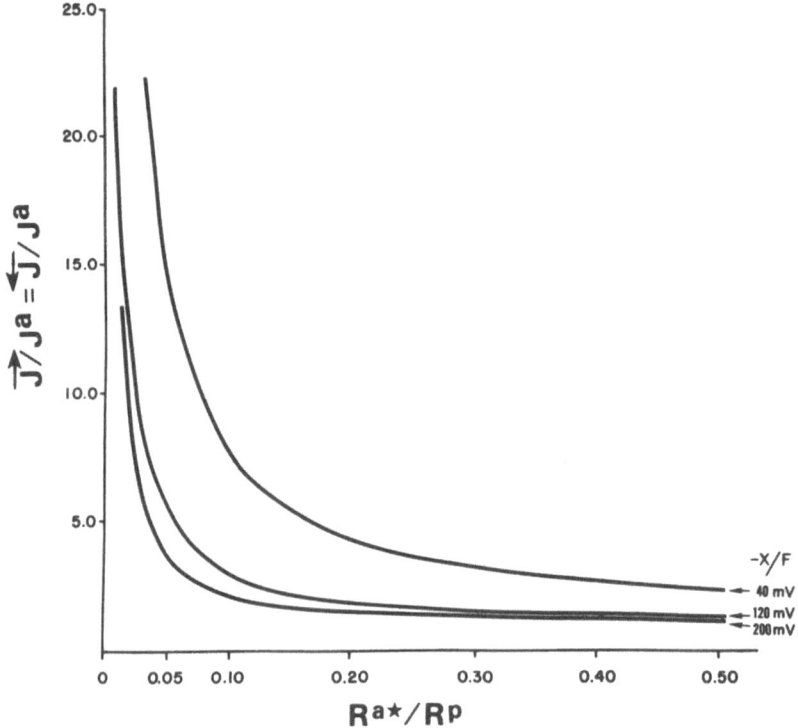

Fig. 9.7. Ratio of unidirectional fluxes to steady-state rate of active transport in symmetrical cell as a function of R^{a*}/R^p and $-X/F$. (From Essig, 1968.)

parent transport may be simple exchange" (Ussing, 1947; Levi and Ussing, 1948). One possible mechanism of exchange which is commonly invoked is the movement of a mobile carrier, facilitating isotopic equilibration across a membrane without consumption of energy. For one example, Minkoff and Damadian, in analyzing the high rates of isotope flux of six solute species in E. coli, have suggested that such exchange diffusion at fixed charge sites constitutes the mechanism by which the cell exchanges tracer for stable isotope. On this basis they exclude membrane pumps dependent on ATP, thereby in their view avoiding a "caloric catastrophe" (Minkoff and Damadian, 1973). The analysis above, however, demonstrates that in principle, high rates of unidirectional flux are not incompatible with membrane pumps, provided only that these pumps permit tracer flux in either direction. The energetic requirements of such mechanisms can only be evaluated when accurate measurements of R^{a*}/R^p and X become available.

9.3.2 Series arrays

Applying Eqs. (9.46) and (9.47) to each element of a series array gives for each element i

$$JR_i^* = RT \ln \frac{J_2 - \rho_{2,i}^{II} J}{J_2 - \rho_{2,i}^{I} J} = RT \ln f_i \tag{9.65}$$

where $\rho_{2,i}^{II}$ and $\rho_{2,i}^{I}$ represent the specific activity of tracer species 2 at the inner and outer borders, respectively, of element i. Since $\rho_{2,i}^{II} = \rho_{2,i+1}^{I}$, summing over elements gives

$$\sum JR_i^* = RT \ln \frac{J_2 - \rho_2^{II} J}{J_2 - \rho_2^{I} J} = JR^* = RT \ln f \tag{9.66}$$

Thus

$$R^* = \sum R_i^* \tag{9.67}$$

and

$$\frac{R^*}{R} = \sum \left(\frac{R_i^*}{R_i}\right)\left(\frac{R_i}{R}\right) = \sum \left(\frac{R^*}{R}\right)_i \left(\frac{R_i}{R}\right) \tag{9.68}$$

so that each element contributes to the overall value of R^*/R in proportion to its fractional contribution to membrane resistance. This result should be compared to Eq. (9.57), the corresponding relation for ω^*/ω for parallel arrays (remembering, however, that in evaluating ω^*/ω we considered the forces operative in each channel to be identical). Furthermore, since $\Sigma \ln f_i = \ln f$,

$$f = \Pi f_i \tag{9.69}$$

This relationship applies irrespective of the presence or absence of isotope interaction. (Compare Eqs. 9.58 and 9.59.)

In analyzing net flux or tracer flux of an ion across the mucosal (m) and serosal (s) series barriers of an epithelial cell, Schultz and Frizzell (1976) have stressed that the overall transepithelial permeability coefficient derived from analysis of the composite membrane (in our terms ω) is not an accurate measure of the permeabilities of the two limiting membranes combined in series, since it may be influenced by conditions that do not affect ω_m or ω_s (that is to say, cellular concentration and electrical potential profiles). Thus they feel that these permeability coefficients are physically meaningful only if the fluxes are restricted to paracellular pathways. While agreeing that it is necessary to be circumspect in inferring mechanistic details from values of permeability coefficients, we take a somewhat different point of view. Our view is that permeability (and resistance) coefficients are indeed physically meaningful parameters, representing the state of the composite membrane as it stands. It must be remembered, however, that as always with phenomenological coefficients, there is no reason to assume *a priori* that ω and R will not be functions of the state of the system under study. Accordingly, although Eq. (9.67) is the analogue of the simple additivity of series resistances familiar from electrical circuit analysis, R^* will depend on concentration profiles across the membrane, despite the formal simplicity of the relationship. These considerations are implicit in the original Kedem-Katchalsky treatment and are elaborated in their analysis of composite membranes.

9.4 Summary

1. Relationships among the net flow, the flux ratio, and the forces promoting transport may be derived from assumptions of broad validity. For permeation by way of identical pathways, in the absence of isotope interaction, measurement of the flux ratio quantifies the forces; measurement of one unidirectional flux (in the absence of net flow) or both unidirectional fluxes (in the presence of net flow) determines R, the resistance to net flow, or ω, the permeability, without knowledge of either driving forces or coupled flows.

2. For homogeneous pathways, deviations of the flux ratio from normal are attributable to coupling of flows of different species, active transport, or isotope interaction. In the presence of isotope in-

teraction, measurement of the tracer exchange resistance R^* or permeability ω^* does not quantify R or ω, and the flux ratio does not quantify the forces.

3. Measurement of the flux ratio at short circuit, a classical method for evaluation of the electromotive force of sodium transport, is unreliable, since f_0 depends not only on E_{Na}, but also on the extent of leak, isotope interaction, and coupling of transport to metabolism.

4. Heterogeneity of parallel arrays and series elements modifies isotope interaction and flux ratios predictably. The observed flux ratio of an active transport system is highly sensitive to parallel leak. The energetic requirements of a pump–leak system cannot be evaluated from measurements of unidirectional flux without highly detailed knowledge of both transport pathways.

10 Kinetics of isotope flows: mechanisms of isotope interaction

Although formal treatments are precise and general, they provide no physical picture of the nature of a transport process. For this purpose it is helpful to examine specific models. A variety of models has been used to analyze the kinetics of isotope flows and abnormalities of the flux ratio. We shall examine a few which are conveniently related to formulations presented above.

10.1 Frictional models

Frictional models have been usefully applied both to the study of transport processes in synthetic membranes (Spiegler, 1958; Mackay and Meares, 1959; Kedem and Katchalsky, 1961) and to analysis of the flux ratio (Ussing, 1952; Hoshiko and Lindley, 1964; Essig, 1966). We here summarize earlier results (Essig, 1966), following the treatment of Kedem and Katchalsky.

In this view it is assumed that in the steady state the thermodynamic force X_i acting on species i is exactly balanced by frictional interaction with particles of other transported species and the membrane. Thus the force of frictional interaction $F_{ij} = -f_{ij}(v_i - v_j)$, where f_{ij} is the frictional coefficient per mole of i, and the v's are average velocities relative to the membrane, given by the ratio (J/c) of the flow to concentration. Considering first the flow of water through a homogeneous membrane m, the resistance to net flow is given by

$$R_w = \frac{f_{wm}\Delta x}{c_w} \quad (10.1)$$

where Δx is the thickness of the membrane. The exchange resistance to water flow, on the other hand, is given by

$$R_w^* = \frac{(f_{wm} + f_{2w})\Delta x}{c_w} \quad (10.2)$$

where f_{2w} is the water–water friction coefficient, so

$$\left(\frac{R^*}{R}\right)_w = 1 + \left(\frac{f_{2w}}{f_{wm}}\right) \quad (10.3)$$

Equation (10.3) shows that with frictional interaction between water molecules, the exchange resistance will exceed the resistance to net flow, and the flux ratio will be abnormal. However, contrary to what might be expected intuitively, the degree of abnormality of the flux ratio will depend not only on the extent of interaction between water molecules but also on the extent of interaction between molecules of water and the membrane. Thus this simple frictional model can account for the observation that isotope interaction is marked in channels of large caliber, in which f_{wm} will be small, but that isotope interaction is small in fine channels, in which f_{wm} will be large.

Analogously, we can consider the effects of interaction between solute particles. Although such effects are likely to be quite small in dilute aqueous solutions, they might well be significant in hydrophobic environments, such as cell plasma membranes. Expressing the possible frictional interaction between solute particles in terms of a frictional coefficient f_{2s},

$$\left(\frac{R^*}{R}\right)_s = 1 + \frac{f_{2s}}{f_{sm} + f_{sw}} \quad (10.4)$$

Again $R^* > R$; as is evident, a simple frictional model cannot explain negative isotope interaction. Of course, the argument above is crude in its neglect of the details of molecular interactions and can be considered to provide only a qualitative picture of membrane function.

10.2 Lattice and carrier models: passive transport

In various systems of biological interest, the electrochemical potential of the transported species may not be continuous at every point. Thus in some cases the rate-limiting factors for transport across very thin membranes may be adsorption or desorption at the surfaces; in these cases the membrane surfaces will not be in equilibrium with the bathing solutions. In other cases, transport may involve movement between lattice sites with discrete changes in chemical potential. Both of the above possibilities are readily dealt with by means of the kinetic formalism introduced by Hill (Chapter 5), since this facilitates the quantitative analysis of a number of familiar transport models. We consider first a few simple examples of passive transport, in the absence of electrical forces, where the solute of interest is moving down a concentration gradient without interaction with flows of other species (Essig, Kedem, and Hill, 1966). For these systems it proves convenient to use permeability rather than resistance coefficients. For our purposes it is useful to direct attention to states near equilibrium, since the relative simplicity of the analysis then permits explicit evaluation of permeability coefficients in terms of kinetic parameters, thereby elucidating requirements for isotope interaction. By this means it is possible to show that discontinuity of $\tilde{\mu}$ does not in itself result in discrepancies between ω and ω^*; rather, as shown above for continuous systems, this is a manifestation of coupling (isotope interaction).

As previously, we consider "membranes" which represent ensembles consisting of a large number of equivalent units, each comprising one or more sites, either fixed or mobile. Each membrane is exposed at its opposite surfaces to two large isothermal baths A and B, containing dilute ideal solutions of a single nonelectrolyte at concentrations c_A and c_B, respectively; the corresponding tracer concentrations are c_A^* and c_B^*. Each membrane unit may exist in any of a number of discrete states, corresponding to the various possible combinations of site vacancy or occupancy by either the abundant or tracer forms of the test solute. As previously, a kinetic description of the system can be given in terms of first-order or pseudo-first-order rate constants for each transition between states.

In principle it is possible to analyze even complex transport mechanisms by the diagram methods applied in Chapter 5 to characterize coupling of transport to metabolism. For the basic models considered here, however, a simpler, more direct approach is feasible. The method is conveniently explained in terms of our first model, consisting of a single row of N identical sites per unit area; the three possible states of this system are shown in Fig. 10.1. Considering first the rate of net flow in the absence of tracer, we designate the rate constant for adsorption from bath i by $\alpha_i = \alpha^0 c_i$, where α^0 incorporates the standard chemical potential of the solute as well as the specific kinetic constant for adsorption for the given solute–membrane system, which is assumed symmetrical. The rate constant for desorption into each bath is β. Representing the fraction of occupied sites by x, adsorption per unit area from bath i occurs at a rate $N\alpha_i(1 - x)$ and desorption into bath i occurs at a rate $N\beta x$, so that in the steady state

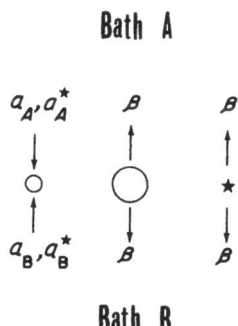

Fig. 10.1. Lattice model of transport system consisting of a single row of identical sites. The large circle and star represent the abundant and tracer forms of the transported species, respectively. (From Essig, Kedem, and Hill, 1966.)

250 Bioenergetics and linear nonequilibrium thermodynamics

$$\frac{dx}{dt} = [\alpha_A(1 - x) - \beta x] + [\alpha_B(1 - x) - \beta x] = 0$$

and

$$x = \frac{\alpha}{\alpha + \beta} \tag{10.5}$$

where $\alpha = (\alpha_A + \alpha_B)/2$. The flow across the membrane is then given by

$$J = N[\alpha_A(1 - x) - \beta x]$$
$$= \frac{N\beta(\alpha_A - \alpha_B)}{2(\alpha + \beta)} = \frac{N\beta\alpha^0(c_A - c_B)}{2(\alpha^0 c + \beta)} \tag{10.6}$$

with c being the arithmetic mean concentration, and the permeability for net flow is

$$\omega = \frac{N\alpha^0\beta}{2\,RT(\alpha + \beta)} = \frac{N\alpha^0\beta}{2\,RT(\alpha^0 c + \beta)} \tag{10.7}$$

Thus permeability shows saturation with increasing mean solute concentration but is independent of the magnitude of Δc.

In order to observe isotope flows, we can add a radioactive form of the test species, which in tracer quantities can be assumed not to affect the distribution of the abundant isotope or the net flow. Given identical kinetic and thermodynamic properties of the tracer and the abundant species, $\alpha^{0*} = \alpha^0$ and $\beta^* = \beta$. Reasoning as above, it is seen that in the steady state

$$\frac{dx^*}{dt} = [\alpha_A^*(1 - x) - \beta x^*] + [\alpha_B^*(1 - x) - \beta x^*] = 0$$

giving with Eq. (10.5):

$$x = \frac{\alpha_A^* + \alpha_B^*}{2(\alpha + \beta)} \tag{10.8}$$

so that

$$J^* = N[\alpha_A^*(1-x) - \beta x^*]$$

$$= \frac{N\beta(\alpha_A^* - \alpha_B^*)}{2(\alpha + \beta)} = \frac{N\beta\alpha^0(c_A^* - c_B^*)}{2(\alpha^0 c + \beta)} \qquad (10.9)$$

Since the tracer flow is unaffected by net flow (being independent of Δc), on comparing Eqs. (10.6) and (10.9), we see that $\omega = \omega^*$. That is, in this case the tracer permeability determined by a self-diffusion experiment (with $\Delta c = 0$) is equal to the permeability to net flow induced by a small concentration difference across the membrane. (Equality of ω and ω^* requires, of course, that in manipulating the concentration difference Δc, the mean concentration c is maintained constant.)

Our next model is more realistic, comprising binding sites at each surface, and possibly within the interior of the membrane as well. Figure 10.2 shows the general n-row model where transport may occur from a site in one row to any empty site in an adjacent row. Here flows depend not only on the rate constants α and β for adsorption and desorption, but also on the rate constant k for transport between sites. For this system a treatment closely analogous to that

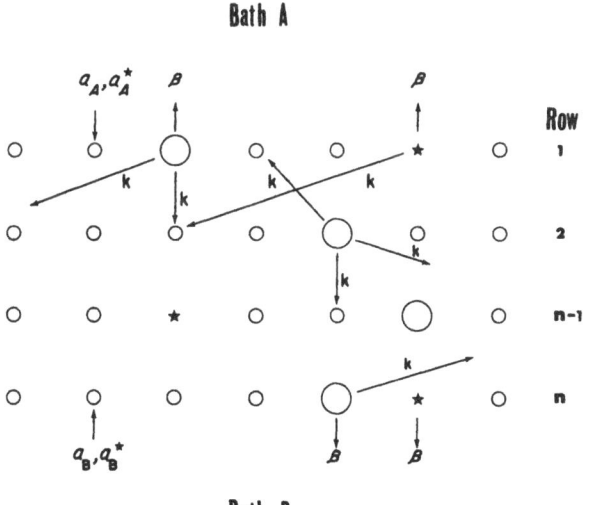

Fig. 10.2. General n-row model with free movement of ligands to empty sites in adjacent rows.

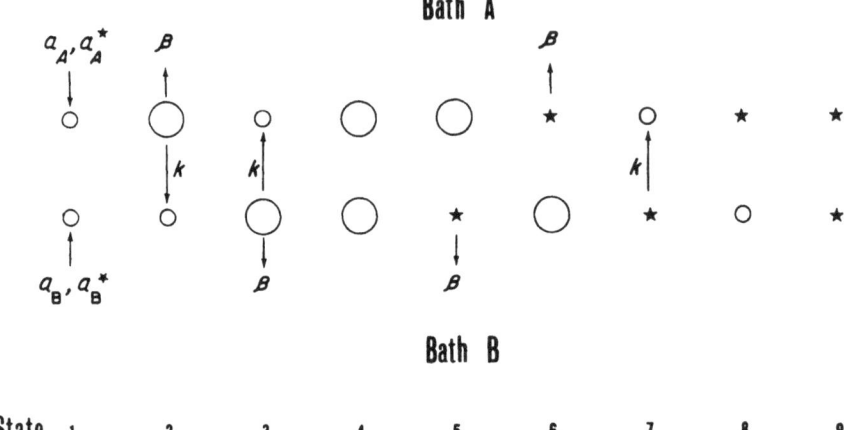

Fig. 10.3. Two-row model with restriction of ligands to individual units.

for the single-row case leads to the result (near equilibrium)

$$\omega = \omega^* = \frac{N\alpha^0\beta k}{RT(\alpha + \beta)[(n - 1)(\alpha + \beta) + 2k]} \qquad (10.10)$$

The equality of ω and ω^* here is a consequence of the freedom of molecules to move between different units. This can be shown by considering a case in which molecular movement is constrained.

For simplicity, we limit our analysis to a two-row lattice, with each pair of sites constituting an independent unit. In this case, in moving from one row to the other, a molecule is restricted to the same unit. The nine possible states of this system are shown in Fig. 10.3. It will be recognized that this system represents one possible type of the classical single-file diffusion model (Hodgkin and Keynes, 1955). Again a treatment analogous to that above is applied, demonstrating, however, that now ω is unequal to ω^*. In this case (again near equilibrium)

$$\omega = \frac{N\alpha^0\beta k}{RT(\alpha + \beta)(\alpha + \beta + 2k)} \qquad (10.11)$$

whereas

$$\omega^* = \frac{N\alpha^0\beta k(\alpha + 2\beta)}{RT(\alpha + \beta)^2(\alpha + 2\beta + 4k)} \qquad (10.12)$$

so that in general ω exceeds ω*, with

$$\frac{\omega^*}{\omega} = 1 - \frac{2\alpha k}{(\alpha + \beta)(\alpha + 2\beta + 4k)} \quad (10.13)$$

It is of interest that as k becomes large relative to α and β, that is, if adsorption and desorption are rate-limiting, Eq. (10.11) becomes identical to Eq. (10.7). Thus, as far as net flow is concerned, two rows with rapid crossing are equivalent to one row. This is not true for tracer flow, however, since even with $k \gg \alpha$ and β, $\omega^* \neq \omega$. On the other hand, single-file movement does not in itself assure that ω and ω^* will differ, since when the surfaces are in near-equilibrium with the adjacent baths ($k \ll \alpha, \beta$) $\omega^* \simeq \omega$. This is also so when the average concentration (and thus α) becomes very small; under these circumstances the crossing of a molecule will only rarely be prevented by occupancy of an adjacent site.

A very different type of behavior is seen with another model commonly invoked in biological studies, of which the simplest possible example is shown in Fig. 10.4. Here a solute can enter the membrane only when combining with a carrier molecule, accessible at one or the other of the two surfaces, and can cross the membrane only when combined with the carrier. (It is immaterial whether the carrier itself actually traverses the membrane or merely by change of its configuration influences solute movement.) In this case, if we assign different rate constants for the unloaded and loaded carrier, k_c and k respectively, the permeability for net flow is

$$\omega = \frac{N\alpha^0 \beta k k_c}{2RT(\alpha + \beta)(\alpha k + \beta k_c + 2 k k_c)} \quad (10.14)$$

whereas the tracer permeability is

$$\omega^* = \frac{N\alpha^0 \beta k}{2RT(\alpha + \beta)(\beta + 2k)} \quad (10.15)$$

so that ω* exceeds ω, with

$$\frac{\omega^*}{\omega} = 1 + \frac{\alpha k}{k_c(\beta + 2k)} \quad (10.16)$$

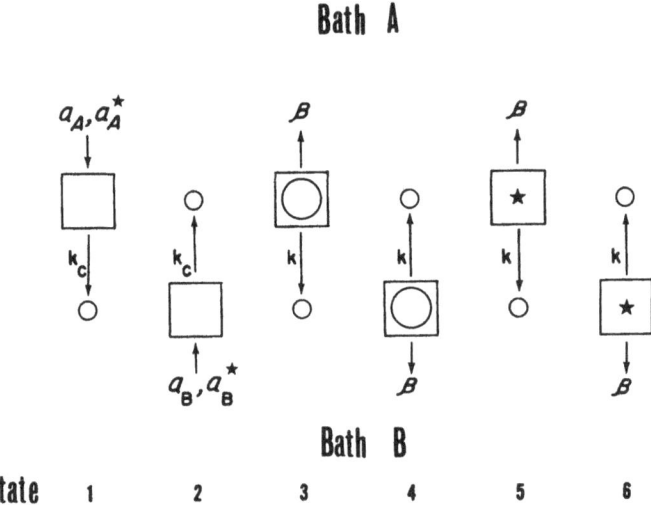

Fig. 10.4. Model in which a substance can cross the membrane only in association with a carrier.

Again it is seen that the extent to which ω^* differs from ω will depend on the conditions of study of the system: as the average solute concentration c, and thus α, becomes small, $\omega^*/\omega \to 1$. Also note that $\omega^* \simeq \omega$ if the rate constant for movement of the unloaded carrier is much greater than that for the loaded carrier ($k_c \gg k$). On the other hand, if k_c becomes much smaller than k, ω^* may become very large; for $k_c \to 0$, there is no movement of the unloaded carrier, and thus no net flow, and $\omega^*/\omega \to \infty$, as in pure exchange diffusion (Levi and Ussing, 1948).

Although the models just discussed provide attractively simple explanations for the commonly observed discrepancies between net flow and tracer permeability coefficients, we stress again that mechanisms other than single-file diffusion or carrier complexes may be responsible. For the two-row lattice of Fig. 10.3, introducing the possibility of a switch between molecules on adjacent occupied sites can convert positive into negative isotope interaction (Essig, Kedem, and

Hill, 1966). More generally, for lattice models the possibility exists that binding sites may change their conformation upon being occupied and that there may be interactions between different sites. Such allosteric cooperativity may induce either positive or negative isotope interaction (Lee et al., 1979). In Sec. 10.4 we shall consider how circulation of flows may result in either positive or negative isotope interaction. In Chapter 11 we shall show an experimental example of still another mechanism of isotope interaction attributable to membrane heterogeneity.

10.3 Carrier model: active transport

Blumenthal and Kedem (1969) have presented an illuminating analysis of the relationship between unidirectional flows and metabolism for a carrier system carrying out active transport. Their treatment, like those immediately above, closely follows that of Hill. For simplicity they consider the transport of a nonelectrolyte. As an example of a plausible carrier they propose an expanding and contracting protein within the membrane (Fig. 10.5). The protein is assumed to have one binding site which shuttles back and forth across the membrane, which interacts with the transported species, and a second binding site restricted to the inner side of the membrane, which is a catalytic site for the interconversion of substrate and product. Figure 10.5a shows the possible states of the model, and Fig. 10.5b shows the possible transitions between states. (We follow Blumenthal and Kedem's convention in denoting the expanded states by even numbers and the contracted states by odd numbers.) The protein can go through three cyclic sequences of states. In cycle a (12431) one molecule of F is transferred inward across the membrane; in cycle b (156821) one molecule of substrate S is transformed into one molecule of product P; and in cycle c (15682431) both processes occur. It is this latter cycle which accounts for coupling between transport and metabolism. If the protein cannot expand or contract except when loaded with the transported species or substrate, that is, if transition 1–2 is impossible, coupling between transport and metabolism is complete; to the extent that this transition can occur, however, coupling is reduced.

In order to analyze the kinetics of isotope flows we can now con-

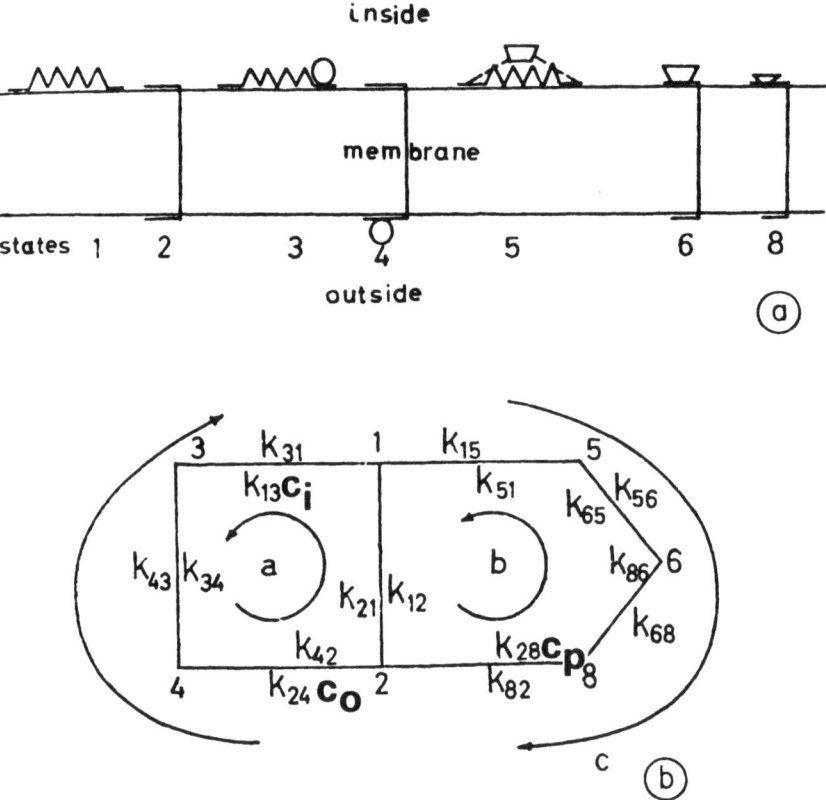

Fig. 10.5. (a) Possible states of the carrier-protein mediating active transport. Quadrangle = substrate; inverted triangle = product; circle = transported species. (b) Transitions between the states and their frequencies. (From Blumenthal and Kedem, 1969.)

sider that two isotopic forms of the transported species are present, with isotope 1 being limited initially to the outside and isotope 2 to the inside of the membrane. This will permit evaluation of the influx $\vec{J}$ and outflux $\overleftarrow{J}$, respectively, the net inward flux $J \equiv \vec{J} - \overleftarrow{J}$, and the flux ratio $f \equiv \vec{J}/\overleftarrow{J}$. With two isotopes, there are now nine instead of seven states (Fig. 10.6a). The protein loaded with the transported species assumes four different states: 3_1 and 4_1 for the contracted and expanded protein-isotope 1 complex, and 3_2 and 4_2 for the contracted and expanded protein-isotope 2 complex. Correspondingly, both cycles a and c are split into two cycles running in parallel (Fig. 10.6b). Since isotopes 1 and 2 are assumed to be thermodynamically and kinetically indistinguishable, except for radioactivity, all rate constants are identical in a_1 and a_2 and in c_1 and c_2; the frequencies of the cycles depend, of course, on the concentrations of species 1 and 2. It is important to note that in addition to splitting the a cycle into two cycles a_1 and a_2, one for each isotope, the presence of two isotopes creates a new cycle a_{12} ($24_1 3_1 13_2 4_2 2$). Cycle a_{12} accomplishes exchange diffusion, the exchange of isotope 1 for isotope 2, without either net transport or reaction. To the extent that the transition between states 1 and 2 is slow relative to that between other states in cycle a, cycle a_{12} will predominate over cycles a_1 and a_2, and coupling between J and $\tilde{J}$ will be tight. Since slow transition between states 1 and 2 is also a condition for tight coupling, it is seen that for this model, isotope interaction and effective active transport go hand in hand.

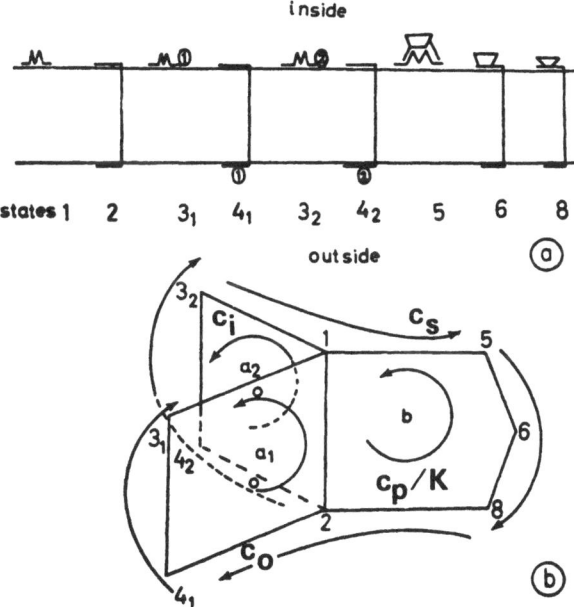

Fig. 10.6. (a) Possible states of the carrier-protein in the presence of two different isotope species (1) and (2). Quadrangle = substrate; inverted triangle = product. (b) Transitions between the states. Only concentration-dependent transition probabilities are indicated.

Evidently, it is possible to characterize all pertinent flows precisely by the proper combination of cycles and application of the diagram methods of Chapters 5 and 6. For our purposes it suffices to summarize certain important results: since we are interested primarily in functions of ratios of flows, we shall follow Blumenthal and Kedem's example in considering only reduced flows j and permeability coefficients ω. Then we have, for transport,

$$j = \left(1 + \frac{R_s}{c_s}\right)\left(\frac{c_o}{c_i} - 1\right) + \left(1 - \frac{c_p}{Kc_s}\right) \qquad (10.17)$$

and for metabolism,

$$j_r = \left(\frac{c_o}{c_i} - 1\right) + \left(1 + \frac{R_F}{c_i}\right)\left(1 - \frac{c_p}{Kc_s}\right) \qquad (10.18)$$

Here c_o and c_i are the concentrations of the transported species at the outside and inside respectively; c_s and c_p are the concentrations of the substrate and product of the metabolic driving reaction, and K its equilibrium constant. R_s is proportional to k_{12}, the rate constant of the transition $1 \to 2$ common to cycles a and b, and to the dissociation constant for the carrier–substrate complex, and it is a function of other rate constants of cycle b; R_F is a corresponding expression for cycle a. When $c_o \simeq c_i$ and $c_p/c_s \simeq K$, $(c_o/c_i - 1) \simeq \ln(c_o/c_i) = X/RT$ and $(1 - c_p/Kc_s) \simeq \ln(Kc_s/c_p) = A/RT$. Therefore, near equilibrium (or, alternatively, constraining X and A to proper pathways), Eqs. (10.17) and (10.18) become standard flow–force relationships of linear NET showing Onsager reciprocity. Referring to Eq. (4.12), we see that the degree of coupling is given by

$$q^2 = \frac{1}{(1 + R_s/c_s)(1 + R_F/c)} \qquad (10.19)$$

where $c \simeq c_o \simeq c_i$. Since both R_s and R_F are proportional to k_{12}, as the rate of transition of unloaded carrier $(1 \to 2)$ increases from very slow to rapid, q^2 falls from ~ 1 toward zero.

For the permeability coefficients, on the other hand,

$$\omega = \frac{c_s + R_s + (R_s/R_F)c}{1 + c/R_F} \qquad (10.20)$$

$$\omega^* = c_s + R_s + (R_s/R_F)c \tag{10.21}$$

so that

$$\frac{\omega^*}{\omega} = 1 + \frac{c}{R_F} \tag{10.22}$$

Thus there is negative isotope interaction unless c is small or R_F is large. The latter will be the case if the dissociation constant of the F-carrier complex is large or the rate of unloaded carrier movement is high.

As discussed in Sec. 9.2.4, it is often considered that $FE_{Na} \equiv (X)_{J=0}$ is given by $RT \ln f_0$. For this model, at short circuit ($c_o = c_i = c$)

$$RT \ln f_0 = RT \ln \frac{c_s + R_s + (R_s/R_F)c}{c_p/K + R_s + (R_s/R_F)c} \tag{10.23}$$

$$= RT \ln \frac{c_s + R_s(\omega^*/\omega)}{c_p/K + R_s(\omega^*/\omega)} \tag{10.24}$$

whereas

$$(X)_{J=0} = RT \ln \frac{c_s + R_s}{c_p/K + R_s} \tag{10.25}$$

so that when metabolism drives transport ($c_s > c_p/K$), $RT \ln f_0 < (X)_{J=0}$. The inequality disappears with loss of isotope interaction and uncoupling of transport and metabolism (see Eq. 10.19). Near equilibrium ($c_s \to c_p/K$), Eq. (10.23) becomes equivalent to the corresponding expression for electrolyte flow, Eq. (9.52).

10.4 Isotope interaction attributable to heterogeneity

The above-described frictional and lattice models provide familiar mechanisms for isotope interaction, intuitively plausible in terms of physical interaction between permeant molecules, either direct or as a consequence of effects of permeant molecules on the transport system (such as blocking of a column or loading of a carrier). It is important to remember, however, that isotope interaction, like other coupling of flows, is defined in phenomenological terms, so that in principle there is no need for direct physical interaction between abundant and tracer species. Indeed, it can be shown that in the absence of such interaction both apparent single-file diffusion and exchange diffusion can come about simply as a result of membrane heterogeneity. Furthermore, in such a case measurements of permeability coefficients may depend significantly on the experimental conditions, for example, whether measurements are carried out in the absence of a hydrostatic pressure difference, as will generally be the case in studies of epithelia mounted in Ussing chambers, or in the absence of volume flow, which will often be the case in the study of symmetrical cells (erythrocytes, muscle, or nerve). This may be appreciated by considering a few examples of the experimental study of coupled nonelectrolyte solute flow. Positive interaction between flows of urea and mannitol was demonstrated by Biber and Curran (1968) in the toad skin and by Franz, Galey, and Van Bruggen (1968) in the frog skin. Similarly, Ussing and Johansen (1969) found that net inward flow of urea enhanced sucrose influx and retarded sucrose efflux in the toad skin. Lief and Essig (1973) found analogous interaction between macroscopic and radioactive tracer urea fluxes in the toad bladder. In all of these cases, for solute–membrane systems with positive reflection coefficients, osmotic water flow would occur in a direction opposite that of net inward solute flow. This would interfere with the observed effects, since it would depress tracer influx and enhance tracer efflux in all pathways. Hence the above-described demonstrations of the influence of macroscopic solute flow on tracer flows have been taken as evidence of direct molecular interaction as the cause of abnormality of the flux ratio.

Although the demonstration of positive coupling of abundant and tracer flows is highly informative, there are significant ambiguities associated with transport measurements at $\Delta p = 0$, in that osmotic water flow interferes with evaluation of the true extent of solute interaction. If it is sufficiently rapid, solvent flow may prevent the demonstration of positive coupling or even lead to an erroneous impression of negative coupling between solute flows. Thus Franz, Galey, and Van Bruggen, in studies of a synthetic membrane, have demonstrated reversal of the direction of net tracer flux on changing

Isotope flows: mechanisms 261

the experimental constraint from zero hydrostatic pressure difference ($\Delta p = 0$) to zero volume flow ($J_v = 0$) and have emphasized the importance of this factor in attempts to evaluate the nature of solute interaction precisely. However, Patlak and Rapoport (1971) have shown that even studies in the absence of net volume flow may lead to ambiguity, owing to the possibility of circulation of volume flows in heteroporous membranes with discrete parallel pathways with different transport parameters.

We shall consider here the influence of such circulation on (phenomenological) isotope interaction (Li and Essig, 1976). In order to demonstrate the effects of circulation in pure form, we assume that there is no direct interaction between the solute flows in any individual pathway. For the ith pathway, let ω_i^* and ω_i be the tracer permeability coefficient and the permeability coefficient for net flow of the test species, respectively, σ_i its reflection coefficient, and L_{pi} the hydraulic conductivity. The parameters of different pathways may differ because of heteroporosity or other factors. For simplicity, we shall normalize all flows relative to unit total membrane area; hence $J = \Sigma_i J_i$ for all flows.

We consider first states in which there is no net volume flow across the membrane, and we begin by analyzing the tracer permeability coefficient, ω^*, determined from a tracer self-exchange experiment. In the absence of a macroscopic concentration difference or a hydrostatic pressure difference across the membrane, there is no macroscopic flow in any pathway, so that

$$\omega^* \equiv \left(\frac{-J^*}{RT\Delta c^*}\right)_{J_i=0} = -\left(\frac{1}{RT\Delta c^*}\right) \sum_i (J_i^*)_{J_{it}=0}$$

$$= -\left(\frac{1}{RT\Delta c^*}\right) \sum (-\omega_i^* RT\Delta c^*) = \sum \omega_i^* \quad (10.26)$$

In the absence of isotope interaction in any pathway, $\omega_i^* \equiv \omega_i$, and we have also

$$\omega^* = \sum \omega_i \quad (10.27)$$

Thus the tracer permeability coefficient is given simply by the sum of the permeability coefficients of the individual pathways.

The permeability coefficient for net flow may be quite different. To determine this quantity, one bathes the two surfaces of the membrane by solutions which are identical except for a concentration difference for the solute of interest. The tendency of the osmotic pressure difference to produce net volume flow is compensated for by applying a hydrostatic pressure difference just adequate to make $J_v = 0$. Then, from Eq. (9.2),

$$\omega \equiv -\left(\frac{J}{RT\Delta c}\right)_{J_v=0} = -\left(\sum \frac{J_i}{RT\Delta c}\right)_{J_v=0}$$

The flows in the individual pathways are given by

$$J_i = -\omega_i RT\Delta c + c(1 - \sigma_i)J_{vi} \qquad (10.28)$$

where c is the logarithmic mean concentration ($c = \Delta c/\Delta \ln c$).

In general the volume flow in any pathway is not zero, but rather

$$J_{vi} = L_{pi}(\sigma_i RT\Delta c - \Delta p) \qquad (10.29)$$

Summing over all pathways and setting $J_v \equiv \sum J_{vi} = 0$ give the value of Δp which would abolish net volume flow:

$$(\Delta p)_{J_v=0} = \frac{\sum \sigma_i L_{pi} RT\Delta c}{\sum L_{pi}} \qquad (10.30)$$

Therefore we have

$$J = \sum J_i = \sum [-\omega_i RT\Delta c + c(1 - \sigma_i)J_{vi}] = \sum (-\omega_i RT\Delta c - c\sigma_i J_{vi})$$

$$= \sum \left\{-\omega_i RT\Delta c - c\sigma_i \left[L_{pi}\left(\sigma_i RT\Delta c - \frac{\sum \sigma_i L_{pi} RT\Delta c}{\sum L_{pi}}\right)\right]\right\} \quad (J_v = 0)$$

and

$$\omega = -\left(\frac{J}{RT\Delta c}\right)_{J_v=0} = \sum \omega_i + c\left[\sum \sigma_i^2 L_{pi} - \frac{\left(\sum \sigma_i L_{pi}\right)^2}{\sum L_{pi}}\right]$$

Comparison with Eq. (10.27) shows that

$$\omega = \omega^* + c\gamma \tag{10.31}$$

or

$$\frac{\omega}{\omega^*} = 1 + \frac{c\gamma}{\omega^*}$$

where

$$\gamma \equiv -\sum (1 - \sigma_i) \frac{J_{vi}}{RT\Delta c} = \sum \sigma_i^2 L_{pi} - \frac{\left(\sum \sigma_i L_{pi}\right)^2}{\sum L_{pi}} \tag{10.32}$$

It is seen that the two coefficients ω and ω^* differ whenever $\gamma \neq 0$. This will occur whenever the σ_i's are not all equal. On the other hand, if σ_i is the same for each pathway, variation in L_{pi} would not cause γ to differ from zero. Since all the L_{pi} are positive, γ must be ≥ 0, irrespective of the values of σ_i. Hence $\omega \geq \omega^*$, that is, the permeability coefficient derived from the measurement of net flow must be equal to or exceed that derived from tracer exchange. In the absence of a net volume flow across the membrane, this discrepancy between ω and ω^* would appear phenomenologically as positive isotope interaction, despite the absence of isotope interaction in the individual pathways.

The magnitude of such isotope interaction cannot, of course, be predicted without detailed knowledge of membrane parameters, which unfortunately is generally unavailable for biological systems of interest. It is helpful, however, to consider the general nature of the relationship between ω and ω^* for a system with two types of parallel channels, which may be considered to represent the cellular (1) and paracellular (2) pathways. For such a system Eq. (10.32) gives

$$\gamma = \frac{(\sigma_1 - \sigma_2)^2 L_{p1} L_{p2}}{L_{p1} + L_{p2}} \tag{10.33}$$

so that with Eq. (10.31)

$$\frac{\omega}{\omega^*} = 1 + \frac{c(\sigma_1 - \sigma_2)^2 L_{p1} L_{p2}}{\omega^*(L_{p1} + L_{p2})} \tag{10.34}$$

The maximal value of this expression will be reached when $L_{p1} = L_{p2}$, $\sigma_1 = 1$ and $\sigma_2 = 0$. Then $\gamma = L_p/2$ and

$$\frac{\omega}{\omega^*} = 1 + \frac{cL_p}{2\omega^*}$$

Substituting now $L_p = \omega_w \overline{V}_w \simeq \omega_w/c_w$, where ω_w, $\overline{V}_w$, and c_w are the permeability coefficient, partial molar volume, and concentration of water, respectively, and the solute mole fraction $x_s = c/c_w$, we have

$$\frac{\omega}{\omega^*} = 1 + \frac{x_s \omega_w}{2\omega^*}$$

so that for a 10-mM solution of solute (for which $x_s = 1.8 \times 10^{-4}$) $\omega/\omega^* \simeq 1 + 10^{-4}\,\omega_w/\omega^*$. On this basis we would not expect much isotope interaction for a nonelectrolyte in a tight epithelium like the toad urinary bladder. In this tissue, in the presence of antidiuretic hormone, Pietras and Wright (1974) found $\omega_w \sim 10^{-3}$ cm · s^{-1}, whereas the poorly permeant nonelectrolyte nicotinamide had a value of $\omega^* = 2.3 \times 10^{-6}$ cm · s^{-1}, so even if the solute concentration had been 10 mM, $10^{-4}\omega_w/\omega^*_{nic} \simeq 0.04 \ll 1$. The effect might possibly be much more significant, however, in a leaky epithelium like the rabbit gallbladder. In that tissue Wright, Smulders, and Tormey (1972) found the osmotic water permeability coefficient $\omega_w \simeq 5 \times 10^{-3}$ cm · s^{-1} whereas Smulders, Tormey, and Wright (1972) found that for sucrose $\omega^* \simeq 4 \times 10^{-6}$ cm · s^{-1}. Since it was felt that because of unstirred layers ω_w was underestimated by at least a factor of 10, whereas ω^*_{suc} was affected only trivially, it is likely that $10^{-4}\,\omega_w/\omega^* \sim 1$. (This can of course be taken as only a rough estimate. However, our simplifying approximations may quite possibly not be unrealistic. As pointed out by Diamond [1979], it is not clear whether 5% or 95% of water flow occurs across the tight junction; it seems reasonable therefore to consider $L_{p1} \simeq L_{p2}$. Also, despite the low value of ω^*_{suc} relative to ω_w, $\sigma_{suc,2}$ may well be nearer to zero than to 1, whereas $\sigma_{suc,1} \simeq 1$. Furthermore, although ω^* is often determined by the use of only tracer concentrations of the test solute, it is not uncommon to employ concentrations $\gg$ 10 mM.)

Often flux ratios and permeability coefficients are estimated by measuring tracer flow without attention to the possible influence of volume flow, or the hydrostatic pressure or concentration differences across the membrane. However, when $\omega \neq \omega^*$, the flux ratio f and the apparent tracer permeability coefficient "ω^*" will vary with the conditions of their measurement. This is easily shown by considering the unidirectional fluxes across the membrane.

Unidirectional fluxes cannot be calculated by the direct application of Eq. (10.28), since this describes net flow. Furthermore, for the case of unidirectional flux of a species, the appropriate mean concentration is not well defined. We can, however, perform a "thought experiment," in which we add tracer quantities of two different isotopes, species a and b, to baths I and II, respectively. With sufficiently large sinks, the tracers will be much diluted on crossing the membrane, and we can conceive that throughout the experiment the total tracer concentration in bath I is $c^{*I} = c_a^I + c_b^I \simeq c_a^I$, and that in bath II is $c^{*II} = c_a^{II} + c_b^{II} \simeq c_b^{II}$. If the tracer concentrations in the two baths are nearly equal, $\Delta c^* = c^{*II} - c^{*I}$ is small and

$$c^* = \frac{\Delta c^*}{\Delta \ln c^*} \simeq (c^{*I} + c^{*II})/2$$

For small Δc^*, and well-defined c^*, in the absence of isotope interaction in the individual pathways, it is valid to apply Eq. (10.28), giving

$$J_i^* = -\omega_i^* RT\Delta c^* + c^*(1 - \sigma_i) J_{vi} \qquad (10.35)$$

that is, in the absence of isotope interaction in the ith pathway, J_i^* is not explicitly dependent on the concentration difference or flow of the abundant species across the membrane; for a given mean concentration c, altering Δc without altering J_{vi} will affect J_i but not J_i^* (see Eq. 10.28). (However, ω_i^*, σ_i, and J_{vi} may, of course, be functions of c.)

Considering first the situation in the absence of net volume flow, adding the J_i^*'s and introducing Eqs. (10.26) and (10.32) gives

$$J^* = -\omega^* RT\Delta c^* - c^*\gamma RT\Delta c \qquad (J_v = 0) \qquad (10.36)$$

Thus it is seen that in contrast to the tracer fluxes in the individual pathways at constant J_{vi}, the total tracer flux is affected by perturbation of Δc, again consistent with the phenomenological interaction of tracer and abundant isotope flows.

The tracer flux J^* must be associated with unidirectional tracer fluxes of the form

$$\vec{J}^* = \omega^* R T c^{*\mathrm{I}} - (c^{*\mathrm{I}}/2)\gamma R T \Delta c + \alpha \qquad (10.37)$$

$$\overleftarrow{J}^* = \omega^* R T c^{*\mathrm{II}} + (c^{*\mathrm{II}}/2)\gamma R T \Delta c + \alpha \qquad (10.38)$$

in order that $J^* \equiv \vec{J}^* - \overleftarrow{J}^*$. Since we deal with tracer fluxes, $\vec{J}^*$ must be proportional to $c^{*\mathrm{I}}$ and independent of $c^{*\mathrm{II}}$, and $\overleftarrow{J}^*$ must be proportional to $c^{*\mathrm{II}}$ and independent of $c^{*\mathrm{I}}$. Hence $\alpha = 0$. Dividing each tracer flux by the specific activity in the appropriate bath gives the unidirectional fluxes

$$\vec{J} = \frac{\vec{J}^*}{c^{*\mathrm{I}}/c^{\mathrm{I}}} = RTc^{\mathrm{I}}\left(\frac{\omega^* - \gamma \Delta c}{2}\right) \qquad (10.39)$$

$$(J_v = 0)$$

$$\overleftarrow{J} = \frac{\overleftarrow{J}^*}{c^{*\mathrm{II}}/c^{\mathrm{II}}} = RTc^{\mathrm{II}}\left(\frac{\omega^* + \gamma \Delta c}{2}\right) \qquad (10.40)$$

Clearly these relationships apply irrespective of the actual concentrations of tracer and abundant isotope employed experimentally, provided that we operate in the range of applicability of Eq. (10.28).

Considering next the situation in the absence of a hydrostatic pressure difference across the membrane, combining Eqs. (10.29) and (10.35) gives

$$J^* = -\omega^* RT\Delta c^* - c^* \theta RT \Delta c \qquad (\Delta p = 0) \qquad (10.41)$$

where

$$\theta = \sum \sigma_i L_{pi}(\sigma_i - 1) \qquad (10.42)$$

The same reasoning as above then leads to the relationships

Isotope flows: mechanisms 267

$$\tilde{J} = RTc^I \left(\omega^* - \frac{\theta \Delta c}{2}\right) \tag{10.43}$$

$$(\Delta p = 0)$$

$$\tilde{J} = RTc^{II} \left(\omega^* + \frac{\theta \Delta c}{2}\right) \tag{10.44}$$

Note that whereas $\gamma \geq 0$, for $0 \leq \sigma_i \leq 1$, $\theta \leq 0$.

Given explicit expressions for the unidirectional fluxes, we can analyze the nature of the flux ratio. For the case of zero volume flow,

$$f = \frac{\tilde{J}}{\tilde{J}} = \frac{c^I(\omega^* - \gamma \Delta c/2)}{c^{II}(\omega^* + \gamma \Delta c/2)} = \frac{c^I}{c^{II}} \frac{(1 - \gamma \Delta c/2\omega^*)}{(1 + \gamma \Delta c/2\omega^*)} \quad (J_v = 0) \tag{10.45}$$

Clearly, for $\gamma \neq 0$, the flux ratio is abnormal, since $f \neq c^I/c^{II}$.[1]

It is also of interest to examine the logarithm of the flux ratio. For sufficiently small Δc, $|\gamma \Delta c/2\omega^*| \ll 1$, and

$$\ln f = \ln \frac{c^I}{c^{II}} + \ln \frac{(1 - \gamma \Delta c/2\omega^*)}{(1 + \gamma \Delta c/2\omega^*)} \simeq \ln \frac{c^I}{c^{II}} - \frac{\gamma \Delta c}{\omega^*} \tag{10.46}$$

Since $\Delta c = c(\Delta \ln c)$,

$$\ln f \simeq \left(1 + \frac{\gamma c}{\omega^*}\right) \ln \left(\frac{c^I}{c^{II}}\right)$$

and with Eq. (10.31)

$$\ln f \simeq \left(\frac{\omega}{\omega^*}\right) \ln \left(\frac{c^I}{c^{II}}\right) = \left(\frac{\omega}{\omega^*}\right) \left(\frac{X}{RT}\right)$$

Hence, for sufficiently small Δc, the general flux ratio relation for homogeneous membranes (Eq. 9.48) continues to apply. Since $\omega \geq \omega^*$, $|RT \ln f| \geq |X|$. Figure 10.7 shows how, in the absence of volume flow, a simple system with two discrete parallel pathways with different values of σ would result in an abnormal flux ratio, as is seen in single-file diffusion.

268 Bioenergetics and linear nonequilibrium thermodynamics

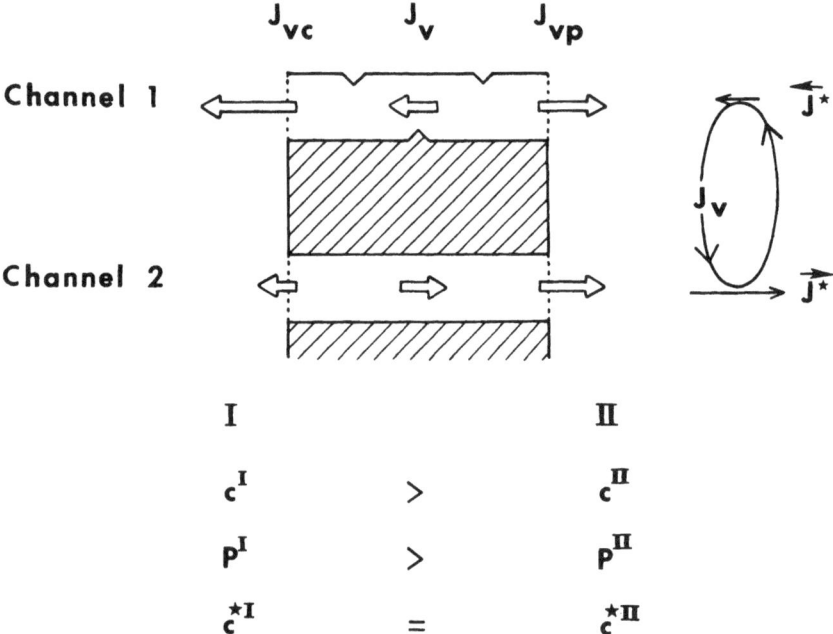

Fig. 10.7. Effect of circulating volume flow on unidirectional fluxes (net volume flow abolished by hydrostatic pressure difference). Channels 1 and 2 are taken as representative of a heterogeneous membrane, with $\sigma_1 > \sigma_2$. (Although the membrane is pictured as heteroporous, differences in σ may equally well be attributable to other factors. For simplicity L_p is assumed the same for all channels.) We consider $c^I > c^{II}$, $p^I > p^{II}$. Volume flow J_{vc} attributable to Δc is from II to I and is of larger absolute magnitude in channel 1 (with higher σ). Flow J_{vp} attributable to Δp is from I to II and equal in both channels. With appropriate setting of Δp, $J_{v_1} < 0$, $J_{v_2} > 0$, and net volume flow $J_v = J_{v1} + J_{v2} = 0$. The effect of the circulating volume flow on tracer solute flux is to markedly enhance $\bar{J}^*$ from I to II through channel 2 (of lower σ) and to slightly enhance $\bar{J}^*$ from II to I through channel 1 (of higher σ); hence with equal initial concentrations of tracer in each bath, net tracer flux $J^* > 0$, and the flux ratio

$$f = \frac{\bar{J}^*/(c^{*I}/c^I)}{\bar{J}^*/(c^{*II}/c^{II})} > \frac{c^I}{c^{II}}$$

as in single-file diffusion. (From Li and Essig, 1976.)

For studies performed in the absence of a hydrostatic pressure difference, an analogous treatment gives

$$f = \frac{\tilde{J}}{\tilde{J}} = \frac{c^{I}}{c^{II}} \frac{(1 - \theta \Delta c/2\omega^*)}{(1 + \theta \Delta c/2\omega^*)} \quad (\Delta p = 0) \quad (10.47)$$

Thus in the absence of a hydrostatic pressure difference, the flux ratio may appear to be abnormal (that is, $f \neq c^{I}/c^{II}$), owing to failure to account for the influence of coupled volume flow.

Comparison of Eqs. (10.45) and (10.47) shows that in principle (for $0 \leq \sigma_i \leq 1$), $f_{J_v=0} > f_{\Delta p=0}$ (and of course $J_{J_v=0} > J_{\Delta p=0}$). The magnitudes of these differences will obviously vary greatly, depending on the membrane and solute system under study. Table 10.1 evaluates γ and θ for a system of two discrete channels for various possible conditions on the γ's and L_p's.

As mentioned above, it is also common for tracer permeabilities to be evaluated from measurements of the quantity $J^*/\Delta c^*$ without consideration of the possible influence of coupled volume flows. Examination of Eq. (10.39) shows that if tracer flow is determined in the direction of influx in the absence of volume flow,

$$"\omega^*" = \frac{J^*}{RTc^{*I}} = \frac{\tilde{J}}{RTc^{I}} = \omega^* - \frac{\gamma \Delta c}{2} \quad (J_v = 0) \quad (10.48)$$

whereas if the corresponding measurement is made in the absence of

Table 10.1. *Properties of heterogeneous membrane comprising parallel channels.*[a]

Condition on σ's	Condition on L_p's	γ $(\sigma_1 - \sigma_2)^2 L_{p1}L_{p2}/(L_{p1} + L_{p2})$	θ $\sigma_1(\sigma_1 - 1)L_{p1} + \sigma_2(\sigma_2 - 1)L_{p2}$
$\sigma_1 = \sigma_2 = \sigma$	$L_{p1} = L_{p2} = L_p$	0	$2\sigma(\sigma - 1)L_p$
$\sigma_1 = \sigma_2 = \sigma$	$L_{p1} \ll L_{p2}$	0	$\sigma(\sigma - 1)L_{p2}$
$\sigma_1 = 1; \sigma_2 = 0$	$L_{p1} = L_{p2} = L_p$	$L_p/2$	0
$\sigma_1 = 1; \sigma_2 = 0$	$L_{p1} \ll L_{p2}$	L_{p1}	0

a. Each membrane is an array of parallel channels of two types, designated 1 and 2. For each set of postulated conditions on the σ's and L_p's, γ and θ are calculated from the theoretical expressions as indicated.

a hydrostatic pressure difference,

$$"\omega^*" = \omega^* - \frac{\theta \Delta c}{2} \quad (\Delta p = 0) \tag{10.49}$$

Again, of course, the significance of ignoring γ and θ will vary greatly with the experimental situation.

Often isotope flows are studied in systems in which opposite surfaces of the membrane are not exposed to identical solutions. Such is the case, for example, for the external and internal surfaces of symmetrical cells, such as erythrocytes, muscle, and nerve cells. Evidently, to the extent that several species permeate the membrane, additional phenomena might be observed. In these cases a complete analysis incorporating the possible influences of multiple flows and forces, active transport, and so on, is quite impractical. However, it is possible and instructive to analyze simplified models of the transport system. We shall therefore examine the case in which the two bathing solutions differ not only in the concentration of the test species, but also in the concentration of a second permeant species, designated by a superscript '. We consider the situation in the absence of a hydrostatic pressure difference, with the concentration difference of the second ("osmotic") species adjusted to make $J_v = 0$. Following the same general approach as above, it can be shown that now

$$\omega = \omega^* + c\epsilon \tag{10.50}$$

where

$$\epsilon = \sum \sigma_i^2 L_{pi} - \frac{(\sum \sigma_i \sigma_i' L_{pi})(\sum \sigma_i L_{pi})}{\sum \sigma_i' L_{pi}} \tag{10.51}$$

(We have here assumed that in any individual pathway there is no isotope interaction and no interaction between the flows of the chemically distinct solute species.) Clearly, in order for the osmotic solute to have effects different from those of an impermeant solute, it is necessary that σ_i' not be identical in all pathways. Otherwise, $\sigma_i' \equiv \sigma'$, and $\epsilon = \gamma$.

More generally, the nature of the relationship between ω and ω^* will depend on whether membrane heterogeneity is more significant for the test solute or the osmotic species. This is most readily seen by considering, for simplicity, that L_{pi} is the same for all pathways. Then $L_{pi} \equiv L_p$, and

$$\epsilon = L_p \left[\sum_i \sigma_i^2 - \frac{\left(\sum_i \sigma_i \sigma_i'\right)\left(\sum_i \sigma_i\right)}{\sum_i \sigma_i'} \right]$$

$$= \frac{L_p}{2 \sum_i \sigma_i'} \left[\sum_i \sum_j (\sigma_i - \sigma_j)(\sigma_i \sigma_j' - \sigma_j \sigma_i') \right] \qquad (10.52)$$

It is clear that with reflection coefficients ≥ 0, we have $L_p/2 \sum_i \sigma_i' > 0$, and thus $\epsilon > 0$ when finite $(\sigma_i - \sigma_j)$ and $(\sigma_i \sigma_j' - \sigma_j \sigma_i')$ are of like sign (for all i, j), and $\epsilon < 0$ when they are of unlike sign. The significance of these relationships is clarified by examining for a representative pair of pathways $\{i, j\}$ the quantity

$$\theta_{ij} \equiv \frac{(\sigma_i - \sigma_j)(\sigma_i \sigma_j' - \sigma_j \sigma_i')}{\sigma_j^2 \sigma_i'} = \left(\frac{\sigma_i}{\sigma_j} - 1\right)\left(\frac{\sigma_i/\sigma_j}{\sigma_i'/\sigma_j'} - 1\right) \qquad (10.53)$$

It can be seen that if heterogeneity of pathways is more significant for the test solute than for the osmotic solute (that is, the σ's differ more than the σ''s), we have for $(\sigma_i/\sigma_j) > 1$, $(\sigma_i/\sigma_j)/(\sigma_i'/\sigma_j') > 1$, and for $(\sigma_i/\sigma_j) < 1$, $(\sigma_i/\sigma_j)/(\sigma_i'/\sigma_j') < 1$, so that $\theta_{ij} > 0$. On the other hand, if heterogeneity is more marked for the osmotic solute than for the test solute, $\theta_{ij} < 0$. With all $\theta_{ij} > 0$ we have $\epsilon > 0$, and with $\theta_{ij} < 0$, $\epsilon < 0$. Thus in the absence of either single-file or carrier mechanisms, we may have either $\omega > \omega^*$ (positive isotope interaction) or $\omega < \omega^*$ (negative isotope interaction), according as membrane heterogeneity is more marked for the test solute or for the osmotic solute, respectively (Fig. 10.8). With a more random pattern of pathway heterogeneity, isotope interaction might be positive, negative, or absent. And of course, the nature of the interaction may vary

272   Bioenergetics and linear nonequilibrium thermodynamics

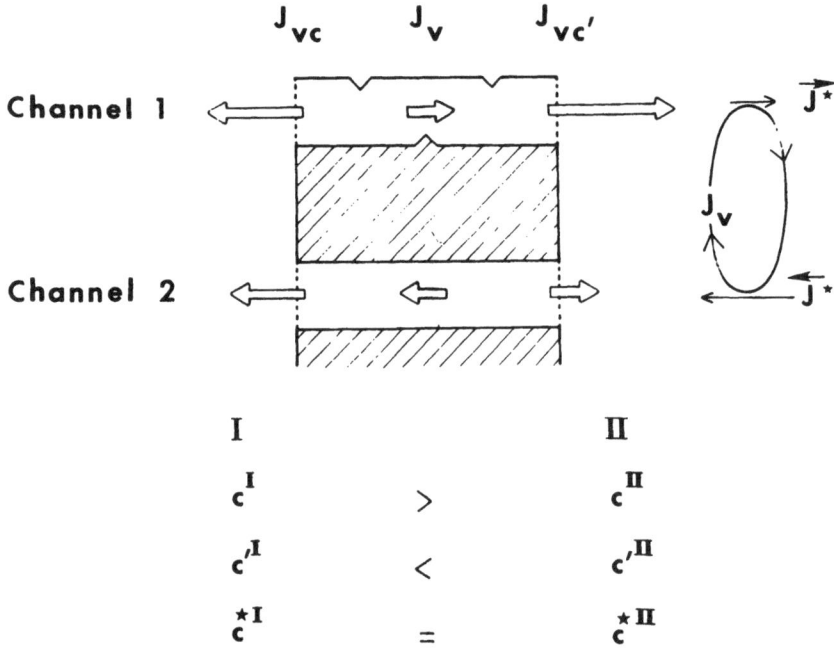

Fig. 10.8. Effect of circulating volume flow on unidirectional fluxes (net volume flow abolished by concentration difference of a permeant "osmotic" solute). Channels 1 and 2 are taken as representative of a heterogeneous membrane such that for the test solute $\sigma_1 > \sigma_2$ and for the osmotic solute $\sigma_1' > \sigma_2'$. For the example pictured it is assumed that membrane heterogeneity is more pronounced for the osmotic solute than for the test solute, that is, $\sigma_1'/\sigma_2' > \sigma_1/\sigma_2$. We consider $c^I > c^{II}$, $c'^I < c'^{II}$. Volume flow J_{vc} attributable to Δc of the test solute is from II to I and of slightly greater magnitude in channel 1. Volume flow $J_{vc'}$ attributable to $\Delta c'$ of the osmotic solute is from I to II and is of substantially greater magnitude in channel 1. With appropriate setting of $\Delta c'$, $J_{v1} > 0$, $J_{v2} < 0$, and net volume flow $J_v = J_{v1} + J_{v2} = 0$. The effect of the clockwise circulating volume flow on tracer solute flux is to slightly enhance $\bar{J}^*$ from I to II through channel 1 (of higher σ) and to markedly enhance $\bar{J}^*$ from II to I through channel 2 (of lower σ); hence with equal initial concentrations of tracer in each bath, net tracer flux $J^* < 0$, and the flux ratio

$$f = \frac{\bar{J}^*/(c^{*I}/c^I)}{\bar{J}^*/(c^{*II}/c^{II})} < \frac{c^I}{c^{II}}$$

as in exchange diffusion. Alternatively, if membrane heterogeneity were more pronounced for the test solute than for the osmotic solute, $f > c^I/c^{II}$, as in single-file diffusion. (From Li and Essig, 1976.)

markedly with change of either the test solute or the osmotic solute employed. In all cases, for sufficiently small flows the general flux ratio relationship of Eq. (9.48) will continue to apply.

10.5 Summary

1. Both positive and negative isotope interaction are explicable by a variety of mechanisms. For passive transport with continuity of the electrochemical potential, frictional interaction can induce positive interaction for both solvent and solutes. For solutes with discontinuity of electrochemical potential, permeation of a lattice along single files will promote positive interaction, and carrier transport will be associated with negative interaction. Circulating volume flow may induce either positive or negative interaction.

2. Analysis of a carrier active-transport model shows that for this system isotope interaction is associated with tight coupling of transport to metabolism; the flux ratio at short circuit underestimates static head.

11 Kinetics of isotope flows: tests and applications of thermodynamic formulation

In Chapters 9 and 10 we considered the general theoretical basis for applying and interpreting isotope kinetic techniques in the study of transport processes. Fundamental relationships were developed which, we hope, will be useful in a wide variety of circumstances. Combining Eqs. (9.47) and (9.48),

$$JR^* = \frac{J}{c\omega^*} = RT \ln f = \frac{R^*}{R}\left(X - \sum R_{0j}J_j\right)$$

$$= \frac{\omega}{\omega^*}\left(X - \sum R_{0j}J_j\right) \quad (11.1)$$

(Here we have written $R_{0j}J_j$ for $\int_0^{\Delta x} \sum r_{0j}J_j dx$, the contribution of coupled flows to the forces promoting transport. As previously, we shall usually employ permeability coefficients rather than resistance coefficients.)

In the present chapter we take a more practical approach. We discuss experimental tests of the theoretical relationships, and we consider how isotope techniques may be utilized to characterize membrane permeability and the forces influencing transport.

Although a great body of experimental work exists which relates unidirectional fluxes and flux ratios to presumed mechanisms of transport, we shall limit ourselves to studies closely relevant to the theoretical formulations considered above. Much caution must be exercised in interpreting tracer flux data in terms of mechanism.

11.1 Validation of theoretical formulations in synthetic membranes

11.1.1 Membranes with incomplete permselectivity and appreciable water permeability

Meares and Ussing (1959a) first tested the validity of the Ussing flux ratio equation by measuring the flux ratios of Na^+ and Cl^- induced by concentration gradients of NaCl and diffusion potentials across a well-characterized cation-exchange resin membrane. The water content of the phenol–sulfonic acid membranes employed (Zeo-Karb 315) was some 75% by weight, and solvent flow was appreciable. The flux ratio was conceived of as the ratio of fluxes of substances a and b, distinguishable only as isotopes of the same chemical substance, placed on opposite sides of a membrane. The theoretical equation which was tested related the flux ratio to the forces responsible for net transport and included, in addition to four terms comprising the electrochemical potential difference of the test species, as considered earlier by Behn and Teorell, a fifth term representing "fluxes of a and b induced by the chemical potential gradients of all other components present." The possible significance of isotope interaction (interaction of fluxes of a and b) was not considered. Good agreement was found between the theoretical and experimental flux ratios for both Na^+ and Cl^-. The solvent drag term was important, whereas the effect of including activity coefficients was small and was considered possibly undesirable. It was concluded that under the conditions employed, the flux ratio equation might be used as a guide in distinguishing active from passive transport.

In a companion study Meares and Ussing (1959b) tested various flux ratio equations in the presence of electrical current flow with appreciable electroosmosis. Theoretical flux ratios calculated from the forces, incorporating an estimate for the contribution of electroosmosis, agreed adequately with experimental flux ratios for the counterion Na⁺, but not for the coion Cl⁻. It was concluded that in the presence of electrical current, the determination of a mass flow correction for the electroosmotic flow introduces some uncertainty in the general application of the equation relating the flux ratio to the forces. However, it was found also that experimental flux ratios for both Na⁺ and Cl⁻ agreed well with theoretical values calculated from the electrical potential difference, the equivalent conductance, and the self-diffusion coefficient of the ion (Meares and Ussing, 1959b, Eq. 18).

In a later, more general study of Zeo-Karb 315 membranes, Meares et al. (1967) studied flux ratios of Na⁺, Br⁻, and Sr²⁺ in a series of mixtures of NaBr and SrBr$_2$ at a total concentration of 0.1 N. The flux ratios were analyzed by means of a formulation based on Eqs. (9.47) and (9.48), as derived earlier by Meares and coworkers. When expressions for coupled flows appropriate to electrical conduction at constant composition were incorporated, it was found that the flux ratio is given (in our terminology) by

$$\ln f = \frac{\Delta x I \tau}{z F c D^*} \qquad (11.2)$$

where Δx is the membrane thickness, I is electrical current density, F is the Faraday, τ is the electrical transport number of the test species, c its concentration in the bathing solutions, and D^* its tracer diffusion coefficient. (This equation will be derived below.) Experimental flux ratios of Br⁻ and Sr²⁺ at intermediate equivalent fractions N_{Na} of sodium in the membrane agreed well with values calculated from Eq. (11.2). At more extreme values of N_{Na}, discrepancies were noted which were unexplained.

In analyzing their results Meares et al. concluded that since the ratio R^*/R does not appear in Eq. (11.2), "No information about isotope interaction can be obtained from flux ratios measured at constant composition." As will be shown below, however, at a given setting of $\Delta\psi$, $(I\tau/cD^*)$ is proportional to R^*/R. Furthermore, the flux ratios might in principle be analyzed by means of Eq. (11.1). Admittedly, it will often not be possible to evaluate the contribution of coupled flows to the forces influencing transport, but if the quantity

$\sum R_{0j}J_j$ is known, the application of Eq. (11.1) will evaluate R^*/R (see Sec. 11.1.2).

Meares and Sutton (1968) later presented a derivation of tracer flux equations in terms of L coefficients rather than R coefficients for greater convenience in relating to conductances and transport numbers. This also led to Eq. (11.2) and to the corresponding relations for forward and backward fluxes of the test substances in terms of the electrical current and the tracer diffusion coefficient. It was stated that although Eq. (11.2) was derived without neglect of tracer interaction, the relations could be derived also from the flux equations given earlier by Meares and Ussing (1959b) "in which the possible effects of isotope interaction were tacitly ignored. In the restricted case of no net concentration gradient, isotope interaction cancels out of the equations for tracer fluxes." In our view a more appropriate statement might be that Eq. (11.2) would be expected to be applicable irrespective of the presence or absence of isotope interaction.

In the latter study Meares and Sutton carried out a large number of precise measurements of counterion and coion tracer fluxes with and against the electric current in NaBr, CsBr, and SrBr$_2$ solutions of constant composition ranging from 0.01–1.00 Eq/L. Theoretical flux ratios were calculated from Eq. (11.2). In general the agreement between observed and theoretical flux ratios was satisfactory except in the most dilute solutions, where discrepancies were attributed to concentration polarization.

The very comprehensive studies of transport in Zeo-Karb 315 membranes have provided much useful information concerning the factors influencing tracer fluxes and flux ratios, both for counterions and coions. However, the use of a membrane which permits appreciable flows of ionic species and solvent, in addition to the flow of the test species of interest, complicates the investigation of the possible contribution of isotope interaction. Zeo-Karb 315 membranes exposed to 0.1 M NaBr have a coion transport number of 0.045 and a water transport number of about 40 moles Eq^{-1}. Accordingly, evaluating the extent of isotope interaction from the application of Eq. (11.1) would necessitate multiple studies to define the magnitude of all the forces influencing net transport. Even for relatively simple systems such studies are laborious.

11.1.2 Membranes with high permselectivity and low water permeability

For these reasons it was of interest to study anion exchange membranes described by Gottlieb and Sollner (1968), which showed tracer exchange rates greater than predicted from their electrical resistance. When such membranes are suitably prepared by adsorption of polyvinyl-benzyl-trimethyl ammonium chloride (PVBT, Dow) on collodion, followed by slow drying, they show nearly ideal permselectivity, with values of cationic transport numbers $\ll 0.01$. Furthermore, osmotic flow and electroosmosis in these membranes are minimal. Hence it appeared likely that there could be little coupling of anion flow with cation or solvent flow and that the discrepancy between tracer exchange rates and electrical resistance was a manifestation of isotope interaction.

In order to examine the influence of isotope interaction, PVBT membranes were studied in a variety of single salt solutions. In preliminary studies the tracer permeability and exchange resistance were evaluated in the absence of net flow according to Eq. (9.45).

$$\omega^* = \frac{1}{cR^*} = -\left(\frac{J^*}{RT\Delta c^*}\right)_{J=0} \quad (\Delta c = 0) \tag{11.3}$$

where J^* represents tracer flow and Δc^* represents the concentration difference of a tracer isotope added to one of the two bathing solutions. Since the membranes were nearly ideally permselective, the phenomenological permeability and resistance coefficients of the anionic test species were conveniently evaluated directly from the electrical resistance in the absence of a concentration difference. Presuming the absence of electroosmosis

$$\frac{1}{c\omega} = R = \frac{X}{J} = \frac{F\Delta\psi}{J} = \frac{-F^2\Delta\psi}{I} = F^2\mathcal{R} \tag{11.4}$$

In conformity with the findings of Gottlieb and Sollner, several membranes showed appreciable discrepancies between tracer flux and electrical measurements such that ω was less than ω^*. These membranes were used to test the validity of the flux ratio formulation (De-Sousa, Li, and Essig, 1971; Li, DeSousa, and Essig, 1974).

It is anticipated that in the absence of isotope interaction,

$$f = \exp\left(\frac{J}{RTc\omega}\right) \quad (11.5)$$

For electrical driving forces, presuming the absence of electroosmosis, we have also from Eq. (9.38),

$$f = \exp\left(\frac{-zF\Delta\psi}{RT}\right) \quad (11.6)$$

More generally, in either the presence or absence of isotope interaction,

$$f = \exp\left(\frac{J}{RTc\omega^*}\right) \quad (11.7)$$

and for electrical forces,

$$f = \exp\left[\left(\frac{\omega}{\omega^*}\right)\left(\frac{-zF\Delta\psi}{RT}\right)\right] \quad (11.8)$$

For concentration driving forces in the absence of isotope interaction, since osmotic permeability in these membranes was demonstrated to be insignificant, we have from (Eq. 9.6),

$$f = \frac{c^I}{c^{II}}$$

whereas in the presence of isotope interaction Eq. (9.48) gives

$$f = \left(\frac{c^I}{c^{II}}\right)^{\omega/\omega^*} \quad (11.9)$$

In principle the equations in terms of flow and in terms of force are equivalent, and either set may be used. However, in experiments employing low bath concentrations (0.01 M) it was found that even low rates of anion flow sufficed to change the bath concentrations appreciably, so that both electrical and concentration driving forces

Table 11.1. Theoretical and observed flux ratios in PVBT-Collodion membranes.

Solutions	$n/n\dagger$[a]	ω/ω^*	$\Delta\psi$ (mV)	f/exp (J/RTcω)	f/exp (J/RTcω^*) (mean±S.E.)
Electrical driving force					
0.10 M KCl	8/4	0.42–0.71	10.0–75.0	0.24–0.91	0.99 ± 0.03
0.03 M KCl	6/3	0.40–0.71	32.0–41.8	0.41–0.72	1.05 ± 0.03
0.01 M KCl	11/5	0.17–0.76	10.0–25.0	0.47–0.88	1.00 ± 0.01
0.10 M KI	2/2	0.74–1.19	25.0–50.0	0.72–1.34	0.94 ± 0.02
0.01 M KI	2/1	0.38–0.44	50.0–75.0	0.17–0.38	1.02 ± 0.05
0.10 M KAc	3/2	0.47–0.72	10.0–25.0	0.60–0.89	0.99 ± 0.01
0.01 M KAc	4/3	0.26–0.56	15.0–50.0	0.26–0.73	0.98 ± 0.00
0.01 M K$_2$SO$_4$	6/2	0.29–0.52	25.0–50.0	0.12–0.31	0.87 ± 0.02
Concentration driving force					
0.10/0.05 M KCl	1/1	0.36	0	0.61	0.95
0.05/0.02 M KCl	6/3	0.40–0.71	0	0.31–0.66	0.88 ± 0.04

Source: Li, DeSousa, and Essig, 1974.
a. n = number of experiments; $n\dagger$ = number of membranes used.

were changing throughout the experiments. Accordingly, since the electrical current and thus J were being measured continuously, it was convenient to analyze the data in terms of Eqs. (11.5) and (11.7). Table 11.1 summarizes the results of studies employing a variety of anions, both monovalent and divalent, and both electrical and concentration driving forces. As is seen, f differed appreciably from exp (J/RTcω) but for the most part agreed satisfactorily with the value predicted by the theoretical expression incorporating the influence of isotope interaction, Eq. (11.7).

Equation (11.7) is easily shown to be equivalent to Eq. (11.2) above, since

$$J = \left(\frac{I\tau}{zF}\right) \qquad (11.10)$$

and from Eq. (11.3)

$$\omega^* = -\left(\frac{J^*}{RT\Delta c^*}\right)_{J=0}$$

where J^* can also be expressed as

$$(J^*)_{J=0} = -D^* \frac{\Delta c^*}{\Delta x} \tag{11.11}$$

Substituting these expressions in Eq. (11.7), we have Eq. (11.2):

$$\ln f = \frac{\Delta x I_T}{zFcD^*}$$

Introducing Eq. (11.1) now gives

$$\frac{\omega}{\omega^*} = \frac{R^*}{R} = \frac{RT(\Delta x I_T/zFcD^*)}{X - \Sigma R_{0j}J_j} \tag{11.12}$$

Hence the inability to evaluate the extent of isotope interaction in experiments utilizing electrical forces at constant composition is not because of cancellation of the contribution of isotope interaction, but because of ignorance of the possible influence of coupled flows J_j of various solute and solvent species. In the experiments employing PVBT membranes, the contribution of coupling of flows of different chemical species was insignificant (the cation transport number and osmotic permeability were trivially small, and the water transport number was <2 moles Eq^{-1}, as against ~40 moles Eq^{-1} in Zeo-Karb 315 membranes in 0.1 M NaBr). In these circumstances it is indeed possible to estimate the extent of isotope interaction.

11.1.3 Coupling of isotope flows

According to the thermodynamic formulation being tested, the discrepancy between ω^* and ω and the abnormality of the flux ratio are manifestations of coupling of isotope flows. The satisfactory agreement between theoretical and experimental flux ratios strongly supports this view, but a more direct demonstration is provided by examining the influence of a net flow on the two unidirectional tracer fluxes. Observations of the effect of an electrical potential difference on tracer fluxes are not completely satisfactory, since perturbation of $\Delta\psi$ alters not only the net flow (and thereby the possible influence of the net flow on tracer flows) but alters also the thermodynamic force

($zF\Delta\psi$) acting on the tracer species. A more convincing demonstration of coupling may be provided by establishing a concentration difference of the abundant species in the absence of an electrical potential difference. This will induce a net flow, but providing that the state of the membrane is unaltered, in the absence of electroosmosis and isotope interaction the tracer fluxes will be unaffected. In the presence of isotope interaction, however, the findings would be quite different. With the negative coupling consistent with the case of $\omega/\omega^* < 1$, a positive net flow ($J > 0$) would decrease the inward tracer flux $\bar{J}^*$ and increase the outward tracer flux $\bar{J}^*$. Table 11.2 shows that both of these effects were demonstrable in PVBT membranes and thus provides a direct demonstration of istope interaction, or coupling of isotope flows.

In interpreting the above findings, one should understand that the demonstration of coupling represents a phenomenological characterization of the system without specific mechanistic implications and that it need not reflect direct interaction at the molecular level. In the interpretation of biological transport studies it is commonplace to attribute negative coupling to countertransport ("exchange diffusion") by means of a carrier shuttling back and forth across the membrane. The unlikelihood of mobile carriers crossing the PVBT membranes studied here suggests that this conventional interpretation is

Table 11.2. Effect of net flow on tracer flows.[a]

Membrane	n	ω/ω^*	$(\bar{J}^*)_{J>0}/(\bar{J}^*)_{J=0}$	$(\bar{J}^*)_{J>0}/(\bar{J}^*)_{J=0}$
F	4	0.71	0.84	1.27
F	4	0.64	0.77	1.66
H	4	0.51	0.77	1.99
H	3	0.49	0.73	1.90
K	4	0.40	0.72	2.21
K	4	0.41	0.70	2.26

Source: Li, DeSousa, and Essig, 1974.

a. n = number of experiments. $\bar{J}^*$ and $\bar{J}^*$ represent inward and outward tracer fluxes respectively. ω/ω^*, $(\bar{J}^*)_{J=0}$, and $(\bar{J}^*)_{J=0}$ were measured with 0.03 M KCl bathing each surface of the membrane. $J > 0$ represents the net flux produced by the use of 0.05 M KCl in the outer compartment and 0.01 M KCl in the inner compartment; $\Delta\psi = 0$.

quite possibly often inappropriate. This possibility is supported also by observations of the transport characteristics of a heterogeneous system comprising parallel pathways of different resistance (Li and Essig, 1977). Although for each element $R^* = R$, the composite array showed marked negative isotope interaction; as in the studies of Table 11.1, flux ratios agreed well with values predicted by Eq. (11.7). Without knowing both the geometrical and electrical properties of a membrane, one cannot derive its transport properties precisely. However, it may be possible to explain the effects of membrane heterogeneity on R^*, R, and the flux ratio in qualitative terms. In the determination of R^* from exchange diffusion in the absence of electrical forces, various pathways will contribute to tracer flow in inverse proportion to their intrinsic resistance, and low resistance pathways will be predominant. In the determination of R with electrical forces, however, we face the difficulty that although the phenomenological treatment assumes a unique transmembrane electrical potential difference, in fact we do not deal with equipotential surfaces. Rather, the transmembrane potential difference will be lower across pathways of low resistance than across pathways of high resistance. Hence the relative contribution to net flow of the low resistance channels may be less than in the case of exchange diffusion, and thus we might expect that introducing channels of low resistance would decrease R less than R^*. Therefore, in the measurement of the flux ratio we would expect the flows in the low-resistance channels to be of major significance, and we would expect f to reflect a value of transmembrane electrical potential close to those across the low resistance channels.

Although the effects described in the table are thus explicable qualitatively, the quantitative agreement with formulations developed for homogenous membranes is perhaps surprising. These results are possibly attributable to the marked differences in resistance of the two types of pathways in our membranes, such that within experimental accuracy J, R^*, $\bar{J}$, and f were all determined essentially by the channels of low resistance. Whether Eq. (11.1) remains valid in membranes with less marked heterogeneity remains to be determined.

11.2 Validation of theoretical formulations in biological membranes: passive transport

11.2.1 General relationships

Chen and Walser (1974) have examined the validity of the above formulations in toad bladder sacs treated with ouabain to eliminate active sodium transport. Because the use of sacs eliminates the edge damage sustained in chambers, the residual transepithelial ion transport in the presence of ouabain presumably represents physiological passive transport. The experiments were analyzed by means of various expressions modified from the treatments of Meares and Sutton and of Krämer and Meares.

The first relation states that the unidirectional flux of a passively transported ion at zero potential difference is the logarithmic mean of the bidirectional fluxes at any other potential. This relation is readily developed in terms of the variables employed in the present treatment. In the absence of net active or passive flows, incorporation of the terminology of Eq. (9.36) into Eq. (11.3) gives

$$RTc\omega^* = -\left(\frac{J^*}{\Delta\rho}\right)_{J=0} = \frac{J_2}{\rho_2^o} = \vec{J}_0 \qquad (11.13)$$

and

$$RTc\omega^* = -\left(\frac{J^*}{\Delta\rho}\right)_{J=0} = \frac{J_3}{\rho_3^i} = \overleftarrow{J}_0 \qquad (11.14)$$

where $\vec{J}_0$ and $\overleftarrow{J}_0$ refer respectively to influx and efflux in the short-circuited state. The introduction of Eq. (11.1) now gives

$$J = \vec{J}_0 \ln f = \overleftarrow{J}_0 \ln f \qquad (11.15)$$

Substituting $J = \vec{J} - \overleftarrow{J}$ and $f = \vec{J}/\overleftarrow{J}$, we obtain the desired expression:

$$\vec{J}_0 = \overleftarrow{J}_0 = \frac{\vec{J} - \overleftarrow{J}}{\ln (\vec{J}/\overleftarrow{J})} \qquad (11.16)$$

Chen and Walser compared the mean of the two unidirectional fluxes at zero potential, $(\tilde{J}_0 + \tilde{J}_0)/2$, with the value predicted from the fluxes at 100 mV, for the case of Na^+, Cl^-, and SO_4^{2-}. The correspondence with Eq. (11.16) was within experimental error.

A second test of the formulation involved the relationship between the flux ratio, the electrical current, and the transport number of the test species. Combining Eqs. (9.36) and 11.10),

$$f = \frac{\tilde{J}}{\tilde{J}} = \frac{J + \tilde{J}}{\tilde{J}} = \frac{I\tau/zF + \tilde{J}}{\tilde{J}} = \frac{I\tau}{zF\tilde{J}} + 1 \tag{11.17}$$

or alternatively,

$$\frac{1}{f} = 1 - \frac{I\tau}{zF\tilde{J}} \tag{11.18}$$

Combining Eqs. (11.10), (11.15), and (11.17) or (11.18) then gives[1]

$$\tau = \frac{zF\tilde{J}_0}{I} \ln\left(\frac{I\tau}{zF\tilde{J}} + 1\right) \tag{11.19}$$

or

$$\tau = -\frac{zF\tilde{J}_0}{I} \ln\left(1 - \frac{I\tau}{zF\tilde{J}}\right) \tag{11.20}$$

Transport numbers calculated by these means agreed closely with those derived directly from measurements of net flux and electrical current at 100 mV.

It was also found useful to introduce the partial conductance g for passive species, defined by

$$J = \frac{I\tau}{zF} = \frac{-g\Delta\psi}{zF} \tag{11.21}$$

Combining Eqs. (11.15) and (11.21),

$$\ln f = \frac{-g\Delta\psi}{zF\tilde{J}_0} \tag{11.22}$$

286   Bioenergetics and linear nonequilibrium thermodynamics

Introducing, in terms of our variables, Chen and Walser's definition

$$Q \equiv \frac{RTg}{(zF)^2 J_0} \tag{11.23}$$

then gives:

$$\ln f = -Q\left(\frac{zF\Delta\psi}{RT}\right) \tag{11.24}$$

where deviation of Q from 1 reflects the factors accounting for deviation from the flux ratio equation of Ussing. The deviations in their study were appreciable: Q for Na^+, Cl^-, and SO_4^{2-} ranged from 0.40 to 0.76.

Q was also expressed as a function of the integral "bulk" diffusion coefficient and the tracer diffusion coefficient. These quantities, denoted D and D^* respectively, are given by

$$D = \frac{-\Delta x J}{c(zF\Delta\psi/RT)} = \frac{\Delta x g RT}{c(zF)^2} \quad (\Delta c = 0) \tag{11.25}$$

and

$$D^* = -\Delta x \left(\frac{J^*}{\Delta c^*}\right)_{J=0} = \frac{\Delta x J_0}{c} \tag{11.26}$$

Combining Eqs. (11.25) and (11.26) with Eq. (11.23),

$$Q = \frac{D}{D^*} \tag{11.27}$$

Thus it was shown that the flux ratio produced by an electrical potential difference is abnormal if and only if the "bulk" diffusion coefficient differs from the tracer diffusion coefficient.

The significance of Chen and Walser's Q may be further clarified by expressing it in terms of the parameters of our Eq. (11.1):

$$RT \ln f = -QzF\Delta\psi = -\frac{\omega}{\omega^*}\left(zF\Delta\psi + \sum R_{\omega j}J_j\right)$$

so that

$$Q = \frac{\omega}{\omega^*}\left(1 + \sum \frac{R_{\omega j} J_j}{zF\Delta\psi}\right) \qquad (11.28)$$

It is seen that Q is the product of two terms, each reflecting a factor contributing to abnormality of the flux ratio: isotope interaction and coupling with the flows of other species. Because they lacked information concerning the possibility of coupling of flows of different species in their system, Chen and Walser drew no conclusions concerning the extent of isotope interaction (or in their terminology, "self interaction"). Irrespective of these effects, their approach permitted the evaluation of transport numbers and partial conductances.

Saito, Lief, and Essig (1974), in studies of toad bladders mounted in chambers, also considered the problems of using tracer isotope techniques to evaluate passive conductance. In general, such attempts are complicated by the need to incorporate the contribution of flows of several species. If it can be shown, however, that passive ionic flows are largely by way of a common pathway, the problem is simplified in that measurements of the flow of one ion may provide information concerning all flows. This appears to be the case in the toad bladder, as indicated by the proportionality of serosal to mucosal ($S \rightarrow M$) ^{42}K flux to that of $S \rightarrow M$ ^{22}Na flux (almost entirely passive), $M \rightarrow S$ ^{36}Cl flux, and $M \rightarrow S$ ^{131}I flux. Given this evidence that passive ionic flows occur largely by way of common pathways and vary proportionally with change in the passive permeability, it is possible to use the measurement of the flow of one species as a means of evaluating also the flows of others. Tracer fluxes can be of great aid in this regard, but their use requires attention to the considerations discussed above. In particular, it cannot be assumed *a priori* in a biological system that the flow of the species of interest will be uninfluenced by flows of other species. This must be tested by experiment. One means of doing so has been alluded to above: in the absence of isotope interaction and coupling of flows of different species, the flux ratio will be "normal," being given by Eq. (9.5) or, if flow is a consequence only of electrical forces, by Eq. (11.6). In practice, however, it is not always convenient to determine the flux ratio; especially when only one tracer form of a given chemical species is

288 Bioenergetics and linear nonequilibrium thermodynamics

available, the determination of two oppositely directed unidirectional fluxes may be difficult or impossible.

11.2.2 Potential dependence of unidirectional flux: passive pathway

In these circumstances it may be more practical to obtain the desired information by measuring only one unidirectional flux, at two settings of the electrical potential. Again writing $\vec{J}$ for the influx and $\overleftarrow{J}$ for the efflux, and combining the net flux and flux ratio of Eqs. (9.36) and (9.37), we have

$$J = (f - 1)\frac{\overleftarrow{J}}{f} \tag{11.29}$$

and

$$J = (f - 1)\overleftarrow{J} \tag{11.30}$$

Equations (11.15) and (11.29) give

$$\frac{\vec{J}}{\vec{J}_0} = f\frac{\ln f}{f - 1} \tag{11.31}$$

Thus, with Eq. (9.5), applicable in the absence of isotope interaction and other coupled flows,

$$\frac{\vec{J}_X}{\vec{J}_0} = \frac{(X/RT)\exp(X/RT)}{\exp(X/RT) - 1} \tag{11.32}$$

or

$$\frac{\vec{J}_{\Delta\psi}}{\vec{J}_0} = \frac{zF\Delta\psi/RT}{\exp(zF\Delta\psi/RT) - 1} \quad (\Delta c = 0) \tag{11.33}$$

Similarly, Eqs. (11.15) and (11.30) give

$$\frac{\overleftarrow{J}}{\overleftarrow{J}_0} = \frac{\ln f}{f - 1} \tag{11.34}$$

and with Eq. (11.6),

$$\frac{J_{\Delta\psi}}{J_0} = \frac{-zF\Delta\psi/RT}{\exp(-zF\Delta\psi/RT) - 1} \quad (\Delta c = 0) \quad (11.35)$$

In Saito et al.'s study of serosal to mucosal Na^+ flux in toad bladders mounted as flat membranes in Ussing-Zerahn-type chambers, the ratio of flows at $\Delta\psi = 50$ mV and $\Delta\psi = 0$ mV was indistinguishable from 2.27, the theoretical ratio derived from the application of Eq. (11.35). This indicates that the flux was almost entirely passive and uninfluenced by isotope interaction, electroosmosis, or coupling with the flows of other ions. For the case of mucosal-to-serosal Cl^- flux, however, the case was different. Here the ratio of flows at $\Delta\psi = 50$ mV and $\Delta\psi = 0$ mV was significantly different from 2.27, the best estimate being about 1.84. Since the studies of Na^+ tracer flux indicated the absence of electroosmosis and coupling of Na^+ and Cl^- flows, and since there is no evidence of active transepithelial Cl^- transport by the Dominican toad bladder, it was presumed that this discrepancy was attributable to isotope interaction. However, this was not tested directly. (In their studies of toad bladders mounted as sacs, Chen and Walser found that both passive Na^+ and Cl^- tracer fluxes responded to change of $\Delta\psi$ to a lesser extent than predicted by Eqs. 11.32–11.35.)

If deviation from Eqs. (11.32)–(11.33) is not in fact attributable to interaction of the flow of Cl^- with that of water or other ions, a development analogous to that above will permit evaluation of the degree of isotope interaction. Thus, in the presence of isotope interaction but in the absence of other coupled flows, introducing Eq. (11.1) into Eq. (11.31) gives

$$\frac{J_X}{J_0} = \frac{[(\omega/\omega^*)(X/RT) \exp[(\omega/\omega^*)(X/RT)]}{\exp[(\omega/\omega^*)(X/RT)] - 1} \quad (11.36)$$

or

$$\frac{J_{\Delta\psi}}{J_0} = \frac{(\omega/\omega^*)(zF\Delta\psi/RT)}{\exp[(\omega/\omega^*)(zF\Delta\psi/RT)] - 1} \quad (\Delta c = 0) \quad (11.37)$$

Similarly, from Eqs. (11.8) and (11.34),

$$\frac{\tilde{J}_{\Delta\psi}}{\tilde{J}_0} = \frac{-(\omega/\omega^*)(zF\Delta\psi/RT)}{\exp\left[-(\omega/\omega^*)(zF\Delta\psi/RT)\right] - 1} \quad (\Delta c = 0) \quad (11.38)$$

(It is seen that for passive flow, both in the presence and in the absence of isotope interaction, $\tilde{J}_0 = J_0$, and $\tilde{J}_{\Delta\psi}/\tilde{J}_0 \equiv \tilde{J}_{-\Delta\psi}/\tilde{J}_0$. This is not the case, of course, for total Na$^+$ flux in the toad bladder, since serosal to mucosal flux is very largely passive, whereas mucosal to serosal sodium flux is predominantly active.)

Eqs. (11.36)–(11.38) cannot be solved explicitly for ω/ω^*, but it is possible to evaluate ω/ω^* graphically (Essig and Li, 1975). This is facilitated by introducing the definition

$$\chi \equiv \frac{\omega}{\omega^*}(-X/RT) \quad (11.39)$$

$$= \frac{\omega}{\omega^*}(zF\Delta\psi/RT) \quad (\Delta c = 0) \quad (11.40)$$

The theoretical relationship between the quantity $\tilde{J}_X/\tilde{J}_{X=0} \equiv \tilde{J}_X/\tilde{J}_0$ ($\equiv \tilde{J}_{-X}/\tilde{J}_0$) and $-(RT/F)\chi$ is shown by the solid line of Fig. 11.1. Thus, considering the case of an electrical force, an experimentally determined value of $\tilde{J}_{\Delta\psi}/\tilde{J}_0(\equiv \tilde{J}_{-\Delta\psi}/\tilde{J}_0)$ is associated graphically with a unique value of χ. Knowing the value of z and $\Delta\psi$ then allows calculation of ω/ω^*: if $\Delta\psi$ is expressed in mV,

$$\frac{\omega}{\omega^*} = \left(\frac{1}{z\Delta\psi}\right)\left(\frac{RT}{F}\right)\chi = \left(\frac{25.3}{z\Delta\psi}\right)\chi \quad (11.41)$$

The validity of these relationships was tested in the above-cited study of anionic tracer fluxes in PVBT membranes permitting little electroosmosis or coupling of flows of different chemical species (Li et al., 1974). The symbols in the figure indicate experimental values of unidirectional fluxes determined at different settings of X. (In most cases $\Delta c = 0$, so $X = -zF\Delta\psi$ and $\chi = (\omega/\omega^*)(zF\Delta\psi/RT)$. In a few cases $\Delta\psi = 0$, so $X = RT \ln (c^I/c^{II})$ and $\chi = (\omega/\omega^*) \ln (c^{II}/c^I)$.) As is seen, the agreement between experiment and theory was good.

Applying this formulation to our findings in the toad bladder, it

Isotope flows: tests and applications 291

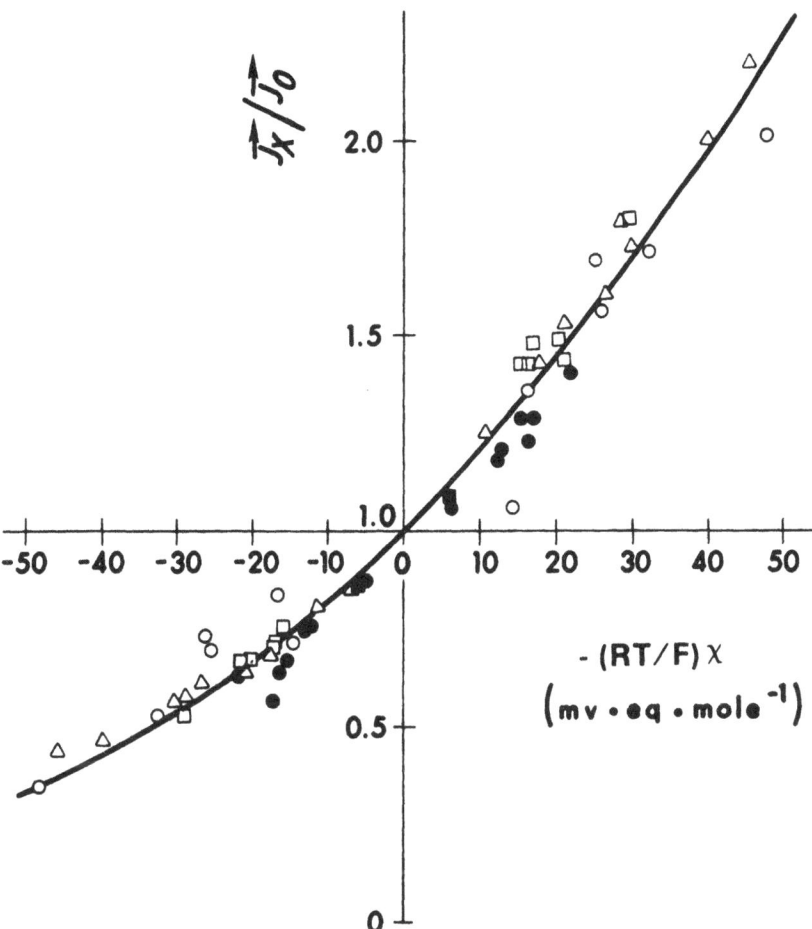

Fig. 11.1. Relationship between $\overrightarrow{J}_x/\overrightarrow{J}_0$ and $-(RT/F)\chi$. Each value of a unidirectional flux represents the mean of three or four determinations. The solid line represents the theoretical relationship of Eq. (11.36). For convenience the abscissa is expressed in terms of electrical units. Theoretically the ordinate $\overrightarrow{J}_x/\overrightarrow{J}_0 = \overleftarrow{J}_{-x}/\overleftarrow{J}_0$; $\overrightarrow{J}_0 = \overleftarrow{J}_0$. (From Li, DeSousa, and Essig, 1974.)

can be seen from Fig. 11.1 that the value of $\overrightarrow{J}_{50}/\overrightarrow{J}_0 = 1.84$ found for Cl$^-$ flux is associated with the value $\chi = -1.4$. Equation (11.42) then gives $(\omega/\omega^*)_{Cl} = [25.3/(-1)(50)](-1.4) \simeq 0.71$. The value of $\overrightarrow{J}_{50}/\overrightarrow{J}_0 = 2.27$ found for Na$^+$ flux is associated with the value $\chi = -2.0$, giving $(\omega/\omega^*)_{Na} = [25.3/(1)(-50)](-2.0) \simeq 1.0$. (These results differ somewhat from those of Chen and Walser with toad bladder sacs, as cited above.)

The calculation of $(\omega/\omega^*)_{Cl}$ and $(\omega/\omega^*)_{Na}$ in the above manner requires, of course, that both Cl^- and Na^+ flux be uninfluenced by the flows of either solvent or other ions. We have presented above our reasons for believing this to be the case in the studies discussed. It should be noted, however, that in contradistinction to the evaluation of the degree of isotope interaction, the partial conductance of an ion can be evaluated unequivocally from two values of the unidirectional flux, irrespective of the nature of the factors influencing the tracer flow. This is seen by considering that in the more general case Eqs. (9.30), (9.45), and (9.48) give

$$f = \frac{\vec{J}}{\overleftarrow{J}} = \exp\left[\left(\frac{\omega}{\omega^*}\right)\left(-zF\Delta\psi - \sum R_{0j}J_j\right)\Big/RT\right] \quad (11.42)$$

where again we have written $\sum R_{0i}J_i$ for $\int_0^{\Delta x} \sum r_{0j}J_j dx$, the contribution of coupled flows to the forces producing net transport. With passive coupled flows proportional to $\Delta\psi$, we may introduce the proportionality constant α, defined by the relation

$$\sum R_{0j}J_j \equiv \alpha z F \Delta\psi \quad (11.43)$$

It is then convenient to introduce a quantity which incorporates all of the factors contributing to abnormality of the flux ratio (Essig and Li, 1975); as is seen from Eq. (11.28), this is equivalent to Chen and Walser's Q:

$$Q = \frac{\omega}{\omega^*}(1 + \alpha) \quad (11.44)$$

Introducing Eqs. (11.42)–(11.44) into Eq. (11.31) gives

$$\frac{\vec{J}_{\Delta\psi}}{\vec{J}_0} = \frac{QzF\Delta\psi/RT}{\exp(QzF\Delta\psi/RT) - 1} \quad (11.45)$$

and again $\vec{J}_{\Delta\psi}/\vec{J}_0 = \overleftarrow{J}_{-\Delta\psi}/\overleftarrow{J}_0$. As for Eqs. (11.36)–(11.38) above, this equation cannot be solved explicitly. Again, however, it can be treated graphically, by introducing the general definition

$$\chi \equiv Qz F\Delta\psi/RT \tag{11.46}$$

(which includes the definition of Eq. 11.40 in the case of $\alpha = 0$). The relation between $\tilde{J}_{\Delta\psi}/\tilde{J}_0$ and χ plotted in Fig. 11.1 will now permit the use of an experimentally measured value of $\tilde{J}_{\Delta\psi}/\tilde{J}_0$ (or $\tilde{J}_{\Delta\psi}/\tilde{J}_0$) to evaluate Q. In general, the interpretation of Q will be equivocal since, as seen from Eq. (11.44), $Q \neq 1$ may result either from isotope interaction ($\omega/\omega^* \neq 1$) or coupling of flows of different species ($\alpha \neq 0$). However, the determination of Q in this manner permits the evaluation of the partial conductance g attributable to an ion. From Eqs. (9.15) and (9.18) we have

$$J = -c\omega \left(zF\Delta\psi + \sum R_{0j}J_j \right)$$
$$= -c\omega^* \left[\left(\frac{\omega}{\omega^*}\right) \left(zF\Delta\psi + \sum R_{0j}J_j \right) \right] = -c\omega^* Q zF\Delta\psi \tag{11.47}$$

But also, for passive flows, Eq. (11.21):

$$J = \frac{-g\Delta\psi}{zF}$$

From Eq. (9.45), $\tilde{J}_0 = RTc\omega^*$, so

$$g = [(zF)^2/RT]Q\tilde{J}_0 \equiv \left(\frac{zF}{\Delta\psi}\right) \tilde{J}_0 \chi \tag{11.48}$$

For $\Delta\psi$ expressed in mV, $\tilde{J}_0$ expressed in mM cm^{-2}sec^{-1}, and g expressed in mmho cm^{-2},

$$g = 9.65 \times 10^7 \, (z/\Delta\psi)\tilde{J}_0\chi \tag{11.49}$$

(In the absence of coupled flows of other species, $Q = \omega/\omega^*$, permitting evaluation of the degree of isotope interaction, as above. In the absence of isotope interaction, the evaluation of χ evaluates the influence of coupled flows of other species.)

11.3 Validation of theoretical formulations in biological membranes: active transport

11.3.1 General relationships

In view of the broad validity of the assumptions underlying the derivation of Eq. (11.1), it might be anticipated that the formulation should be applicable in a wide variety of experimental circumstances. In particular, one would hope that it could be employed to analyze the kinetics of mechanisms of active transport. In attempting to employ the flux ratio relationship for this purpose, however, one should appreciate that in the derivation of Eq. (11.1) the contribution of the flow of metabolism to the forces promoting transport was introduced only by formal analogy to those of coupled solute and solvent flows and that it differs fundamentally in its nature, in that whereas material flows J_j are vectorial and the corresponding local phenomenological resistance coefficients are scalar, the flow of metabolism J_r (such as the rate of oxygen consumption) is a scalar quantity, and hence the coefficient r_{ir} must be a vector. Therefore, although incorporation of the force $r_{ir}J_r$ appears justified intuitively, it is important to test the flux ratio relationship experimentally in the presence of active transport, as previously for passive transport.

One such test has been carried out by Chen and Walser (1975) in studies of active sodium transport in toad bladder sacs. In order to evaluate the active components of sodium flux as a function of the transepithelial electrical potential difference, unidirectional fluxes were measured before and after the addition of sufficient ouabain to eliminate active transport. Since it was shown that ouabain does not affect passive serosal to mucosal sodium flux in the toad bladder, it was considered that the ouabain-inhibitable fluxes are via the active pathway. By means of this approach it was found that whereas at $\Delta\psi = 100$ mV ouabain-inhibitable $S \to M$ sodium flux was not demonstrable, at $\Delta\psi = 150$ mV it averaged (in electrical units) 0.073 ± 0.001 μA/mg (n = 15). Thus sodium transport through the active pathway is bidirectional. Although such bidirectionality is implicit in the very concept of quantification of flux ratios in the active pathway, we are not aware of its experimental demonstration prior to that of Chen and Walser.

If indeed Eq. (11.1) applies in the presence of active transport,

Eq. (11.13) remains valid, giving

$$J^a = J^a{}_{J^a=0} \ln f^a \equiv \bar{J}^a_E \ln f^a = \bar{\bar{J}}^a_E \ln f^a \qquad (11.50)$$

and

$$\bar{J}^a_E = \bar{\bar{J}}^a_E = \frac{\overrightarrow{J}^a - \overleftarrow{J}^a}{\ln(\overrightarrow{J}^a/\overleftarrow{J}^a)} \qquad (11.51)$$

where the superscript a refers to the active pathway; since we consider here only sodium flux. we omit the subscript. (In the case of active transport $\bar{J}^a_{J^a=0} = \bar{\bar{J}}^a_{J^a=0}$ is not given by $\bar{J}^a_{\Delta\bar{\mu}=0}$, but rather by $\bar{J}^a_{\Delta\bar{\mu}=zFE} \equiv \bar{J}^a_E$, where E, the electromotive force of sodium transport, is the value of $\Delta\psi$ necessary to reduce the rate of net active sodium transport J^a to zero.) An experimentally convenient means of testing Eq. (11.1) is now provided by introducing the fundamental relationship of the equivalent circuit formulation (Ussing, 1960), in our terms given by

$$I^a = FJ^a = F(\overrightarrow{J}^a - \overleftarrow{J}^a) = \kappa^a_{Na}(E - \Delta\psi) \qquad (11.52)$$

Combining Eqs. (11.50) and (11.52) gives:

$$F\overrightarrow{J}^a = \frac{FJ^a}{\exp(J^a/\bar{J}^a_E) - 1} = \frac{\kappa^a_{Na}(E - \Delta\psi)}{\exp[\kappa^a_{Na}(E - \Delta\psi)/F\bar{J}^a_E - 1] - 1} \qquad (11.53)$$

and

$$F\overleftarrow{J}^a = \frac{FJ^a}{1 - \exp(-J^a/\bar{J}^a_E)} = \frac{\kappa^a_{Na}(E - \Delta\psi)}{1 - \exp[-\kappa^a_{Na}(E - \Delta\psi)/F\bar{J}^a_E]} \qquad (11.54)$$

Table 11.3 presents the experimental results of Chen and Walser (1975), comparing steady-state active sodium fluxes observed at $\Delta\psi = 0$, 100, and 150 mV with those predicted from the relationships above. As is seen, the agreement between experiment and theory is very good. With an average value of $E = 154$ mV in their membranes, the values of serosal to mucosal sodium flux at $\Delta\psi = 0$ and 100 mV were predictably immeasurably small.

The validity of the general flux ratio relationship in the presence

Table 11.3. Theoretical and observed electrical potential-dependence of toad urinary bladder active sodium fluxes.[a]

	J^a (μA/mg)		$\overleftarrow{J}^a$ (μA/mg)	
$\Delta\psi$ (mV)	Theoretical	Observed	Theoretical	Observed
0	2×10^{-6}	0	1.50	1.56 ± 0.19
100	3×10^{-3}	0	0.55	0.54 ± 0.06
150	0.076	0.073 ± 0.001	0.15	0.14 ± 0.03

Source: Chen and Walser (1975).
a. J^a and $\overleftarrow{J}^a$ here represent serosal to mucosal components and mucosal to serosal components, respectively, of active sodium flux, which in Chen and Walser's system contributes about 80% of the current attributable to active transport. Theoretical values of J^a and $\overleftarrow{J}^a$ were calculated from Eqs. (11.53) and (11.54) (their Eqs. 18 and 19).

of active transport has also been tested in toad bladders mounted as planar membranes (Wolff and Essig, 1977). Using bicarbonate-Ringer's solution, only Na^+ contributes significantly to active transport, and passive Na^+ tracer flow is uninfluenced by isotope interaction or coupled flows with other chemical species. Therefore, measurements of tissue conductance and sodium influx in the presence and absence of active transport provide a convenient means of determining the potential-dependence of both influx and efflux in the active pathway. In conformity with the findings of Chen and Walser, it was found that as $\Delta\psi$ approached E, active efflux $\overleftarrow{J}^a$ became demonstrable. When $\Delta\psi$ exceeded E, $\overleftarrow{J}^a$ exceeded $\overrightarrow{J}^a$, so that net flux J^a was negative. Again, experimental values of $\overleftarrow{J}^a$ agreed well with values predicted theoretically.

11.3.2 Potential dependence of flux ratio: active pathway

Also of interest to Chen and Walser was the extent to which the unidirectional fluxes and flux ratio of the active sodium transport pathway are influenced by "abnormal" values of the parameter Q^a ("the ratio of the bulk diffusion coefficient to the tracer diffusion coefficient in this pathway"), that is, the values of $Q^a \neq 1$. In order to evaluate this quantity, the flux ratio equation employed earlier for the analysis of passive transport, Eq. (11.24), was modified by incorporating the contribution of the electromotive force of the sodium pump, giving

$$\ln f^a = -Q^a zF(\Delta\psi - E)/RT \quad (11.55)$$

By applying this and the above relations to their data, they calculated Q^a to be 2.54 ± 0.34, considered to be characteristic of single-file diffusion.

The use of Eq. (11.55) to analyze active fluxes is intuitively reasonable in view of experimentally observed linear relationships between $\ln f$ and $\Delta\psi$ in various systems, and satisfaction of the condition that $f^a_{\Delta\psi=E} = 1$. It is useful, however, to examine the nature and significance of the parameter Q^a in terms of the pertinent thermodynamic variables (Essig and Lang, 1975). This may be done by equating the flux ratio relations of Eqs. (11.1) and (11.55). It is convenient to consider the situation in which there is no concentration difference across the membrane, so that in the absence of coupled flows of species other than sodium, as appears to be the case for Dominican toad bladders mounted in chambers,

$$Q^a zF(\Delta\psi - E) = \left(\frac{\omega}{\omega^*}\right)^a (zF\Delta\psi + R_{or}J_r) \quad (11.56)$$

Setting $\Delta\psi = E$ shows that

$$R_{or} = \frac{-zFE}{J_{rE}} \quad (11.57)$$

so that $Q^a(\Delta\psi - E) = (\omega/\omega^*)^a[\Delta\psi - (E/J_{rE})J_r]$. Setting $\Delta\psi = 0$ now gives

$$Q^a = \left(\frac{\omega}{\omega^*}\right)^a \left(\frac{J_{r0}}{J_{rE}}\right) \quad (11.58)$$

Q^a may also be expressed in terms of the degree of coupling q; since for linear systems with constant affinity, q^a is unique and given by $J_{rE}/J_{r0} = 1 - q^2$ (see Eq. 7.39),

$$Q^a = \left(\frac{\omega}{\omega^*}\right)^a \frac{1}{(1-q^2)} \quad (11.59)$$

The above development shows that Q^a reflects not only the extent of isotope interaction but also the coupling of transport to metabolism. (If the flow of sodium were coupled with that of other species, as suggested by Chen and Walser as a possibility in their preparation, Q^a would of course reflect this coupling as well.) In the absence of coupled material flows, since $J_{r0} > J_{rE}(q^2 > 0)$, Q^a must overestimate the magnitude of $(\omega/\omega^*)^a$. Clearly Q^a cannot be considered to be a reliable index of the extent of isotope interaction. A corollary is that in the absence of isotope interaction, that is, if $\omega/\omega^* = 1$, Q^a would not equal 1. Therefore, as was pointed out in Sec. 9.2.4, even in this situation it is not to be expected that the flux ratio will obey the simple classic relationship, $RT \ln f = F(E - \Delta\psi)$, even if it were possible entirely to eliminate flux through parallel passive pathways. These considerations are pertinent to interpretation of the finding that in short-circuited frog skins with only minimal edge damage, $RT \ln f$ agrees well with values of E estimated by other means (Helman, O'Neil, and Fisher, 1975). This agreement seems likely to be fortuitous, reflecting the combined effects of a value of $Q > 1$ in the active pathway, which acts to raise f, and residual small fluxes via passive pathways, which act to lower f. It would be useful to test this interpretation by determining Q^a in the frog skin.

11.3.3 Isotope interaction in the active transport pathway

As discussed above, for Dominican toad bladders mounted in chambers, there appear to be no coupled flows other than those of metabolism and active sodium transport. Hence the application of Eq. (11.58) should permit evaluation of the extent of isotope interaction in the active pathway. We cannot yet evaluate this quantity with precision, since the pertinent variables have not yet been evaluated simultaneously in the same tissues. However, it has been possible to estimate ω/ω^* roughly. Combining Chen and Walser's value for Q^a in toad bladder sacs with Lang's value of $J_{rE}/J_{r0} = 1 - q^2$ measured in chambers gave a mean value of $(\omega/\omega^*)^a = 0.82$ (Essig and Lang, 1975). The use of a presumably more accurate value of $q = 0.86$ (Lang, Caplan, and Essig, 1977) and a value of $Q^a = 2.07$ for toad bladders mounted in chambers (Wolff and Essig, 1977) gives $(\omega/\omega^*)^a = 0.54$. Labarca, Canessa, and Leaf (1977) have suggested that coupling of transport and oxidative metabolism in the toad bladder is very tight; if so, values of $(\omega/\omega^*)^a$ would be markedly lower than indicated above. Clearly none of these values can be accepted with confidence. However, they suggest the existence of negative isotope interaction in the active pathway. Although without further information this finding cannot be taken as evidence for any specific transport mechanism, it is clearly consistent with the commonly accepted Na-K ATPase carrier mechanism for active sodium transport, as would not be the case for findings suggestive of single-file diffusion. It is important to emphasize that in order to rule out this latter possibility, suggested on the basis of a value of $Q^a > 1$, it is necessary to take account explicitly of all coupled flows, including that of metabolism.

In commenting on the above formulation, Chen and Walser have pointed out that it is possible that K flux and Na flux may interact in association with the function of Na-K ATPase and that such interaction might possibly influence the magnitude of $(\omega/\omega^*)^a$. This is of course so. However, as we have emphasized previously, it must be appreciated that ω and ω^* are phenomenological coefficients, and the term "isotope interaction" is accordingly a phenomenological characterization and not a description of molecular events. Thus $(\omega/\omega^*)^a$ describes the behavior of the system as it stands, whatever the nature of the factors which make it so.

11.4 Summary

1. The use of synthetic membranes with high permselectivity and low water permeability permits transport studies in which the pertinent forces are well defined. The unidirectional fluxes and flux ratio are then predicted by a theoretical formulation incorporating the influence of isotope interaction.

2. The demonstration of isotope interaction has no specific mechanistic implications: negative coupling of isotope flows (countertransport, exchange diffusion) is demonstrable in synthetic systems lacking carrier mechanisms for transmembrane transport. Negative isotope interaction can be induced by combining homogeneous membrane elements in a heterogeneous parallel array.

3. Theoretical relationships for unidirectional fluxes and flux ratios are obeyed also in epithelia, both for passive and active transepithelial transport.

4. The voltage-dependence of the unidirectional flux of an ion quantifies its partial conductance.

5. Knowledge of the interaction of diverse flows and/or metabolism allows also evaluation of the degree of isotope interaction. The value of R^*/R (or ω/ω^*) may then be determined from the voltage-dependence of either the flux ratio or the unidirectional flux, as convenient.

6. The above formulations have been applied experimentally in studies of the toad urinary bladder. Measurements of serosal to mucosal tracer fluxes permit evaluation of the passive, presumably paracellular conductance.

7. Even with corrections for passive flows, as predicted previously on theoretical grounds, measurement of the flux ratio at short circuit does not quantify the electromotive force of sodium transport E_{Na}.

8. Combined measurements of the voltage-dependence of Na tracer flux and oxidative metabolism permit evaluation of the extent of isotope interaction in the active pathway. Best current estimates in the toad urinary bladder indicate that $(\omega/\omega^*)^a < 1$, in conformity with the commonly accepted ATPase carrier mechanism for active Na^+ transport.

12 Muscular contraction

Mechanochemical energy conversion in muscle is perhaps the most impressive of all examples of biological energy conversion, since it is the closest to our immediate experience. Not only do we perceive the output directly, but we sense the level of input quite vividly as well. Wilkie (1967) has remarked, "The efficiency of their muscles is clearly a matter of the greatest importance to animals and men." This may be so, although the matter of *prime* importance is surely the degree of coupling of their muscles. In isometric and (essentially) unloaded contractions the efficiency is actually zero, yet such contractions occur frequently: the former to maintain a given mechanical situation against change, the latter to secure quick movement. The emphasis on coupling obviously anticipates a nonequilibrium thermodynamic approach; it does not alter the significance of efficiency measurements in well-characterized states. Measurements of this kind have played a central role in muscle physiology (see for example, Hill, 1939, 1964b), but until quite recently they were based entirely on simultaneous observations of work and heat.[1]

It is clear from our earlier considerations that efficiency may be represented by the ratio (output)/(dissipation + output). The dissipation of the system is by no means synonymous with the heat it produces; this would be the case only under very special circumstances, to be discussed below. Consequently the commonly used efficiency ratio (work)/(heat + work), sometimes termed the mechanical efficiency, must be interpreted with caution. Wilkie (1960) has suggested that this ratio could be converted into a valid measure of efficiency by multiplying the denominator by a correction factor $Y = \Delta G/\Delta H$, where ΔH is the enthalpy change per mole of reaction. In terms of the relationship between the affinity and the Gibbs free energy discussed in Chapter 2, and subject to the considerations mentioned there, $Y = -A/\Delta H$. Unfortunately, the factor Y is not necessarily a constant. It could be if A were constant; however, A may vary, and furthermore ΔH is not independent of changes in the activities of reactants and products except in ideal solutions.

Although the application of NET to muscular contraction has not progressed as far or as fast as it has in transport studies, we shall outline the treatment in the following sections and compare it with some of the standard approaches. As in other areas, NET appears to be an invaluable tool in constructing a logically consistent description of the energetics of the system, and it permits a variety of hitherto puzzling phenomena to fall into place in an intuitively appealing way. Furthermore, the force–velocity characteristic of contracting muscle raises important issues in relation to regulation and control. These issues can be dealt with in the context of linear NET, and the results, whether or not they prove relevant in the case of muscle, are certainly suggestive and of general significance.

In this chapter we will not make explicit use of the diagram method, even though muscular contraction may be considered a special case of active transport: according to the sliding filament hypothesis, metabolic energy is expended to transport actin filaments relative to myosin filaments against mechanical tension. However, two complicating factors severely restrict our ability to employ the method in the usual way. First, actin is not a "small molecule" ligand comparable to a hydrogen or sodium ion but is actually part of the macromolecular machine. Secondly, stationarity in the *strictest* sense occurs in muscle only during an isometric contraction. Despite these difficulties, the concept of basic free energy levels (Chapter 5) can be exploited in a rather powerful manner to elucidate the quantitative features of specific molecular models. A consideration of such muscle models is outside the scope of this book, but the interested reader is referred to the substantial series of studies by T. L. Hill and collaborators (see, for example, Hill, 1974, 1975, 1977; Eisenberg and Hill, 1978; Eisenberg, Hill, and Chen, 1980).

12.1 Nonequilibrium thermodynamic analysis of muscle

12.1.1 The dissipation function

We can arrive at the dissipation function for muscle by an approach similar to that used for "discontinuous systems" in the study of transport problems (Chapter 2). Figure 12.1 is a diagrammatic representation of a striated muscle, showing a single fibril bearing the load. Ernst (1963) has discussed in detail the proposition that the myofibril is the smallest functional unit of muscle. Adopting this assumption, we shall consider the fibril, or the total assembly of filaments within a fibril, to be the essential working element or "black

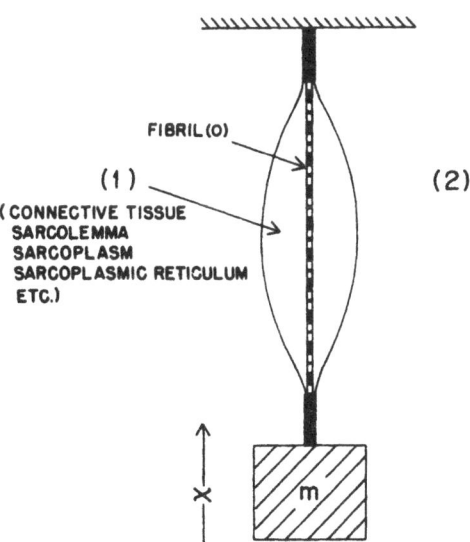

Fig. 12.1. The muscle as an energy converter. The single fibril shown represents the "black box" or working element. It raises a mass m through a distance x.

box" of the system, and for *small* steady tetanic contractions shall suppose it to be in a stationary state (the experimental evidence supporting this will be summarized later). We restrict our considerations solely to the *steady-state phase of tetanic contraction*, that is, the initial phase of both constant velocity and constant tension (including zero velocity, or isometric contraction, as one limiting case, and zero tension, or unloaded contraction, as the other). These considerations explicitly exclude twitches.

Attached to the fibril (0) is a reservoir of reagents, compartment (1), from which it derives its input. The region denoted as compartment (2) is the external world, which includes the mass m. Then, for an infinitesimal contraction of the fibril at a temperature T during the stationary state, the Gibbs equation for compartment (1) is:

$$dU^{(1)} = TdS^{(1)} - pdV^{(1)} + \sum_i \mu_i^{(1)} dn_i^{(1)} \qquad (12.1)$$

where p and V represent pressure and volume, and U and S internal energy and entropy, as usual. The corresponding equation for compartment (2), representing now the change in total energy of the external world as a consequence of the contraction, is:

$$dU^{(2)} = TdS^{(2)} - pdV^{(2)} + \sum_i \mu_i^{(2)} dn_i^{(2)} + mgdx \qquad (12.2)$$

where x is the distance through which the mass m is raised, and g is the acceleration of gravity. The summation term accounts for transfer of matter between compartments (2) and (1), for example, oxygen consumption. Since steady contractions only are considered, no inertial terms are involved; the potential energy change brought about in compartment (2) is included in $dU^{(2)}$ but that in compartment (1) is regarded as negligible (that is, the muscle is regarded as weightless). Summing these two equations and applying the first law, we obtain

$$TdS^{\text{total}} \equiv Td_tS = -mgdx - \sum_i \mu_i^{(1)} dn_i^{(1)} - \sum_i \mu_i^{(2)} dn_i^{(2)} \qquad (12.3)$$

where, as before, d_tS is the "internal entropy production," or total creation of entropy as a consequence of the irreversible processes taking place. By definition the dissipation is then given by

$$\Phi \equiv \frac{Td_iS}{dt} = -mg\dot{x} + \sum_j A_j v_j \qquad (12.4)$$

and as we have seen before, this is positive-definite, although any individual term may be negative ($\dot{x} = dx/dt$). In writing Eq. (12.4), we have introduced the reaction parameters described in Chapter 2. The summation results from the two summations in Eq. (12.3) on introducing affinities and reaction velocities and refers to a set of independent chemical reactions (diffusional processes which occur within compartment 1 are readily included in this formalism). We now assume that it is possible to identify among the chemical reactions j a *unique* reaction k which is coupled to the mechanical process of contraction and that this reaction is not coupled directly to any of the remaining reactions. This is equivalent to separating Φ into two positive-definite parts as follows:

$$\Phi = \Phi_{\text{mech}} + \Phi_{\text{chem}} \qquad (12.5)$$

where in the stationary state

$$\Phi_{\text{mech}} = -mg\dot{x} + A_k v_k \qquad (12.6)$$

(the mechanochemical part of the dissipation), and

$$\Phi_{\text{chem}} = \sum_j A_j v_j \qquad (j \neq k) \qquad (12.7)$$

(the purely chemical part of the dissipation, which does not include the reaction k). The absence of direct coupling between reaction k and the rest of the reactions j implies that reaction k will not advance if its own affinity is zero and no mechanical force is applied, even though other reactions may have finite affinities and velocities. However, *indirect* effects of some or all of the processes included in Φ_{chem} on the reaction k, and thus on Φ_{mech}, are not excluded. For example, the reactions j may occur in such a way as to maintain A_k at a fixed value.

Now it is clear that the fibril cannot be in a truly stationary state while it is shortening. Nevertheless, the treatment above will be a good approximation if the fibril is "quasi-stationary," in other words, if its state parameters change very slowly in comparison with those of compartment (1). This is very reminiscent of the situation in epithelial tissues, as discussed in Chapter 7. There is a considerable body of evidence to support the assumption of quasi-stationarity; indeed, this assumption is quite generally made in discussions of tetanic contraction. We shall summarize the evidence briefly under several headings.

Mechanical evidence for stationarity. In isometric tetanic contractions, tension develops to a maximum value, which thereafter remains constant for periods on the order of seconds in the case of frog and toad sartorii at 0°C (Gasser and Hill, 1924; Hill, 1949). In isotonic contractions, after an initial short period of adjustment to the load, the velocity of contraction becomes astonishingly constant and may be maintained for 60–70% of the total shortening the muscle undergoes (Sandow, 1961). This is all the more remarkable in view of the fact that isometric tension is a function of length, decreasing fairly rapidly at lengths greater or shorter than "rest" length.

Thermal evidence for stationarity. Since initial velocity of contraction under isotonic conditions is constant, the well-known fact that "shortening heat" is directly proportional to distance shortened (Hill, 1938; Abbott, 1951) reflects the constant rate at which heat is produced during shortening. During isometric contractions the heat production also reaches a constant rate (Hill, 1938), originally referred to as the "maintenance heat" production but now termed the "stable heat" (Carlson and Wilkie, 1974). Like isometric tension, this is a function of length, with a maximum at the rest length (Abbott, 1951).

Chemical evidence for stationarity. It has been shown that a linear relationship exists between time and ATP breakdown in isometric contractions of frog sartorius (Infante, Klaupiks, and Davies, 1964b; Maréchal, 1964). Again there is a length dependence, breakdown being maximum at rest length. This was established (in the presence of iodoacetate) by measuring the rate of formation of inorganic phosphate and also by measuring the consumption of ATP directly in sartorii treated with fluorodinitrobenzene. A similar result was obtained for phosphorylcreatine usage in isometric contractions of rectus abdominis (Infante, Klaupiks, and Davies, 1964a). Further striking evidence of stationarity is the direct proportionality found between phosphorylcreatine breakdown and mechanical work performed at constant load, using frog rectus abdominis muscles which contracted once or twice to different extents and for different times at 0°C (Cain, Infante, and Davies, 1962). This result will be discussed further below.

Structural evidence for stationarity. The assumption that the state parameters of a myofibril change relatively slowly during steady contraction does not seem incompatible with what is known of its structure. For example, one might expect the configurational entropy of the fibril to undergo a marked change as it shortens. But one of the implications of the sliding filament mechanism is that such changes are minimal. The filaments themselves appear to be rigid (Elliott, Lowy, and Millman, 1965; Huxley, Brown, and Holmes, 1965), and the "entropy of mixing" of actin and myosin when a sarcomere contracts cannot be large. For short contractions, cross-sectional volume elements in the interacting regions of the A-band do not have strongly time-dependent characteristics, although changes in lattice spacing do occur (Elliott, Lowy, and Millman, 1965).

12.1.2 The phenomenological equations for muscle: linearity and incomplete coupling

Theoretical considerations. In the description of muscular contraction to be considered here it is essential to write linear phenomenological relations with constant coefficients. Since chemical reactions are linear only under special circumstances, we need to examine the requirements for linearity rather closely. As discussed in Chapters 6 and 7, it has been pointed out that when the affinity of a given chemical reaction is large, the reaction may often be split into a considerable number of elementary reactions. In the stationary state the velocities of all the elementary reactions become identical. If the system is open, but only the initial reactants and final products can be exchanged with the environment, the stationarity condition above includes transport processes, that is, influx of reactants and outflux of products; this is in fact the condition of minimum entropy production for a given value of the external affinity (Prigogine, 1961). The transport processes may be linear over a fairly wide range of forces (differences in electrochemical potential) as compared with the reactions. Kinetic schemes for enzymatic reactions invariably consist of a set of consecutive steps. The mechanochemically coupled hydrolysis of ATP in muscle apparently involves an elaborate series of intermediate states, commencing with binding of ATP to the H-meromyosin cross bridges and followed by various interactions and conformational changes before ADP and P_i are finally released (Carlson and Wilkie, 1974). The reaction term in Φ_{mech}, Eq. (12.6), therefore consists in general of a series of terms arising from the sequence of elementary reaction steps as follows:

$$\Phi_{mech} = -mg\dot{x} + A_\alpha v_\alpha + \sum_{\rho=1}^{r} A_\rho v_\rho + A_\omega v_\omega \qquad (12.8)$$

where the total affinity, A_k, is given by

$$A_k = A_\alpha + \sum_{\rho=1}^{r} A_\rho + A_\omega \qquad (12.9)$$

Here the process α may refer to transfer of reactant from cytoplasm or local compartment to the myosin filament, ω may refer to transfer of products from the filament, and ρ indicates a series of r intermediate steps. In the stationary state

$$v_\alpha = v_1 = v_2 = \cdots = v = v_\omega = v_k \qquad (12.10)$$

and Φ_{mech} collapses to its original form. Our requirement for linearity would be realizable if the individual affinities A_ρ were sufficiently small or in an appropriate range, *provided that factors which change the A_ρ's did not alter the pertinent kinetic parameters significantly*. This, of course, is on *a priori* grounds unlikely in a far-from-equilibrium system. Nevertheless, in order to proceed, we shall invoke the concept of a proper pathway (Chapter 6) for the mechanochemical process (which corresponds to Eq. 12.6). The well-known nonlinearity manifested by the force–velocity relation will be associated entirely with the chemical process (which corresponds to Eq. 12.7).

These considerations clearly do not imply that $\dot{x}$ bears a fixed relation to v_k—that the ratio $\dot{x}/v_k$ is constant under all conditions. If it were, the muscle would be fully coupled ($q = 1$); in this case (and *only* in this case) the reaction would be brought to a halt in an isometric contraction and reversed in all stretches during tetanus. On the other hand, if the muscle is incompletely coupled, the reaction could in principle still be brought to a halt and ultimately reversed, but only by a tension greater than isometric. (The smaller the value of q, the larger the increase in tension above isometric needed for this purpose.) Tensions insufficiently greater than isometric would only slow down the reaction. It was shown by Abbott, Aubert, and Hill (1950) that all of the work done on a muscle when it is stretched while being stimulated may be "absorbed" (that is, does not appear as heat). Infante, Klaupiks, and Davies (1964c) showed that stretching an activated fluorodinitrobenzene-poisoned frog sartorius caused a reduction in the breakdown of ATP by as much as 90% "even though" the tension developed was about 70% greater than the isometric tension at rest length (no net resynthesis was observed). These authors also demonstrated a steady breakdown of ATP during isometric contractions at 0°C. We will return to a discussion of this matter later; however, on the basis of these observations we conclude that muscle is an incompletely coupled energy converter.

Before we write down phenomenological equations for muscle as an energy converter, it is useful to rewrite Eq. (12.6), the mechanochemical dissipation function, in a more familiar form using the customary notation for the output force and flow introduced by A. V. Hill:

$$\Phi_{mech} = V(-P) + J_r A \qquad (12.11)$$

where P is the tension and V the initial constant velocity in tetanic shortening; J_r is the velocity of the coupled metabolic reaction, and A its affinity as usual. It is convenient to consider $-P$ as the equivalent of the conventional output force X_1, taking the direction of contraction as positive, since the spontaneous effect of P is to *stretch* the muscle. Then P_0, the "isometric tension" or tension developed when the muscle is prevented from shortening, is directly analogous to the emf of an electrochemical cell, while V_m, the velocity of contraction when the load is zero, is analogous to the short-circuit current. If V is expressed in m/sec, P must be expressed in newtons for dimensional consistency. The resistive load, or mechanical resistance $R_L = P/V$, may vary from zero to infinity. Equation (12.11) represents a two-term dissipation function, the implication of which is that the system has two degrees of freedom. Thus the phenomenological relations corresponding to Eq. (12.11) are:

$$V = L_c(-P) + L_{cr}A \tag{12.12}$$

$$J_r = L_{cr}(-P) + L_r A \tag{12.13}$$

where Onsager symmetry has been assumed (the subscripts c and r indicate contraction and reaction). Equations (12.12) and (12.13) contain vectorial cross coefficients; consequently it is implied that coupling takes place in an anisotropic medium. In this respect the treatment is no different from that of Kedem for active transport (Chapter 3).

The above phenomenological relations may, of course, also be written in inverse form:

$$-P = R_c V + R_{cr} J_r \tag{12.14}$$

$$A = R_{cr} V + R_r J_r \tag{12.15}$$

Comparison with experiment. Equation (12.14) can be expressed in the following way:

$$J_r = \left(\frac{-1}{R_{cr}}\right) P + \left(\frac{-R_c}{R_{cr}}\right) V \tag{12.16}$$

Using the subscripts o and m to refer to isometric and unloaded contraction, or static head and level flow, respectively,[2] we readily see that Eq. (12.16) is equivalent to

$$J_r = \left(\frac{J_{ro}}{P_o}\right) P + \left(\frac{J_{rm}}{V_m}\right) V \tag{12.17}$$

Bornhorst and Minardi (1969) drew attention to the constancy of J_{ro}/P_o, for reasonable length changes, in both frog rectus abdominis and frog sartorius muscles (Infante, Klaupiks, and Davies, 1964a, b) and suggested that this is consistent with the assumption that the phenomenological coefficients are constant with changes in muscle length. There is also good evidence for constancy of J_{rm}/V_m and stationarity of contraction in frog rectus abdominis—the direct propor-

Muscular contraction 311

tionality found between phosphocreatine breakdown and mechanical work performed at constant load (Cain, Infante, and Davies, 1962). Constant load implies constant velocity in a stationary contraction; thus for contractions of different duration at constant load, Eq. (12.17) may be rewritten in the form

$$J_r = \left(\frac{J_{ro}}{P_oV} + \frac{J_{rm}}{V_mP}\right) PV$$

and multiplying both sides by the duration of contraction gives the result of Cain, Infante, and Davies. Further evidence for the constancy of J_{rm}/V_m has been adduced by Oplatka (1972). These findings are consistent with the validity of linear equations with constant coefficients. In support of this view, additional experimental data of Cain and coworkers were analyzed by Bornhorst and Minardi. Integrating Eq. (12.17) explicitly for isotonic contractions, Bornhorst and Minardi obtained

$$-\Delta PC = \left(\frac{J_{ro}}{P_o}\right) Pt + \left(\frac{J_{rm}}{V_m}\right) \Delta l \qquad (12.18)$$

where ΔPC represents the change in phosphocreatine content, Δl is the change in muscle length, and t is the duration of the contraction. Figure 12.2 is a comparison of the mechanochemical data of Cain, Infante, and Davies (1962) with Eq. (12.18). The data refer to "constant shortening" contractions of frog rectus abdominis under varying loads; the load P in Eq. (12.18) has been replaced by $W/\Delta l$, where W is the reported external work done by the muscle. As can be seen, the linearity of the plot is impressive. Values of R_{cr} and R_c are readily obtained from the slope and the intercept, respectively (it should be remembered that the resistive cross coefficient R_{cr} must be negative in a positively coupled system).

For isometric contractions, Eq. (12.16) becomes simply

$$J_{ro} = \left(\frac{-1}{R_{cr}}\right) P_o \qquad (12.19)$$

Paul, Peterson, and Caplan (1973, 1974) have examined this relation in vascular smooth muscle (VSM) using techniques derived from

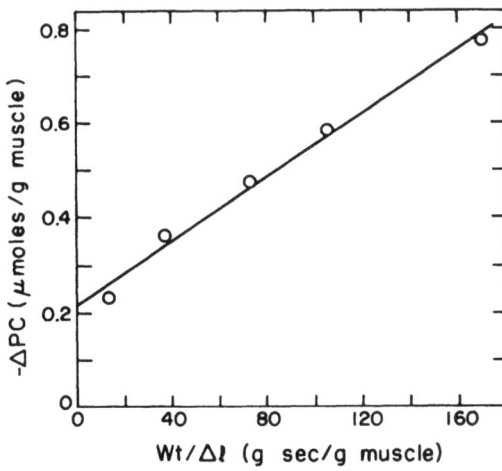

Fig. 12.2. Changes in phosphocreatine content (ΔPC) of frog *rectus abdominis* in "constant shortening" contractions under varying loads. W is the external work done, Δl the distance contracted. (Data of Cain, Infante, and Davies, 1962, according to Bornhorst and Minardi, 1969.)

similar studies in epithelia (Vieira, Caplan, and Essig, 1972). From an experimental point of view VSM offers considerable advantages in mechanochemical studies, since it appears to be one of the slowest of all muscles (Bárány, 1967; Rüegg, 1971), and in contrast to skeletal muscle preparations it contains no large pools of high-energy phosphates (Lundholm and Mohme-Lundholm, 1962; Daemers-Lambert, 1964). Daemers-Lambert and Roland (1967) have demonstrated that ATP, through glycolytic and oxidative phosphorylation, is synthesized continuously during contraction and that a steady state is reached. In the steady state, when the overall concentration of ATP is constant, the rate of ATP hydrolysis by the muscle is equal to the rate of ATP production by intermediary metabolism. The contribution of aerobic glycolysis to the total ATP production is small (Paul, Peterson, and Caplan, 1973, 1974). Thus monitoring of the oxygen consumption rate in these tissues should be a valid measure of the driving chemical reaction. A polarographic method, using Clark-type oxygen electrodes, was developed for this purpose. Figure 12.3 shows simultaneous measurements of J_{ro} and P_o in bovine mesenteric vein under progressively varying stimulation with different stimulants. The linear behavior seen here was observed repeatedly.

Clearly Eq. (12.19) refers only to suprabasal values of J_{ro}; note that the difference between basal J_{ro} as determined directly—prior to stimulation—and the extrapolated intercept of the J_{ro}–P_o regression line is only 4% of the basal value of J_{ro} (and independent of the stimulant used). This negligible difference implies that any contribution to suprabasal J_{ro} from *noncontractile* as well as contractile processes must be proportional to the force generated. If such a contribution is present, it may arise from regulatory functions of intermediary metabolism.

The invariance of the J_{ro}–P_o relation under varying pharmacological stimulus strongly suggests that the slope of the curve indeed represents a measure of an intrinsic mechanochemical coupling coef-

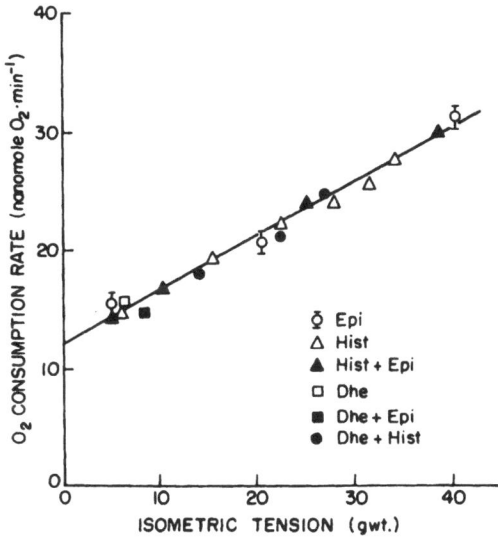

Fig. 12.3. A plot of J_{ro} versus graded P_o for bovine mesenteric vein, using varying levels of the adrenergic stimulant epinephrine (Epi)—norepinephrine shows identical behavior—and the nonadrenergic stimulant histamine (Hist). The effect of the adrenergic α-blocking agent dihydroergotamine (Dhe) is also shown. The regression line drawn is for all points; regression lines for categorized subsets of points are not statistically different. ▲ represents stimulation with maximal concentrations of both Hist and Epi and subsequent dilutions. ● represents maximal stimulation with Hist of the tissue when blocked with Dhe, and subsequent dilutions. Error bars represent ±1 S.E. (From Paul, Peterson, and Caplan, 1974.)

314 Bioenergetics and linear nonequilibrium thermodynamics

ficient. One interpretation of these experiments might be that the value of A is being altered at varying levels of stimulus; however, other interpretations can be made, as will be discussed later.

12.2 Regulation of energy conversion

12.2.1 The Hill equation

A conceptual model. A characteristic force–velocity relation for muscle, which describes accurately the mechanical behavior of muscles of all types from many different species, was discovered experimentally by A. V. Hill (1938). It is generally written:

$$(P + a)(V + b) = (P_o + a)b = (V_m + b)a \qquad (12.20)$$

The quantities a and b are purely mechanical constants (Hill, 1964a), although originally a was identified with the "shortening heat" per unit length of contraction. It follows that the ratios a/P_o and b/V_m are equal; we shall denote them by θ. The most remarkable property of this relation, a property which sets it apart from all other proposed force–velocity relations, is its symmetry (Caplan, 1968a, b). Thus in this respect it contrasts sharply with an earlier relation proposed by Fenn and Marsh (1935):

$$P = P_o \exp(-aV) - kV \qquad (12.21)$$

where a and k are constants, as well as with the somewhat similar exponential expressions proposed by Polissar (1952) and Aubert (1956). The symmetry of Eq. (12.20) emerges clearly if we normalize both tension and velocity. We then obtain:

$$\left(\frac{P}{P_o} + \theta\right)\left(\frac{V}{V_m} + \theta\right) = (1 + \theta)\,\theta \qquad (12.22)$$

It is seen that the functional form of $P(V)$ is identical to that of $V(P)$, a situation reminiscent of the phenomenological relations of NET (the significance of this will become apparent later). The normalized equation represents a family of hyperbolae cutting both axes at unity. The curvature of any given hyperbola depends on the quantity θ, which was singled out by Hill (1938) as an index of performance for the muscle. Typically θ has a value between 0.2 and 0.3, and the curvature is quite high. Although Eq. (12.20) is set aside from the other relations by virtue of both its symmetry and its algebraic simplicity, all were put forward on a largely empirical basis. The choice among them has been described succinctly by Hill (1965): "The real question as between the four equations is, which is the most useful? Useful for what? Helping to make a complicated mixture of observed facts more digestible, pointing out where to look for new ones and providing a 'take-off' for the next jump into the unknown." In fact, the ubiquitous use of Eq. (12.20) in the literature indicates that for most workers in this field the choice was made long ago. We have, therefore, good reason to take seriously a remark made by Pringle (1961) that it is "worth enquiring if any conceptual model does really attach to the equation."

The nonlinearity of the force–velocity relation suggests the possibility that a regulatory process is at work and that the muscle is able to adjust and match itself to its load. (In this respect, an important property of the Hill equation is that at sufficiently low values of θ the power output remains essentially constant over a wide range of loads, thus "buffering" the rate of doing work against changes, sudden or otherwise, which may occur in the magnitude of the load.) The concept of self-regulation in muscle is by no means new; it was known in principle by Heidenhain as early as 1864 (Hill, 1965) and was originally proposed explicitly by Fenn (1923). Essentially the same idea has been used by Pringle (1961) in postulating a property of the contractile component called "activation," which is increased by tension and which controls the velocity of shortening at a given tension. In phenomenological terms, regulation could be effected by two possible mechanisms, acting either singly or in concert. On the one hand, the phenomenological coefficients might be strong functions of tension or velocity; this would imply a fundamental nonlinearity of the system. On the other hand, the input—in particular the affinity of the driving reaction—might be altered in response to changes in tension.

An analysis of systems of the latter type was carried out by considering a linear energy converter in series with a regulator that modifies the input derived from a primary energy source (Caplan, 1966).

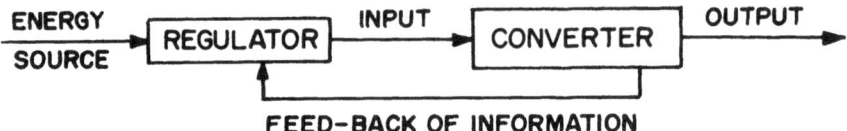

Fig. 12.4. Diagram of a self-regulated energy converter. (From Caplan, 1966.)

The output of the regulator is the input to the converter; nonlinearity is confined to the regulator. Such a scheme is shown diagrammatically in Fig. 12.4; it may be remarked that the stability characteristics of systems of this type have been subjected to extensive study (see, for example, Walter, 1972; Stucki, 1978). It will be noticed that in the scheme under consideration, the feedback to the regulator comes from the converter, not from an arbitrary point beyond the output "terminals." This is an essential property of the system: the regulator has no source of information about the load other than the performance of the converter itself. It was found (see the appendix to this chapter) that if the energy converter is incompletely coupled, operates between specified static-head and level-flow limits, and gives a unique adaptive response to any imposed load, then its behavior can always be described by the same general expression, which may be regarded as the canonical regulator equation. In its simplest form this expression reduces to the Hill force–velocity relation. This interpretation leads to a rather striking conclusion: the degree of coupling of the converter is given by

$$q = \frac{1}{\sqrt{1+\theta}} \qquad (12.23)$$

In this view Eq. (12.22) has only one adjustable parameter: the degree of coupling, which may be determined from the curvature of the force–velocity curve. Hill (1938) showed that θ is usually about 0.25 in frog sartorius. This gives a degree of coupling of about 90%, corresponding to a maximum efficiency of about 40%. The definition of the mechanical (better termed mechanochemical) efficiency follows, of course, immediately from Eq. (12.11):

$$\eta_{\text{mech}} = \frac{VP}{J_r A} \qquad (12.24)$$

As defined above, the efficiency of a given contraction in the steady state may in principle be evaluated experimentally without recourse to heat measurements with their attendant assumptions and approximations. In smooth muscle at least, as shown by the work of Paul and colleagues, three of the parameters should be directly measurable. The fourth, the affinity, requires special consideration but, as will be seen, may also lend itself to experimental evaluation by nondestructive methods.

Experimental evaluation of the affinity in muscle. In contrast with the active transport systems we have examined earlier, the problem here is that we are specifically allowing for the possibility that the affinity may vary in response to the operation of a regulator. Nevertheless, two approaches may be considered which are analogous to methods described in Chapter 7 for determining the affinity in epithelia. These are summarized by the following two operational equations, which are readily derived from (12.12) and (12.13) or (12.14) and (12.15):

$$A_m = -\frac{V_m}{(\partial J_r/\partial P)_{A_m}} \quad \text{(very small loads, in the range } V \text{ close to } V_m\text{)} \quad (12.25a)$$

$$A_o = \frac{P_o}{(\partial J_r/\partial V)_{A_o}} \quad \text{(near maximal loads, in the range } P \text{ close to } P_o\text{)} \quad (12.25b)$$

Thus Eq. (12.25a) would require estimating the variation of, say, oxygen consumption with tension in a range of very light loads over which the affinity corresponding to the operation of the muscle in the unloaded state might not be expected to vary appreciably, while Eq. (12.25b) would similarly evaluate the affinity in a range of operation close to isometric contraction. Experiments of this type have not yet been done and would be of tremendous interest in indicating whether indeed there is considerable variation in the affinity between level flow and static head. However, these values of the affinity, A_m and A_o, refer only to the two limiting states. In any steady state other than the two referred to by Eqs. (12.25a) and (12.25b), the affinity may be determined in principle by the more general formula also derivable from the phenomenological equations:

$$A = \frac{P}{(\partial J_r/\partial V)_A} - \frac{V}{(\partial J_r/\partial P)_A} \quad \text{(small variations of load around a given state)} \quad (12.26)$$

318 Bioenergetics and linear nonequilibrium thermodynamics

Here it is required that oxygen consumption rate and velocity of contraction be determined simultaneously for sufficiently small variations of load around a given state characterized by the corresponding values of P and V. If A remains essentially constant for such small perturbations, and if the sensitivities of the measuring devices are adequate, it should prove possible to evaluate A.

12.2.2 The muscles as an autonomic energy converter

"Affinity" regulation: regulation based on energetic factors. According to the view taken above (Fig. 12.4), the force–velocity relation (Eq. 12.20) contains only a single adjustable parameter when normalized—the degree of coupling of the working element of the muscle. The tighter the coupling, the greater the curvature of the Hill hyperbolae; this is readily seen on the normalized plots shown in Fig. 12.5. As mentioned earlier, a characteristic of the Hill equation is that at *high* degrees of coupling the power output remains essentially constant over a considerable range of load resistance. This is roughly the case in muscle, as was pointed out by Fenn and Marsh (1935). It was also mentioned that in frog sartorius θ has a mean value of about 0.25, which corresponds to a q of 0.89 and an η_{max} of 38%. On the basis of work and heat measurements, using a Levin-Wyman ergometer, which maintains constant shortening speed, Hill found the

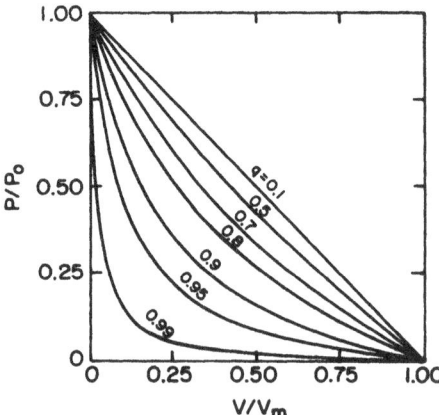

Fig. 12.5. Normalized Hill plots at different degrees of coupling q. (From Caplan, 1966.)

maximum "mechanical efficiency" of frog sartorius to be 39.4% (Hill, 1939; Hill and Woledge, 1962). Later he obtained a maximum value of about 45% for maintained isotonic contractions (Hill, 1964b). This corresponds to a q of 0.92 and a θ of 0.17; usually in these sartorii the range of θ extends from 0.2 to 0.3.

Caution must be exercised, however, in interpreting the measurements quoted above, since they refer to the ratio $PV/(\dot{Q} + PV)$ where $\dot{Q}$ is the observed rate of heat production of the muscle. On the other hand, remembering Eq. (12.6) or (12.11), the efficiency is the ratio $PV/(\Phi_{mech} + PV)$. As Φ_{mech} and $\dot{Q}$ are not identical, the two ratios do not coincide. The former *may* be a reasonable approximation to the latter, providing (1) the heat production associated with Φ_{chem} is negligible or can be corrected for, (2) the muscle is virtually a closed system during the experiment, and (3) the absolute value of $dS^{(1)}/dt$ is negligibly small in comparison with that of $dS^{(2)}/dt$, since d_iS/dt is given by the sum of these quantities (see Eqs. 12.1–12.4).

The *inverse* Hill equation, relating the affinity and velocity of the driving reaction, follows from the phenomenological equations and Eqs. (12.22) and (12.23) (Caplan, 1968a):

$$\left(\frac{A}{A_m} - \frac{J_r - J_{ro}}{J_{rm}}\right)\left(\frac{J_r}{J_{ro}} - \frac{A - A_m}{A_o}\right) = 1 \qquad (12.27)$$

As before, the subscripts o and m refer to isometric and unloaded tetanic contractions, respectively. The degree of coupling is implicit in this equation, since the constants are related to one another through q^2:

$$\frac{A_m}{A_o} \cdot \frac{J_{ro}}{J_{rm}} = 1 - q^2 \qquad (12.28)$$

Equation (12.27) may demand rather large variations in the affinity, especially if the degree of coupling is high. The implication is that the regulator brings about large changes in the activities of the reactants or products (or both) of the driving reaction. Clearly this could only take place locally. The possibility that substantial changes occur in the local concentrations or activities of adenine nucleotides, that is, that they participate in muscular contraction in a "bound" or compartmentalized form, has been raised by several workers (Mommaerts, Brady, and Abbott, 1961; Seraydarian, Mommaerts, and Wallner, 1962; Yagi and Mase, 1962, 1965). Other instances have also been cited in which metabolic control by compartmentation of ATP, ADP, and phosphate appears to be important (Maitra and Chance, 1965).

The mechanism of the type of regulation envisaged here is best appreciated in terms of a scheme such as the one used to discuss active transport of sodium in the appendix to Chapter 7. Those steps in the reaction sequence which are indicated as being directly coupled to the sodium pump are now to be considered coupled to the contractile mechanism: this is the linear energy converter. The remaining steps constitute the nonlinear regulator; the activities of intermediates in these steps are assumed to be sensitive to tension or velocity. Hydrolysis of the main phosphocreatine pool corresponds to the whole (or a large part) of the sequences shown, while intermediate steps may correspond to local compartmentation. Under stationary conditions, stationary-state coupling applies to the whole chain of reactions, that is, to both regulator and converter. The total affinity for the change $aA + bB \rightarrow yY + zZ$ is regarded as constant (fixed, for example, by the phosphocreatine pool), but that portion of the affinity responsible for the change $mM + nN \rightarrow pP + qQ$—the immediate driving reaction—is determined by the regulator according to the mechanical load. When the load is changed, a corresponding change occurs in some or all of the chemical potentials of the intermediates in the series C to X. If the affinity across the converter sequence should increase, then the affinity across the regulator sequences must decrease (and vice versa). A possible explanation of how the force exerted by the applied load could influence the affinity of the driving reaction, which hinges on the reversible binding of calcium ions to tropomyosin, has been put forward by Oplatka (1972).

It should be stressed that the mode of regulation discussed here is very different from that which has been considered, perhaps implicitly, in the discussion of epithelial active transport in Chapter 7. In the epithelial system the function of the "regulator" is to maintain the affinity of some intermediary pool of reactants, possibly ATP, ADP, and P_i, *constant:* this is the conclusion to which the available experimental evidence points. In muscle the requirements are different, and it seems quite probable that the affinity varies to match the load.

Detailed criticisms of this approach have been offered by Wilkie and Woledge (1967). In support of it they found that not only is the correct force–velocity relation obtained, but an experimentally observed relation between maximum efficiency and maximum normalized power output (for muscles of several species) appears to be similar to the predicted one. However, they also concluded that the relation between the rate of the driving reaction and the load is unlike the predicted one, and that the predicted variations in free energy of the driving reaction are so large that they could not actually occur. Caplan (1968b) pointed out that these objections rest on a number of questionable assumptions, including (1) that the velocity of the driving reaction during steady-state shortening can be inferred from the observed rates of heat and work production and the experimental estimates of the enthalpy of creatine phosphate breakdown, and (2) that the affinity of the driving reaction depends on the concentrations of the reactants and products as determined by chemical analysis of the whole muscle. Bornhorst and Minardi (1969) were able to evaluate the affinity of the driving reaction from the data of Cain, Infante, and Davies (1962) on frog rectus abdominis with the following result: $A_o = 10.6$ kcal/mole ATP, $A_m = 7.5$ kcal/mole ATP. The maximum value is within the range expected by Wilkie and Woledge (1967). (Wilson et al., 1974, found an affinity of 11.4 kcal/mole ATP for ATP utilization under conditions obtaining in isolated rat liver cells.) More recently Wilkie and coworkers (Gilbert, Kretzschmar, and Wilkie, 1973; Curtin et al., 1974) concluded that in tetanic contractions of frog sartorius muscle the rate of phosphocreatine splitting at any instant is generally *not* proportional to the rate of production of (heat + work) at that instant; the former, therefore, cannot be inferred directly from the latter. On the other hand, these investigators showed that the splitting of phosphocreatine is specifically linked to the performance of work, and its breakdown is sufficient to account for the work evolved. The improper inference occurs not only in the critique of Wilkie and Woledge (1967), but also in a number of other theoretical studies, notably (as pointed out by Wilkie and coworkers) those of A. F. Huxley (1957), Bornhorst and Minardi (1970a), and Chaplain and Frommelt (1971).

The chemical reaction scheme outlined earlier as a mechanism for

322   Bioenergetics and linear nonequilibrium thermodynamics

regulation implies that if we write A_{tot} for the total fixed affinity of the overall reaction sequence, then

$$A_{tot} - A = A_{reg} \tag{12.29}$$

where A is the affinity across the converter sequence, that is, the input to the linear converter, and A_{reg} is the combined affinity across the nonlinear regulator sequences. If we write

$$A_{reg} = R_{reg} J_r \tag{12.30}$$

where the nonlinearity is expressed in the fact that R_{reg} is implicitly, if not explicitly, a function of J_r, we may replace Eq. (12.15) by

$$A_{tot} = R_{cr} V + (R_r + R_{reg}) J_r \tag{12.31}$$

The degree of coupling of the *overall* system (regulator plus converter) in some condition of operation is then

$$q_{tot} = \frac{-R_{cr}}{\sqrt{R_c(R_r + R_{reg})}} \tag{12.32}$$

and clearly $q_{tot} < q$, the degree of coupling of the converter, given as usual by

$$q = \frac{-R_{cr}}{\sqrt{R_c R_r}} \tag{12.33}$$

Thus for the overall system, in this view, a reduced degree of coupling is the price paid for regulation. The utilization of energy for purely control purposes in bioenergetic pathways is, however, now understood to be a rather general phenomenon (Hess, 1975). One could perhaps view the overall system as an energy converter with a variable degree of coupling, and look upon q_{tot} as the quantity which adjusts itself to the load. This view would not necessarily require q to be constant, but its constancy is an essential property of the class of energy converters considered here. It is worth noting that if R_{reg} were constant, so that the overall system behaved linearly, regulation of a kind would still occur and A would increase with the load. (Since J_r would then necessarily decrease with increasing load as the system moved from level flow to static head, A_{reg} would fall according to Eq. 12.30. This would lead to a rise in A according to Eq. 12.29 as long as A_{tot} remained constant.) However, such a system would be unnecessarily inefficient, since it requires extremely low values of q_{tot} to achieve anything like a uniform output over a wide range of loads.

"Stoichiometry" regulation: Regulation based on kinetic factors. The regulatory scheme described above has the apparent disadvantage that the regulator is dissipative. Thus the efficiency of the overall system, calculated on the basis of the bulk affinity, must always be lower than that of the energy converter alone. This, as noted, may be the price paid for regulation. However, in an attempt to formulate the problem without this wastage, Bornhorst and Minardi (1970a, b) put forward a phenomenological model operating at constant affinity, regulation being achieved by means of variable coefficients. The variation of the coefficients was related to the sliding-filament model of contracting muscle by assuming, *inter alia,* that (1) the cross bridges, considered as the fundamental subunits, are linear energy converters with constant phenomenological coefficients (this is also assumed in the affinity regulation model), (2) the number of active cross bridges varies with velocity or load, and (3) the Hill force–velocity relation is valid.

Bornhorst and Minardi pointed out that the coefficients of the overall system bear a simple relationship to those of the individual cross bridges when the latter vary in number. If n denotes the number of "activated bridges" in a given stationary state, and in accordance with Eqs. (12.12) and (12.13) L_c^i, L_{cr}^i, and L_r^i are the coefficients of an individual bridge, then (since $V = V^i$, $P = np^i$, $J_r = nJ_r^i$, and $A = A^i$)

$$L_c = L_c^i/n$$

$$L_{cr} = L_{cr}^i$$

$$L_r = nL_r^i \tag{12.34}$$

Providing n has the appropriate force or velocity dependence, this system will follow the Hill equation. In terms of velocity, this dependence is

$$\frac{n}{n_o} = \frac{a/P_o}{V/V_m + a/P_o} \tag{12.35}$$

No *a priori* basis is given for Eq. (12.35). It is readily seen that the following relations hold:

$$q = q^i \tag{12.36}$$

$$Z = Z^i/n \tag{12.37}$$

where $Z = \sqrt{L_c/L_r}$ as usual (see Chapter 4). Thus the degree of coupling of the overall system is constant, whereas Z varies. (Since qZ is a measure of the apparent stoichiometry of the system V_m/J_{rm}, "stoichiometry regulation" seems to be an apposite description of this process.) In contrast, the affinity regulation considered earlier is equivalent to having a *variable degree of coupling* for the overall system (regulator plus converter). It is interesting to note that any energy converter for which q is constant but Z varies with the force or flow ratio will still have a maximum efficiency given by $q^2/(1 + \sqrt{1 - q^2})^2$.

In the stoichiometry regulation model, the degree of coupling is no longer simply related to θ, being given instead by

$$q = \frac{1}{\sqrt{1 + (AJ_{ro}/P_o V_m)}} \tag{12.38}$$

The predicted variation of reaction rate with tension is also significantly different from that predicted by the affinity regulation model. In the case of the isometric contractions studied by Paul and colleagues (Fig. 12.3), the stoichiometry regulation model suggests that the effect of stimulant is to alter n_o rather than A, since from Eqs. (12.12), (12.34), and (12.35)

$$P_o = \left(\frac{L_{cr}^i}{L_c^i}\right) n_o A \tag{12.39}$$

To establish the validity of either model requires an analysis of chemical data over a wide range of conditions. As a start in this direction Bornhorst and Minardi (1970b) have extended both models to take explicit account of the effects of length variation.

Bornhorst and Minardi (1970a) observed that both types of regulation can be represented in terms of a simple electrical analogue. Affinity regulation corresponds to a battery driving a motor in series with a variable resistor, while stoichiometry regulation corresponds to a battery driving a large number of motors in parallel, all of which are geared to the same mechanical load. In the latter case each of the motors may be switched in or out of circuit at will, control being dependent on the number of switches closed at any instant. The closing of the switches (which is equivalent to establishing active cross bridges) appears to be nondissipative in this analogue, but this is illusory since no real process can take place without accompanying dissipation. However, the dissipation caused by stoichiometry regulation may well be, in principle, much less than that caused by affinity regulation.

12.2.3 Relationship of the phenomenological coefficients to molecular parameters

It has been known for a long time that muscle maintains a fairly uniform output power over a wide range of loads. However, for very small or very large loads the output power cannot be of any consequence. In the important limiting states, isometric and unloaded contraction, the fact that output and efficiency are both zero is of no significance. We have commented on this earlier in our general discussion of static head and level flow; what is important in these states for muscle is the ability to maintain tension or to bring about rapid contraction. Indeed it can be shown that if the Hill force–velocity relation reflects affinity regulation, then both the tension and velocity efficacies (compare Eqs. 6.23 and 6.24) remain high over appreciable ranges (Essig and Caplan, 1970).

These considerations have a direct bearing on the results of studies of the mechanochemistry of vertebrate skeletal muscle by Oplatka (1972). Oplatka finds that the equations of state for these muscles can be written in terms of three universal molecular constants: (1) the average molecular contractile force $\bar{\phi}_0$ developed during the lifetime of an active complex in an *isometric* contraction, (2) the relative movement Δl_m of half a thick filament (molecular "jump" or step distance) following the splitting by each of its subfragment-1 subunits of one ATP molecule in an *unloaded* contraction, and (3) the ratio of reaction velocities under unloaded and isometric conditions (the significance of this ratio was also stressed by Caplan, 1968a, and Bornhorst and Minardi, 1969). Writing the phenomenological equations for a single active unit in terms of resistance coefficients (compare Eqs. 12.14 and 12.15), Oplatka found that

$$R_{cr} = -\Delta l_m R_c = -\left(\frac{\bar{\phi}_o}{\bar{a}_o}\right) R_r \qquad (12.40)$$

where $\bar{a}_o$ is the affinity per molecule of ATP hydrolyzed in an isometric contraction. Oplatka pointed out that $\bar{a}_o$ must be practically the same for different striated muscles, and therefore the degree of coupling should be another universal molecular constant:

$$q = \sqrt{\frac{\bar{\phi}_o \Delta l_m}{\bar{a}_o}} \qquad (12.41)$$

Substituting values derived from the literature for the molecular parameters, Oplatka calculated q to be approximately 0.9, in agreement with the estimate made by Caplan (1966) on the basis of Eq. (12.23).

Equation (12.40) may also be used to find the phenomenological stoichiometry Z, which should be yet another universal molecular constant:

$$Z = \sqrt{\frac{\bar{a}_o \Delta l_m}{\bar{\phi}_o}} \qquad (12.42)$$

As expected, the apparent stoichiometry in an unloaded contraction is $qZ = \Delta l_m$.

12.3 Some general considerations

12.3.1 The usefulness of NET

The classical approach to the energetics of muscular contraction has always been, and still is, expressed in terms of classical thermodynamics. Carlson and Wilkie (1974) specifically request their readers to note that neither of the laws of classical thermodynamics restricts application to systems at equilibrium, and they go on to point out that the two laws apply to all systems whether they are close to equilibrium or not. They further point out that this generally is the basis of the usefulness of thermodynamics in biology, for in living systems we are almost always concerned with transformations of chemical energy by processes that are not reversible, under conditions that are far from equilibrium.

This observation is perfectly correct, but in fact it underlines the *necessity* for treating the energetics of the contractile system in terms of NET, which of course is as soundly rooted in the first and second laws as classical thermodynamics, being essentially a powerful extension of it designed specifically to deal with systems away from equilibrium. Instead of considering only the initial and final states corresponding to some change in a system, NET deals explicitly with the *process* of change itself. In contrast to the assumption, implicit in the classical thermodynamic treatment of muscle, that the system is closed, no such assumption is suggested or required by NET, which is particularly well adapted to the study of open systems.

However, perhaps the most important contribution of NET is its ability to represent the muscle as a two-flow system. This information cannot be expressed in a classical treatment precisely because it deals with states rather than processes. For example, the classical treatment of the electrochemical cell invariably assumes complete coupling between electrical current and chemical reaction; in other words, the treatment is essentially that of a one-flow system (compare Eq. 6.20). The phenomenological equations (12.12) and (12.13), or (12.14) and (12.15), constitute in our view the simplest possible description of the system consistent with the two degrees of freedom it possesses.

Below are listed four mechanochemical characteristics that have given rise to conceptual difficulties from time to time, but in terms of the NET description are readily explicable.

1. *Variation of efficiency with velocity*. It has been stated that operationally, the variation of efficiency with velocity means there is a variable coupling coefficient between the work and ATP splitting by the mechanical generator (Kushmerick and Davies, 1969). However, as seen in Chapter 4, such variation is in fact not unexpected but will

occur in a predictable manner in a system of fixed degree of coupling as defined here.

2. *Energy expenditure in isometric contractions.* It has been claimed that there may be a significant amount of work being done *internally* during a brief isometric contraction (Kushmerick and Davies, 1969), this being suggested by the demonstration (Infante, Klaupiks, and Davies, 1964b) that ATP utilization is mainly associated with external work done in isotonic contractions, and by the extrapolation of this observation to isometric contractions, assuming a "unitary concept" of the mechanism of mechanochemical coupling. However, in a thermodynamic sense the notion of internal work is inherently unsound, since work is a mode of energy transfer *between* systems and is identifiable by the fact that it may always be expressed in terms of raising a weight. In an incompletely coupled system, static head is associated with pure dissipation, and no work is done. The system expends energy simply in order to *maintain* this state.

3. *Slowdown of ATP hydrolysis: active stretch.* Curtin and Davies (1973, 1975) showed that at low velocities of stretching, the rate of ATP breakdown was very much lower than during isometric contraction. Furthermore, they concluded that during stretching, crossbridge links can form and maintain a force over their range of movement without ATP breakdown. They therefore argued that "any theory of muscle contraction must be able to account for the ability of the muscle to develop tension even greater than P_0 and to maintain it over long distances of stretching with very little breakdown of ATP" (Curtin and Davies, 1973). However, Eqs. (12.12) and (12.13) show clearly that if the muscle is stretched under tetanic stimulation, that is, if V is caused to become negative, then $P > P_0$ (we may assume that under these conditions $A \simeq A_0$ since, as shown in the appendix to this chapter, affinity regulation implies that A_0 is a limiting value), and $J_r < J_{r0}$. In principle, if the phenomenological coefficients and the affinity were to remain essentially constant under such strenuous conditions, J_r would be reduced to zero at an appropriately high value of P and at still higher values would become negative—that is, ATP would be synthesized. This extreme behavior has never been observed.

4. *The "catch" mechanism.* Certain molluscan muscles are able to "set" (see, for example, Carlson and Wilkie, 1974) while maintaining tension so that they behave like a stretched elastic body and can con-

tinue to exert tension between fixed points for a very small expenditure of energy. The energetics are therefore consistent with the notion that in this type of muscle the contractile system is either fully or extremely highly coupled (Caplan, 1968b). In this view no additional catch mechanism is necessary: a completely coupled muscle would consume no metabolic energy (above basal) during a steady-state isometric contraction.

12.3.2 Implications of self-regulation

The regulator which we here suppose to be present in muscle appears to correspond exactly to the idea Fenn had in mind when he wrote, many years ago, "There is another regulatory mechanism within each individual fibre which, within certain limits, is able to adapt the energy output to the work done . . . The other mechanism is independent of the nervous system and works merely by virtue of the fundamental nature of the muscle machine, whatever that may be . . . The energy liberated . . . can be modified by the nature of the load which the muscle discovers it must lift" (Fenn, 1923). The processes involved in regulation are accounted for in Φ_{chem}, Eq. (12.7). Small changes in affinity correspond to quite large changes in activity, and these can be envisaged only in terms of *local* activities at the reaction sites. Alberty (1968) has pointed out the strong influence of both magnesium and calcium on the standard affinity for ATP hydrolysis, and in addition a possible pH effect cannot be excluded. In view of the important role played by calcium in excitation, it is quite conceivable that the local affinity could be regulated in some way by the calcium level, in accordance with the suggestion of Oplatka (1972). This view is not inconsistent with the theory of Pringle (1967). (It has been suggested by Cohen and Longley, 1966, that reversible binding of bivalent cations by tropomyosin may have a regulatory function in muscle contraction.) Pertinent discussions of feedback regulation at the molecular level and of the interaction between metabolite-modulated enzymes have been given by Atkinson (1965) and Hess (1975).

If the considerations which lead to Eq. (12.23) are really the basis of the observed mechanochemical behavior of muscle, it is an example of a regulated linear energy converter with constant coefficients. Alternatively, as discussed, one might consider muscle to be an energy converter with variable coefficients. However, the difference between these two descriptions may well be found to lie only in the choice of the "black box" or working element; it would certainly be surprising if a system conforming to the second description and characterized by the Hill equation could not be reinterpreted in terms of the first. This question could be resolved by measurement of A according to Eqs. (12.25)–(12.27).

The idea of self-adaptive regulation requires that the velocity of shortening depend uniquely on the tension, an assumption which is well established by mechanical observations, as for example in the elegant method described by Macpherson (1953) for determining the force–velocity relation from the initial stages of two isometric contractions. Another method for determining the force–velocity relation, which has not been used (to the best of our knowledge), would be to allow the muscle to contract against a series of purely dissipative mechanical resistances, or dashpots. The regulator would then be called upon to respond directly to the mechanically resistive load $R_L = P/V$, instead of to P alone as in an isotonic contraction or to V alone as in a Levin-Wyman ergometer. The ability to respond in such a way as to give rise to the appropriate Hill curve is a necessary consequence of the model of Fig. 12.4, as shown in the appendix to this chapter. Among the relations which describe the chemical behavior of the system, the linear dependence of tension on both velocity of contraction and velocity of reaction (Eq. 12.14) should be readily testable. This relation has already been partially tested in the work of Paul, Peterson, and Caplan (1973, 1974); it remains to be tested for isotonic contractions. If linearity can be established, it is possible, in principle, to determine all the parameters in Eqs. (12.14) and (12.15), as q is known from the force–velocity curve.

12.4 Conclusions

The muscle fibril may be regarded as a working element characterized by a dissipation function of two terms: mechanical power output and chemical power input. There is a considerable body of evidence that during at least the initial phase of isometric or isotonic tetanic contraction the fibril attains an essentially stationary state. (Such stationarity probably occurs frequently *in vivo*, as in the demonstration by Wilkie, 1950, of the Hill force–velocity relation in human muscle.)

The mechanochemically coupled hydrolysis of ATP in muscle may well comprise a lengthy series of intermediate steps, each of low affinity. Some of the steps must represent transport of reactants and products into and out of local "compartments." For this reason the application of linear phenomenological relations to the initial phase of tetanic contraction may possibly not be unjustified. In any case, in order to proceed further we are led to invoke the concept of a proper pathway for the mechanochemical process. This allows us to treat the fibril as a linear energy converter. Its degree of coupling (Chapter 4) must be less than unity, as shown, for example, by the continued breakdown of ATP during isometric tetanic contraction.

The muscle may in principle belong to the class of simple autonomic (self-regulated) energy converters (Caplan, 1966), since the mechanochemical process may be linear and apparently satisfies the requirements of stationarity and incomplete coupling. If so, the Hill force–velocity relation contains only a *single* adjustable parameter—the degree of coupling.

The type of regulation which characterizes this class of energy converters may be described as affinity regulation, based on energetic factors. This may be contrasted with stoichiometry regulation, based on kinetic factors.

The degree of coupling q and the phenomenological stoichiometry Z of the mechanochemical process may both be expressed in terms of three molecular parameters of the mechanochemical process: the average molecular contractile force developed during the lifetime of an active complex in an isometric contraction, the relative movement of half a thick filament following the splitting by each of its subfragment-1 subunits of one ATP molecule in an unloaded contraction, and the affinity per molecule of ATP hydrolyzed in an isometric contraction.

The chemical input of the muscle may be expected to follow the inverse Hill equation (Eq. 12.27). This demands a considerable variation in the affinity of the reaction if the degree of coupling is high. The corresponding activity changes could only be brought about *locally* by the regulator.

Values of the affinity at any given tension are in principle determinable, in smooth muscle, from simultaneous measurements of tension, oxygen consumption rate, and rate of contraction. Such values should provide a useful comparison with chemical estimates on whole muscle (see, for example, Kushmerick and Davies, 1969; Curtin et al., 1974) in which the assumptions involved are at some remove from direct experiment.

12.5 Summary

1. The dissipation function for a contracting muscle is derived by considering the fibrils as working elements. It appears to be divisible into two separate positive-definite parts, one relating to mechanochemical processes and the other to purely chemical processes. Evidence is presented that for the purposes at hand the fibrils may be regarded as if they were in a stationary state during contraction.

2. Expressing the "mechanochemical" dissipation in the form $\Phi_{\text{mech}} = V(-P) + J_r A$ where, in the customary notation introduced by A. V. Hill, P is the tension and V the initial constant velocity in tetanic shortening (the remaining symbols have their usual significance), the phenomenological relations are readily written in terms of either L- or R-coefficients. Experimental evidence is presented in support of the assumed linearity of these equations. The relationship of the phenomenological coefficients to molecular parameters is briefly discussed.

3. The characteristic hyperbolic force–velocity relation for muscle (the Hill equation) is shown to lead to a conceptual model based on the notion of self-regulation (a theoretical analysis of self-regulated linear energy conversion is given in the appendix). In this view the Hill equation has only one adjustable parameter: the degree of coupling, which may be determined from the curvature of the force–velocity characteristic. Thus for frog sartorius one finds a degree of coupling of about 90%, corresponding to a maximum efficiency of about 40%.

4. In contrast to the active transport systems previously examined, the affinity in muscle is expected to vary with different loads in response to the operation of a regulator. Methods for experimentally evaluating the affinity in a given state are suggested.

5. The nature of the regulatory process is examined, and several significant implications of self-regulation are pointed out. Regula-

tion on the basis of energetic factors (affinity regulation) is compared with regulation on the basis of kinetic factors (stoichiometry regulation). At present the validity of either model is not established.

6. The usefulness of a nonequilibrium thermodynamic approach to the understanding of muscle is discussed in terms of the following mechanochemical characteristics: variation of efficiency with velocity, energy expenditure in isometric contractions, slowdown of ATP hydrolysis during active stretch, and the economical use of energy by "catch" mechanisms.

7. An interpretation of both the degree of coupling and the phenomenological stoichiometry of the mechanochemical process is given in terms of molecular parameters.

Appendix: Derivation of the Hill equation for self-regulated linear energy converters

Optimal load

The subject of energy conversion in linear systems—systems which can be described by the linear phenomenological relations of nonequilibrium thermodynamics—has been discussed in Chapter 4, where it was shown that the degree of coupling is of considerable utility in describing such systems. In most cases it can in principle be measured, and this enables one to calculate without further information the maximum efficiency of the system, as well as the efficiency at maximum output. The latter quantity is frequently of greater interest and has the same value whether the system is driven by a fixed force (as a "constant-voltage source") or by a fixed flow (as a "constant-current source"). The power output of a reasonably tightly coupled converter in these two modes of operation is greatest at very different values of the load: the fixed-force mode favors low loads while the fixed-flow mode favors high.

When a linear energy converter is required to produce high output power over a wide range of loads, operation in one of the above modes is not worthwhile. Instead, it pays to equip the converter with a device which regulates the input in response to the load and thus provides for better "matching." Servomechanisms of this kind are well known in engineering practice; for example, heat engines in-

334 Bioenergetics and linear nonequilibrium thermodynamics

variably incorporate automatic governors. In biology, regulatory mechanisms are known to be essential. We call a converter equipped with such a device self-regulated; our purpose is to examine the input–output characteristics of this type of system.

Linear energy converters of the type we have been considering clearly possess two degrees of freedom; that is, any two externally controlled parameters determine a stationary state. One of these is commonly the load resistance, which in itself is sufficient to fix the actual operating efficiency. A generalized load resistance is defined (in the general terms used in Chapter 4) by

$$R_L = \frac{-X_1}{J_1} \qquad (12.\text{A}1)$$

(This expression is positive in the driving region.) In order to achieve maximum efficiency or maximum output power, the load must be suitably matched to the converter. For example, for efficiency the optimal load is given by

$$R_L = R_L^\eta = \frac{1}{L_{11}\sqrt{1-q^2}} \qquad (12.\text{A}2)$$

The quantity $(1/L_{11})$ is the generalized "internal resistance" of the system when X_2 is constant, that is, when the converter is analogous to a voltage source. It has been shown (Kedem and Caplan, 1965) that if we denote by m_R the ratio between load resistance and internal resistance, that is,

$$m_R = R_L L_{11} \qquad (12.\text{A}3)$$

then the efficiency is given by

$$\eta = \frac{q^2}{(1+m_R)\left(1 - q^2 + \dfrac{1}{m_R}\right)} \qquad (12.\text{A}4)$$

Combining Eqs. (12.A2), (12.A3), and (12.A4), we can express η in terms of the ratio R_L/R_L^o:

$$\eta = \frac{q^2}{(1-q^2)} \cdot \frac{1}{\left(1 + \dfrac{R_L}{R_L^o \sqrt{1-q^2}}\right)\left(1 + \dfrac{R_L^o}{R_L \sqrt{1-q^2}}\right)} \quad (12.A5)$$

Equation (12.A5) can also be written as a quadratic in R_L, hence each value of η corresponds to two values of R_L. At $\eta_{\max}$ they coincide.

Input-output diagrams

It is often illuminating to consider graphical representations of the energy converter. The simplest is the output diagram shown in Fig. 12.6, where a number of possible paths across the driving region are plotted as loci in a two-dimensional "output space" (to avoid negative values, $-X_1$ is plotted against J_1). This type of diagram differs from the conventional indicator diagram used to describe the operation of a machine working in cycles. Each point here represents a stationary state, and the corresponding output power is given by the

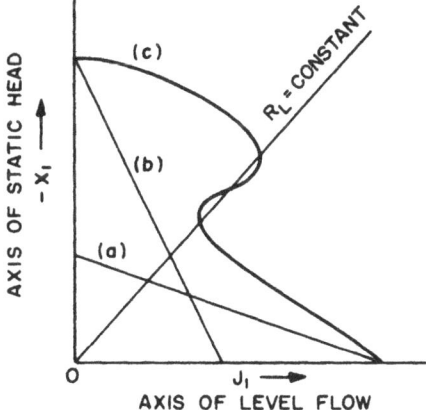

Fig. 12.6. Output diagram, showing some possible loci in output space. An analogous diagram is obtained for input space by plotting the corresponding values of X_2 against J_2. Lines of constant R_L (load lines) radiate from the origin. (From Caplan, 1966.)

area of the rectangle enclosed by the axes, the origin, and the point. A locus merely represents a series of neighboring stationary states through which the system may pass as a result of successive alterations in some external parameter, such as R_L. Its limits occur at the boundaries of the driving region, where R_L is zero or infinity.

At this stage we have no *a priori* grounds for restricting a system to any given locus. In principle all points in output space are available, and the description of a certain path implies knowledge of the way in which the system is, or was, *controlled*. The output diagram therefore gives information about modes of operation of a particular system. In itself, however, it does not distinguish between systems of different q. In contrast, all possible loci of a given system coincide if efficiency is plotted against the ratio of forces (or flows), or if the force ratio is plotted against the flow ratio, since, as shown in Chapter 4, such diagrams distinguish *only* between systems of different q.

A more sophisticated representation is the input–output diagram shown in Fig. 12.7. It arises from the fact that the phenomenological equations, being linear, describe an affine transformation. Thus output space can be mapped onto input space, and vice versa, the characteristic of the transformation being that straight lines remain straight, and parallel lines parallel. This operation simply places every point in output space over its "image" in input space, and hence any curve on the diagram represents a locus in both spaces simultaneously. (Essentially it is a two-dimensional projection of a four-dimensional plot.) The input–output diagram conveys information about both the system itself and its mode of operation; indeed, with suitable scales attached it may be of practical value in operating such converters or in giving a concise graphical description of their behavior. As coupling tends to zero, output space shrinks toward a single straight line; that is, the axes of static head and level flow eventually merge; as coupling tends to completion, the slope of the static head axis tends to infinity. The value of q^2 has a straightforward geometrical significance: any horizontal or vertical line drawn on the diagram is divided in the ratio $q^2/(1 - q^2)$ by the axes. Thus a fraction q^2 of the total length of either of the lines (a) and (b) in Fig. 12.7 lies between the output-space axes. The area outside output space (that is, outside the driving region) represents regions in

Muscular contraction 337

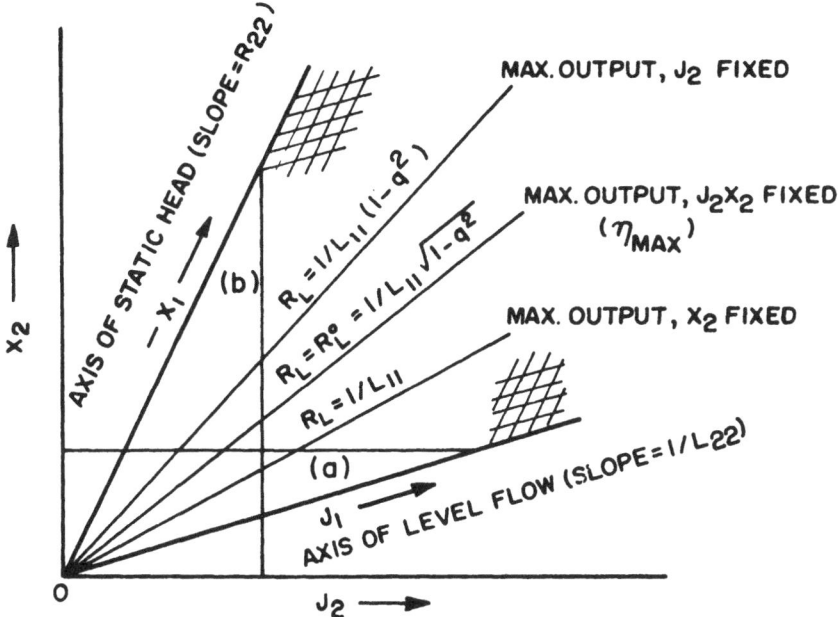

Fig. 12.7. Input–output diagram, showing two of the loci, (a) and (b), which appear in Fig. 12.6. Output space is now mapped onto input space (portions of the transformed output coordinate system are indicated). The three load lines specify optimal matching for different purposes. (From Caplan, 1966.)

which both terms in the dissipation function are positive, and consequently no energy conversion in either direction can take place.

We observed that the control of any two parameters, in other words, the application of two restrictions, completely specifies a stationary state. Generally one restriction is applied on the output side by the nature of the load, while the other is applied on the input side by the nature of the feed. Two common classes of machines are those in which the input is restricted by fixing either the driving force X_2 ("constant voltage source") or the driving flow J_2 ("constant current source"). Typical loci of such machines are given by the lines (a) and (b), respectively, in Figs. 12.6 and 12.7; any given stationary state on a locus specified in this way is determined by a single output restriction. This may be applied by fixing the value of X_1, J_1, or R_L. Each of these possibilities may be represented by a straight line on the input-output diagram that intersects the locus determined by the input restriction at a point corresponding to the stationary state. It is seen that none of the output restrictions mentioned above is capable of determining a locus *across* the driving region.

Three load lines appear in Fig. 12.7, representing three different optimally matched loads. The load at which efficiency is at a maximum is independent of input restrictions, but if maximum output is required, the input restriction becomes important. If X_2 is fixed, the load resistance must match the internal resistance, while if J_2 is fixed, the load conductance must match the internal conductance (Kedem and Caplan, 1965). These are the values shown on the diagram. The ratio of the two loads maximizing output is $(1 - q^2)$ and is thus far from unity if the degree of coupling is high. Indeed, energy converters of the constant-voltage-source type should have low internal resistance, while those of the constant-current-source type should have low internal conductance. The former are consequently ill adapted for operation into loads of high R_L whenever the maintenance of high output power is essential. Similarly, the latter are ill adapted for loads of low R_L.

Self-regulation of energy converters

In order to achieve a high or more uniform output-power distribution over a relatively wide range of loads, other modes of operation may be introduced in place of those described by the loci (a) and (b). It is clear that in discussing operation at fixed X_2 or fixed J_2 we have already postulated the existence of some kind of regulator controlling the input. We now consider the characteristics of a feedback system in which the regulator controls the input *in response to the load*, as indicated in Fig. 12.4. The problem is to find out whether the choice of loci which a system of this kind could be made to follow is limited in any way, and if so to ascertain the nature of the limitation.

On physical grounds it is obvious that the input-output space within which the regulator can function must be bounded on all sides. Thus for any real converter there is evidently a peak input power for each value of the load which cannot be exceeded without damage to the system. It is to be expected that the regulator would adjust output force to a maximum when $R_L = \infty$, so that the system could push hardest when faced with maximum resistance, and it would adjust output flow to a maximum when $R_L = 0$, for a similar reason. The corresponding bounded input–output diagram is shown in Fig. 12.8, in which limiting values of the flows and forces reached at static head are denoted by the superscript s, while limiting values reached at level flow are denoted by the superscript l. It will be realized that although the driving region may be bounded by the lines $-X_1 = -X_1^s$ and $J_1 = J_1^l$, it is equally likely to be bounded by the lines $X_2 = X_2^s$ and $J_2 = J_2^l$. However, the precise nature of the boundary across the driving region is less significant than the two points at which it cuts the static head and level flow axes. At these two points $-X_1$ or J_1 must exceed all other values in the driving region determined by the regulator, in order not to violate a requirement for stability described below. This is also true of any given locus, whether or not static head and level flow lie at the actual working limits specified by the physical nature of the converter. Consequently we regard $-X_1 = -X_1^s$ and $J_1 = J_1^l$ as limits set by the *regulator* to safeguard the converter. It is important to realize that $-X_1^s$ and J_1^l do not just represent static head and level flow for some family

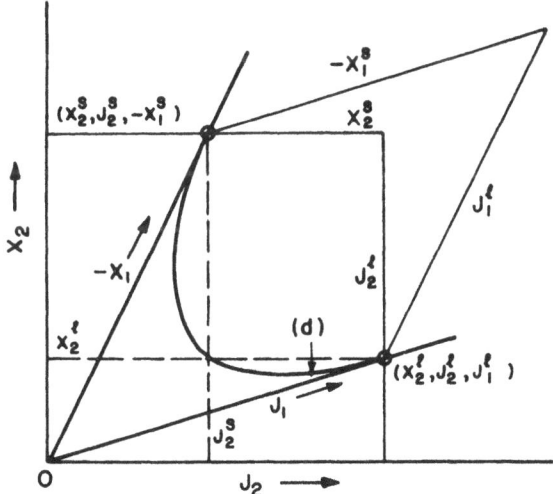

Fig. 12.8. Bounded input-output space, showing a self-regulated locus (d). The peak values of the output force and flow $-X_1^s$ and J_1^l will be reached only at static head and level flow, respectively. The boundaries may equally well be given by X_2^s and J_2^l. (From Caplan, 1966.)

of loci. They represent primarily the maximum values of $-X_1$ and J_1 allowed to the system, and their association with static head and level flow, although justifiable on general grounds, is necessitated by considerations of stability. Nothing in the following analysis excludes the possibilities $X_2^s < X_2^l$, or $J_2^l < J_2^s$, although these situations seem improbable. Note that knowledge of the magnitudes of all the flows and forces in both boundary states, or indeed in any two stationary states within the driving region, gives more than sufficient information to evaluate all the phenomenological coefficients if they are constants.

Once the limits are given within which the regulator is to operate, the problem becomes a matter of investigating the class of permissible loci which run between these limits. We commence by considering the nature of the information about the load which is fed back to the regulator. This depends on the nature of the restriction applied at the output end. In engineering practice an energy converter usually works into a dissipative load; that is, R_L is fixed and the regulator must choose a point on the R_L line. In biology it is more likely to be $-X_1$ that is fixed, or possibly J_1 (in muscle it may be any one of the three). A general requirement is that the system should produce stable regulation whatever the nature of the output restriction applied. It is evident that all the regulator does is impose the input restriction necessary to specify a stationary state. This may be expressed in a number of ways, but it is advantageous to put it explicitly in terms of the output flow and force. For example, one could write simply

$$J_1 = \psi(-X_1) \tag{12.A6}$$

and proceed to study the class of functions ψ which fit the requirement that J_1 should be a *single-valued* function of $-X_1$ with appropriate limits. A further requirement which severely narrows the range of possibilities is that the inverse function

$$-X_1 = \psi^{-1}(J_1) \tag{12.A7}$$

should also be single-valued, which ensures in addition that both $-X_1$ and J_1 will be single-valued functions of R_L. The all-important condition that the functions be single-valued is necessary for stable regulation, since any output restriction applied can then give rise to one specific stationary state only. It will be observed that the functions ψ and ψ^{-1} cannot be single-valued if the values J_1^l or $-X_1^s$ are reached or exceeded elsewhere within the driving region.

Now ψ and ψ^{-1} evidently characterize the performance of the regulator and can be regarded as expressions of the regulator function. However, as functions of output, they may also include the properties of the converter. Since an output restriction is sensed, regulator functions written in terms of input are not considered more fundamental; in any case, one type is readily transformed into the other. In biology the distinction between the two black boxes is unlikely to be clear, even if it exists, and it may be impossible to isolate the input to the converter. (The combined behavior of regulator and converter would then suggest the presence of a single nonlinear black box.) It is preferable, therefore, to represent the regulator in terms of output functions. We assume that it is possible to separate a "programming" function, which can be altered arbitrarily to suit circumstances, from certain elements which are essential to the basic operation of the regulator. More precisely, the assumption is that a canonical expression can be found for the regulator function which takes either of the following forms:

$$J_1 = \psi[-X_1, -X_1^s, J_1^l, q^2, h(R_L)] \tag{12.A8}$$

$$-X_1 = \psi^{-1}[J_1, -X_1^s, J_1^l, q^2, h(R_L)] \tag{12.A9}$$

where $h(R_L)$ is one possible representation of the arbitrary function. Here $-X_1$ and J_1 appear as single-valued implicit functions of one another within the driving region; $h(R_L)$ is a function of both. It is important to bear in mind that an equation of the form of (12.A8) or (12.A9) represents a locus, between fixed boundary points, characterized primarily by the programming function $h(R_L)$.

Included among the essential elements in Eqs. (12.A8) and (12.A9) is the degree of coupling. The regulator function need not necessarily contain the phenomenological coefficients of the converter in an explicit way. But if the performance of the converter is to be judged by the regulator in accordance with an energetic criterion, the degree of coupling can hardly be left out of account. A further consideration is that in response to a restriction (say on $-X_1$) the regulator causes the converter to pass through a series of quasi-stationary states until the appropriate final state is reached. The information fed back may consist of quantities such as $(\partial J_1/\partial X_2)_{-X_1}$ and $(\partial J_2/\partial X_2)_{-X_1}$. From these, all the phenomenological coefficients can be evaluated if q^2 is given. Absolute ratios may also be sensed by the regulator in any given stationary state. In fully coupled systems, however, there are fewer potential sources of information of this kind. The differences between partially and fully coupled converters are profound, as is pointed out in Chapter 4, and we exclude the latter from consideration here.

It is worth emphasizing that single-valuedness of J_1 as a function of $-X_1$ does not necessarily imply the inverse, nor does it imply single-valuedness of either quantity as a function of R_L. This is clearly seen by examining locus (c) in Fig. 12.6. One approach to the problem of covering all conditions of single-valuedness simultaneously is to express output power as a single-valued function of the load, that is, as a function of the form

$$-J_1 X_1 = \phi[R_L, -X_1^s, J_1^l, q^2, h(R_L)] \tag{12.A10}$$

Indeed the fundamental issue in designing a regulator is the nature of ϕ. The class of loci described by Eq. (12.A10) includes all loci for which both ψ and ψ^{-1} are single-valued functions. This class may equally well be described by the following parametric equations, where f and g again represent single-valued functions if ϕ is single-valued:

$$J_1 = f[R_L, -X_1^s, J_1^l, q^2, h(R_L)]$$

$$-X_1 = g[R_L, -X_1^s, J_1^l, q^2, h(R_L)] \tag{12.A11}$$

These two functions are simply related by

$$R_L = \frac{g(R_L)}{f(R_L)} \tag{12.A12}$$

but we shall consider them separately because we are interested in their limits. These are

$$R_L \to \infty: f \to 0, g \to -X_1^s$$

$$R_L \to 0: f \to J_1^l, g \to 0$$

Muscular contraction 343

From this it is clear that the functions f and g may be replaced by normalized functions $\bar{f}$ and $\bar{g}$:

$$J_1 = J_1^l \bar{f}[R_L, -X_1^s, J_1^l, q^2, h(R_L)]$$

$$-X_1 = -X_1^s \bar{g}[R_L, -X_1^s, J_1^l, q^2, h(R_L)] \qquad (12.A13)$$

It is seen that ϕ is a product of two single-valued functions, and that Eq. (12.A10) can be put in the form

$$-J_1 X_1 = -J_1^l X_1^s \bar{f}[R_L, -X_1^s, J_1^l, q^2, h(R_L)]$$
$$\cdot \bar{g}[R_L, -X_1^s, J_1^l, q^2, h(R_L)] \qquad (12.A14)$$

The problem has become one of finding an explicit canonical expression corresponding to Eq. (12.A14). We are helped at this point by the following observation. The quantity $-J_1^l X_1^s$ which appears in Eq. (12.A14) is related uniquely to the input in a certain series of stationary states falling within the limits of the driving region. The relationship is

$$-J_1^l X_1^s = J_2^s X_2^l \frac{q^2}{1-q^2} \qquad (12.A15)$$

and it applies to all states in which the input is equal to $J_2^s X_2^l$. These lie on the hyperbola in input space $\alpha\beta$ shown in Fig. 12.9. For *this* set of stationary states we can write, from Eqs. (12.A5) and (12.A15), remembering the definition of η,

$$-J_1 X_1 = \frac{-J_1^l X_1^s}{\left(1 + \dfrac{R_L}{R_L^o \sqrt{1-q^2}}\right)\left(1 + \dfrac{R_L^o}{R_L \sqrt{1-q^2}}\right)} \qquad (12.A16)$$

Equation (12.A16) indicates immediately how the desired function can be constructed; the right-hand side is nothing other than (input) × (efficiency). Now Eq. (12.A14) embraces all possible loci running between the given limits, and therefore also loci which include stationary states from the set described by Eq. (12.A16). In other words, among the loci described by Eq. (12.A14) are those which meet or intersect the curve $\alpha\beta$. At all such points the right-hand side of Eq. (12.A14) must become identical to the right-hand

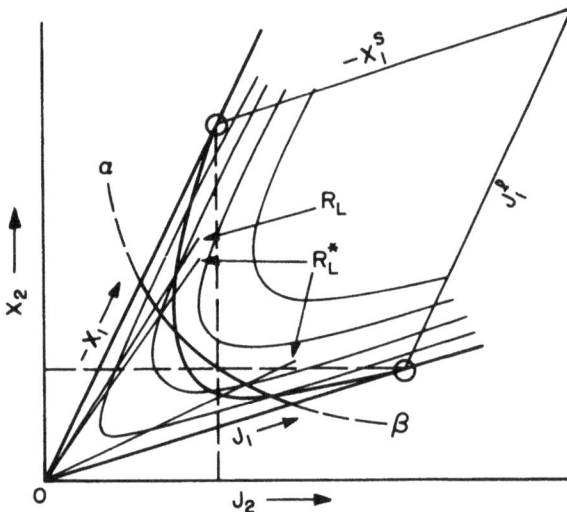

Fig. 12.9. Relationship between peak values of output force and flow and the series of stationary states $\alpha\beta$ described by Eq. (12.A16) (see text). The two load lines marked R_L^* characterize a single constant-output curve (a hyperbola in output space), which passes through their points of intersection with $\alpha\beta$. They are associated with the load line R_L which intersects the self-regulated locus on the same constant-output curve. (From Caplan, 1966.)

side of Eq. (12.A16). This means that every choice of limits for the output flow and force is associated with a unique set of stationary states in which Eq. (12.A14) merely expresses the relationship among output, input, and load resistance. In these states the canonical form of Eq. (12.A14) reduces to a characteristic form involving a simple transfer function, the efficiency, and therefore one would expect that in all other states it must include the essential mathematical structure of the latter.

Each stationary state on the curve $\alpha\beta$ corresponds to a different value of R_L. We can, therefore, characterize any hyperbola in *output* space by a pair of conjugate values of R_L $[R_L^a R_L^b = (R_L^*)^2]$ associated with the points of intersection of the output hyperbola with $\alpha\beta$. Even hyperbolae which fail to intersect $\alpha\beta$—those which represent relatively large constant outputs—are associated with complex or negative conjugate values of R_L, owing to the quadratic nature of Eq. (12.A16). We denote any value of R_L which characterizes a level of output in this way by R_L^*. Consider the representative locus tra-

Muscular contraction   345

versing the driving region between fixed limits shown in Fig. 12.9. It is seen that at each point the locus is intersected by an output hyperbola, so that the actual value of R_L at a point is associated with two values of R_L^*, either of which specify the output. As R_L varies between zero and infinity, one value of R_L^* varies between zero and infinity in the same sense, and the other varies in the opposite sense. A region of complex or negative values of R_L^* may intervene, or even a discontinuity such that R_L^* jumps between two conjugate values at some value of R_L. In the latter case the locus cuts $\alpha\beta$, passes below it, and touches some hyperbola (which represents maximum output for that locus) tangentially at the R_L value in question. Thus R_L^* can be expressed as a function of R_L for any locus. The function may be single-valued or may have "conjugate" discontinuities. Accordingly, we write for any locus, from Eq. (12.A16),

$$-J_1 X_1 = \frac{-J_1^l X_1^s}{\left(1 + \dfrac{R_L^*}{R_L^o \sqrt{1-q^2}}\right)\left(1 + \dfrac{R_L^o}{R_L^* \sqrt{1-q^2}}\right)} \quad (12.\text{A}17)$$

where as a consequence of Eq. (12.A14) R_L^* represents a function of the form

$$R_L^* [R_L, -X_1^s, J_1^l, q^2, h(R_L)]$$

which reduces to R_L, or to $(R_L^o)^2/R_L$, wherever the locus intersects or touches the curve $\alpha\beta$. The limits of R_L^*, although identical to those of R_L, need not necessarily coincide with them. It turns out that whichever way the limits are assumed to lie, Eq. (12.A17) gives rise directly to the two functions $\bar{f}$ and $\bar{g}$ of Eqs. (12.A13) with exactly the required properties. For example, if the limits of R_L^* do coincide with those of R_L, we obtain

$$J_1 = \frac{J_1^l}{1 + \dfrac{R_L^*}{R_L^o \sqrt{1-q^2}}} \cdot \frac{1}{h}$$

$$-X_1 = \frac{-X_1^s}{1 + \dfrac{R_L^o}{R_L^* \sqrt{1-q^2}}} \cdot h \quad (12.\text{A}18)$$

346 Bioenergetics and linear nonequilibrium thermodynamics

where the quantity h, a function of R_L, must be introduced for generality. If it were not, the function R_L^* would be fully defined by Eqs. (12.A1) and (12.A18), and the latter would be the parametric equations of a single locus. Thus h represents the arbitrary part of R_L^*, and we identify it with our previous programming function. It is dimensionless with limits of unity when R_L reaches either zero or infinity. If h is a constant, it must be unity; it can never become zero or infinite, since this would imply zero output at a finite load and also that ψ and ψ^{-1} were not single-valued. The other choice of limits for R_L^* simply interchanges the denominators in Eqs. (12.A18).

Since the right-hand side of Eq. (12.A17) is the function ϕ of Eq. (12.A10), we are now in a position to look for the functions ψ and ψ^{-1} of Eqs. (12.A8) and (12.A9). When Eqs. (12.A18) are combined by addition, and Eqs. (12.A15) and (12.A17) are reintroduced, the parametric function R_L^* is eliminated and the following result appears:

$$\frac{X_1}{X_1^s h} + \frac{J_1 h}{J_1^l} = 1 + \frac{J_1 X_1}{J_2^s X_2^l} \qquad (12.A19)$$

We shall now, for convenience, define two constants:

$$a = \frac{J_2^s X_2^l}{J_1^l} = -X_1^s \left(\frac{1-q^2}{q^2}\right)$$

$$b = \frac{J_2^s X_2^l}{-X_1^s} = J_1^l \left(\frac{1-q^2}{q^2}\right) \qquad (12.A20)$$

Thus

$$q^2 = \frac{-X_1^s}{-X_1^s + a} = \frac{J_1^l}{J_1^l + b} \qquad (12.A21)$$

From Eqs. (12.A19) and (12.A20) we obtain

$$J_1 = b \left(\frac{-X_1^s + X_1/h}{-X_1 + ah}\right)$$

$$-X_1 = a \left(\frac{J_1^l - J_1 h}{J_1 + b/h}\right) \qquad (12.A22)$$

Equations (12.A22) correspond directly to Eqs. (12.A8) and (12.A9).

The regulator program

It is useful to consider the cybernetic interpretation of the equations derived above. To cause a given converter to follow an arbitrarily chosen locus, it is evidently necessary to build a suitable "program" into the regulator. Our contention is that $h(R_L)$ represents the element of choice in designing such a program. For example, it can be used to ensure maximum output at a selected value of the load resistance. Ashby (1963) has pointed out that the act of designing a regulator is an act of communication between the designer (who may be a gene) and the regulator itself. Information describing the regulatory function must pass between them. The constraints of the present problem are such that when $h(R_L)$ is a constant it can only be unity, and its informational content for the program is zero. We shall term this the unmodulated h-function. By modulating the h-function, the designer is able to choose and modify the characteristics of the regulator, although within limits. In so doing he introduces information into the program over and above the minimum required for the regulator to function at all.

The most interesting feature of this analysis is that the locus of the unmodulated h-function corresponds exactly to the force–velocity relation for muscle discovered experimentally by Hill (1938). The Hill equation, Eq. (12.20), may be written in either of the forms

$$V = b \left(\frac{P_o - P}{P + a} \right)$$

$$P = a \left(\frac{V_m - V}{V + b} \right) \qquad (12.A23)$$

On setting h equal to unity, one sees that Eqs. (12.A22) can be identified immediately with Eqs. (12.A23). Rewriting Eqs. (12.A21) in this notation, we obtain

$$q = \sqrt{\frac{P_o}{P_o + a}} = \sqrt{\frac{V_m}{V_m + b}} \qquad (12.A24)$$

(Compare Eq. 12.23.)

13 Energy coupling in mitochondria, chloroplasts, and halophilic bacteria

It is now generally accepted that both in mitochondria and in chloroplasts, coupling occurs among three main processes, all of which are in principle reversible: electron transport, ADP phosphorylation, and proton translocation across the coupling membrane. As Rottenberg (1978) has pointed out, a structural analogy can be drawn between these organelles, although the polarities of the coupling membranes are reversed. Thus many of the differences in ion transport and phosphorylation observed between mitochondria and chloroplasts are due simply to the fact that in mitochondria, proton pumping takes place into the medium, while in chloroplasts it takes place into the thylakoid space. The nature of the coupling in both cases is as yet not fully understood, but a strong belief exists that both mechanisms depend on proton circulation and will eventually prove to be essentially identical. The crucial role of proton circulation in the coupling mechanism has been underscored by the discovery of the light-driven proton pump which drives phosphorylation under anaerobic conditions in certain halophilic bacteria.

13.1 Oxidative phosphorylation and photophosphorylation

Oxidative phosphorylation in mitochondria (and bacteria) is the process by which oxidation of Krebs-cycle intermediates (or other reducing agents), catalyzed by the respiratory system, results in synthesis of ATP (Fig. 13.1). Most thermodynamic treatments have been based on the determination of a "stoichiometric ratio," the ratio between the rates of phosphorylation and oxidation (P/O ratio). Normally, the *maximal* experimental ratio, rounded off to an integer, is considered to represent the stoichiometry which would obtain ideally (for example, 3 with NAD-linked substrates, 2 with succinate); taken together with other evidence it is assumed to indicate the number of coupling sites (Slater, 1966). The free energies of the reactions are compared, and the efficiency of the process computed. The second law of thermodynamics then provides a criterion for the self-consistency of the data. Thus if n is the presumed stoichiometric ratio, it is necessary that

$$-\Delta G_{ox} \geq n\Delta G_{phos} \qquad (13.1)$$

or in other words, the efficiency obeys the relation

$$-n\Delta G_{phos}/\Delta G_{ox} \leq 1 \qquad (13.2)$$

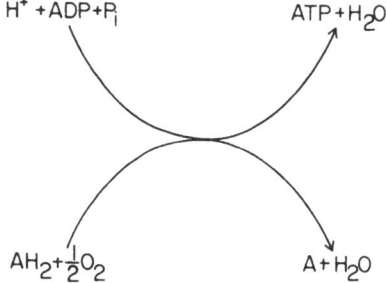

Fig. 13.1. General scheme of oxidative phosphorylation.

This consideration is also used to examine the possibility of coupling in different regions of the respiratory chain. Thus, assuming $n = 1$ at a given site, the oxidoreduction potentials (free energies) are compared with the phosphate potential (Lehninger, 1964; Slater, 1966). On thermodynamic grounds, however, n need not necessarily equal 1 at a given site; it is conceivable that oxidative phosphorylation might occur at sites not ordinarily considered, with values of n less than 1. For a detailed account of the thermodynamics of oxidation-reduction reactions (both equilibrium and nonequilibrium), the reader is referred to the excellent review by Walz (1979).

As was pointed out by Rottenberg, Caplan, and Essig (1967), the above considerations are unambiguous only if there is complete coupling. Since this is obviously not the case,[1] stoichiometric ratios cannot be discussed without reference to the state of the system; otherwise, such calculations only accidentally reflect its physical parameters. It is noteworthy that the determinations of n, ΔG_{ox}, and ΔG_{phos} are seldom performed simultaneously, and calculations are based on extrapolated and estimated values not necessarily referring to the same state.

Similarly, there are inadequacies in the usual energetic analyses of data obtained in state 4 (oxidation in the absence of phosphorylation owing to a limiting amount of ADP; Chance and Williams, 1956) and state 3 (steady-state phosphorylation after the addition of ADP). The classical treatment calculates the "respiratory control ratio," that is, the ratio of the oxidation rate in state 3 to that in state 4 (Chance and Williams, 1956), but although this is taken to be an indicator of the tightness of coupling, it does not make use of all relevant information. It is clear that a system in state 4, in which the P/O ratio is zero, is at static head and accordingly reflects the operation of the energy conversion mechanism just as well as a system in state 3.

Photophosphorylation in chloroplasts (and chromatophores) shows strong parallels to oxidative phosphorylation. Thus when a suspension of broken or Class II chloroplasts (those in which the envelope is broken, giving free access of metabolites and reagents to the mostly intact grana membranes) is illuminated in a medium containing ADP and P_i, ATP is synthesized (Arnon, Allen, and Whatley, 1954). As in mitochondria, this synthesis is coupled to electron transport, which here occurs primarily from H_2O to NADP, driven by the radiant energy absorbed in two distinct photosystems. En route, the electrons are caused to flow spontaneously down a chain of redox enzymes similar to the respiratory chain, resulting likewise in the generation of ATP. However, the observed P/2e$^-$ ratio seldom exceeds 1 (Avron, 1971).

A more immediate form of photophosphorylation is manifested by the extreme halophile *Halobacterium halobium*, an obligate aerobe which can survive at low oxygen levels (and eventually even in the absence of oxygen), providing it is exposed to light. Under these conditions it synthesizes patches containing a purple pigment scattered throughout the cell membrane. The pigment replaces the electron transport system by acting as a proton pump driven directly by the light.

Rottenberg (1978) points out that the near-identity between uncouplers of mitochondria and uncouplers of chloroplasts suggests a similar mechanism of coupling. We have not as yet commented on the coupling mechanism in oxidative phosphorylation. Although one should bear in mind that the coupling mechanisms in mitochondria and in chloroplasts could ultimately turn out to be quite different, this seems unlikely; a discussion of mechanism need not therefore refer explicitly to either system. However, for convenience as well as clarity, our discussion as far as these organelles are concerned will focus on mitochondria. Although thermodynamic considerations per se are independent of mechanisms, a particular molecular model often carries thermodynamic implications which can be tested experimentally. Most of the many models and theories that have been proposed for oxidative phosphorylation fall into two categories: the "chemical" hypothesis and the "chemiosmotic" hypothesis.

The chemical hypothesis postulates direct chemical coupling: oxidation results in the formation of a high-energy intermediate which drives phosphorylation (Slater, 1966). Since it is known that mitochondria can maintain concentration gradients of ions (including protons) (Lehninger, Carafoli, and Rossi, 1967), this hypothesis is usually extended to include coupling of ion transport to breakdown of the intermediate. The chemiosmotic hypothesis, on the other hand, postulates that oxidation is coupled to ejection of protons from the mitochondria (Mitchell, 1961, 1966). The resultant electrochemical potential gradient of the protons drives phosphorylation in the membrane by a reversal of an ATPase-proton pump. Cation concentration gradients are a direct result of the electrical membrane potential produced by proton transport.

352 Bioenergetics and linear nonequilibrium thermodynamics

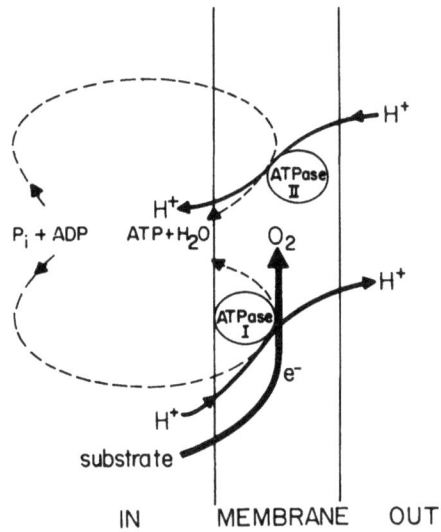

Fig. 13.2. A model for the coupling of oxidation, phosphorylation, and proton translocation, showing parallel coupling of oxidation and phosphorylation via two mechanisms. (From Rottenberg, Caplan, and Essig, 1967.)

Both theories link ion transport to oxidative phosphorylation, but whereas in the chemical theory coupling results from the sharing of a high-energy intermediate, in the chemiosmotic theory ion flows are intrinsic to the very mechanism of oxidative phosphorylation. The thermodynamic implications of the two alternatives are quite different. However, as was discussed by Rottenberg, Caplan, and Essig (1967), the mechanisms are not mutually exclusive, and it is possible that both are operative, as shown in Fig. 13.2, where a "parallel" coupling model is illustrated. In order to examine these matters further, we must consider the models in greater detail.

13.2 Models of energy coupling in oxidative phosphorylation

13.2.1 Possible coupling modes

In Fig. 13.3 all possible coupling modes are considered. J_P and J_O refer to the rates of phosphorylation and oxidation, respectively, while J_H

represents the outward flux of protons across the coupling membrane. X represents a hypothetical intermediate which may couple all three processes: by convention, $\tilde{X}$ denotes its so-called high-energy state. However, X and $\tilde{X}$ may also represent particular conformational states of one or more enzymes of the system. All six of the "flows" or processes shown are understood to be reversible under appropriate conditions. The six possible coupling modes indicated by encircling the corresponding flows are enumerated below:

A. Positive coupling between oxidation and phosphorylation.
B. Negative coupling between phosphorylation and outward proton translocation.
C. Positive coupling between oxidation and outward proton translocation.
D. Positive coupling between oxidation and synthesis of a high-energy intermediate.
E. Positive coupling between phosphorylation and breakdown of a high-energy intermediate.
F. Positive coupling between outward proton translocation and breakdown of a high-energy intermediate.

Reversibility of the flows means, for example, that in coupling $\dot{B}$, which we have designated as negative in accordance with the po-

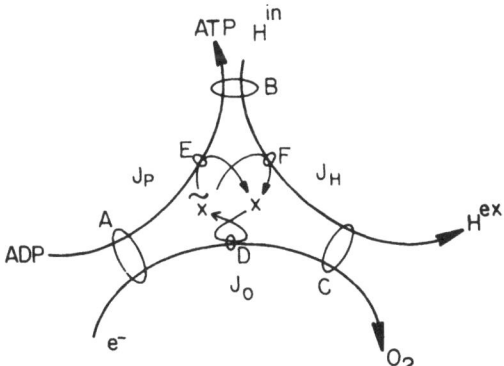

Fig. 13.3. Possible modes of coupling in oxidative phosphorylation (see text). In the case of the processes J_O, J_P, and J_H the arrows indicate our convention for positive direction of flow. The antiparallel character of the flows at B signifies negative coupling in accordance with this sign convention.

354   Bioenergetics and linear nonequilibrium thermodynamics

larity convention adopted, hydrolysis of ATP would tend to drive protons out according to the direction of the arrow, while an influx of protons would tend to produce phosphorylation. In coupling F, an influx of protons would lead to an accumulation of $\bar{X}$ at the expense of X. Couplings B, C, and F all represent reversible proton pumps; that is, they give rise to active transport of protons.

It will be realized that, from a purely phenomenological point of view, as long as the high-energy intermediate cannot be experimentally identified, coupling A is equivalent to the combination of couplings DE, B is equivalent to EF, and C to DF.

13.2.2 The chemical hypothesis

In its most general form, the chemical hypothesis includes reversible coupling of proton transport to the high-energy intermediate reaction (Slater, 1971). Thus the couplings assumed to be present are D, E, and F. In this view the breakdown pathway for $\bar{X}$ coupled to proton transport is essentially a side reaction. To the extent that there is incomplete coupling at F, or leakage of protons across the coupling membrane, or loss of $\bar{X}$ from the site of phosphorylation by side reactions or by diffusion, the overall coupling DE will be partially or even substantially decoupled. On the other hand, the proton electrochemical potential difference across the coupling membrane ($\Delta\bar{\mu}_H$) could serve as an important energy store, since its capacity in this regard may be very much greater than that of the active intermediate. Active transport of other ions would, of course, serve the same purpose, providing their rates of leakage across the membrane were sufficiently low.

13.2.3 The chemiosmotic hypothesis

As postulated by Mitchell (1961, 1966), the chemiosmotic hypothesis assumes the presence of couplings B and C only. These are associated with physically separate and distinct sites on the coupling membrane: C at site I and B at site II, as illustrated in Fig. 13.4. Sites I and II are equivalent, respectively, to Mitchell's proton-translocating oxidoreduction chain and to his proton-translocating ATPase system. Thus J_P and J_O are coupled only through circulation of H^+, rather than through a high-energy intermediate (although of course the formal-

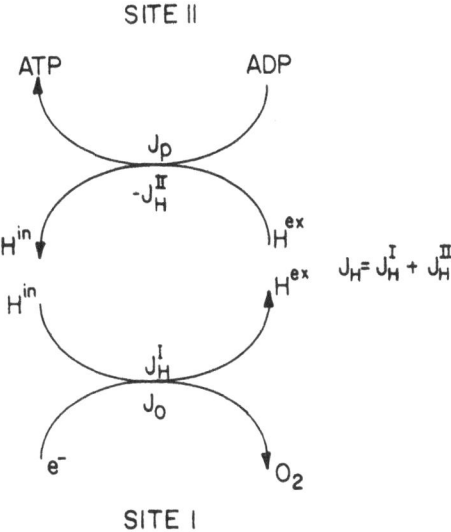

Fig. 13.4. A schematic representation of the chemiosmotic hypothesis. (From Caplan and Essig, 1969.)

ism does not exclude possible participation of a high-energy intermediate at either site).

Although the essence of the chemiosmotic hypothesis is contained in Fig. 13.4, Mitchell includes in his model a rather specific arrangement of the electron transport chain. This involves three "loops" from NADH to O_2, or two from succinate. The loops constitute the coupling sites at which the proton and electron transports are linked, and a stoichiometry is assumed of $2H^+/2e^-$ in each loop: two protons are taken up at the inner face of the membrane and extruded at the outer. This leads to H^+/O ratios of 6 for NADH and 4 for succinate. With Mitchell's additional assumption that the stoichiometry of the proton-translocating ATPase system is $2H^+/ATP$, the overall P/O ratio becomes 3 for NADH and 2 for succinate, providing no leaks or independent proton-coupled transport processes occur. Recent studies, however, suggest that the H^+/O and H^+/ATP ratios given above should be approximately doubled, leaving the P/O ratio essentially unchanged (a detailed discussion of this question is given by Rottenberg, 1979).

The mechanism of uncoupling by such classical uncouplers as

DNP (2,4-dinitrophenol), FCCP (carbonyl cyanide p-trifluoromethoxyphenylhydrazone), and so on, which are postulated to be membrane-soluble proton carriers (Mitchell, 1961, 1966; Bakker, van den Heuvel, and van Dam, 1974), appears to be a straightforward dissipation of the "protonmotive force" ($\Delta\bar{\mu}_H/F$). The uncoupling effects of various ionophores evidently result from the dissipation of either $\Delta\psi$ or ΔpH as a consequence of facilitated or exchange diffusion (Chance and Montal, 1971). Complete or near-complete uncoupling of oxidative phosphorylation may be brought about, which seems inconsistent with the chemical hypothesis. Although these uncoupling phenomena are taken to favor the chemiosmotic hypothesis, Rottenberg (1978) has taken the view that the chemical hypothesis (as described above) cannot be distinguished from the chemiosmotic on this basis, on the grounds that a dissipation of the protonmotive force would bring the high-energy intermediate to equilibrium with its hydrolysis product. This seems to be a rather extreme position, and hard to justify. The coupling at F constitutes an $\tilde{X}$-driven proton pump, and although it would clearly run faster at level flow than under other conditions, there is no reason to suppose it could run so fast as to essentially deplete its energy supply.

13.2.4 The parallel coupling hypothesis

In order to stress the mutual nonexclusivity of the hypotheses just discussed, Rottenberg, Caplan, and Essig (1967) suggested a parallel coupling model in which the two mechanisms are combined. This assumes the presence of coupling A (or DE), as in the chemical hypothesis, as well as couplings B and C at two separate sites, as in the chemiosmotic hypothesis (Fig. 13.2). Experimental evidence in support of this model will be commented on below.

13.3 Nonequilibrium thermodynamic analysis of oxidative phosphorylation and ion flows

13.3.1 The dissipation function

A good deal of the effort to distinguish between models of oxidative phosphorylation has centered on attempts to isolate high-energy intermediates. However, NET considerations lead to criteria which do not depend on the isolation of an intermediate; neither is it necessary to observe a component of H$^+$ flow used for phosphorylation (Caplan and Essig, 1969). Although it is necessary to determine $\Delta\bar{\mu}_H$ (and possibly $\Delta\bar{\mu}_K$), the other parameters are evaluated externally. In principle it is possible to envisage conditions in which J_H may be maintained nonzero and constant for brief periods by appropriately setting $\Delta\bar{\mu}_H$. For this system, under isothermal conditions, it seems reasonable to write

$$\Phi = J_P A_P + J_H \Delta\bar{\mu}_H + J_O A_O \qquad (13.3)$$

As above, the subscripts P, H, and O refer, respectively, to phosphorylation, net H$^+$ flow, and substrate oxidation. Despite the complexity of the system, Eq. (13.3) is adequate; we consider only steady states in which flows such as those of Ca^{++}, Mg^{++}, and H$_2$O will have come to a halt because their conjugate forces, that is, their electrochemical potential gradients, are not experimentally controlled.[2] Generation and breakdown of a possible high-energy intermediate would probably take place on a time scale far smaller than that required for the establishment or relaxation of a steady-state protonmotive force. In any case, this reaction constitutes in principle a typical *cyclic* coupling process as discussed by Katchalsky and Spangler (1968); in the steady state such processes are expressed only implicitly in the phenomenological coefficients, not explicitly in the dissipation function. A demonstration that in the present instance a cyclic coupling process (involving a high-energy intermediate) would be expressed implicitly in the phenomenological coefficients can be found in the review by Rottenberg (1978).

The dissipation function we have written above is ambiguous as to whether the affinities are to be measured internally or externally. Furthermore, since the flows may be nonconservative, that is, a substance may be metabolized within the composite membrane, the rate at which substances enter at one surface may not equal the rate at which they leave at the other. Hence there is also ambiguity as to the definition of the flows. Clearly, it would be very convenient to evaluate the affinities and the reaction rates externally, since internal activities are difficult to determine. This is found to be appropriate, as shown below, providing J_H is taken as the rate of decrease of H$^+$

358 Bioenergetics and linear nonequilibrium thermodynamics

content inside the mitochondrion, and $\Delta\bar{\mu}_H$ is defined as $\bar{\mu}_H^{in} - \bar{\mu}_H^{ex}$. (In the absence of a reaction involving H$^+$ in the interior of the mitochondrion, J_H is simply the net outward flux across the inner surface of the membrane.)

Consider the mitochondrial oxidation of a substrate to a product. The rate of dissipation of free energy isothermally when the membrane is in a stationary state is

$$\Phi = -\sum_j (\dot{n}_j^{in}\bar{\mu}_j^{in} + \dot{n}_j^{ex}\bar{\mu}_j^{ex}) \tag{13.4}$$

in which $\bar{\mu}_j$ and n_j represent electrochemical potential and number of moles of species j ($\dot{n}_j = dn_j/dt$). The species j include substrate, product, O$_2$, CO$_2$, H$_2$O, ATP, ADP, P$_i$, H$^+$, and possibly additional components such as K$^+$, Ca^{++}, and Mg^{++}. If we introduce the definition $\Delta\bar{\mu}_j = \bar{\mu}_j^{in} - \bar{\mu}_j^{ex}$, Eq. (13.4) can be written

$$\Phi = -\sum_j [\dot{n}_j^{in}\Delta\bar{\mu}_j + (\dot{n}_j^{in} + \dot{n}_j^{ex})\bar{\mu}_j^{ex}] \tag{13.5}$$

Now, designating the rth metabolic reaction rate by J_r and the corresponding stoichiometric coefficients by ν_{jr},

$$(\dot{n}_j^{in} + \dot{n}_j^{ex}) = \sum_r (\dot{n}_{jr}^{in} + \dot{n}_{jr}^{ex}) = \sum_r \nu_{jr} J_r$$

and hence Eq. (13.5) becomes

$$\Phi = -\sum_j \left(\dot{n}_j^{in}\Delta\bar{\mu}_j + \sum_r \nu_{jr} J_r \bar{\mu}_j^{ex} \right) \tag{13.6}$$

It is immaterial whether the reactions take place entirely within the mitochondrial membrane or to some extent within the interior of the mitochondrion as well, but we exclude the possibility of reactions occurring externally. The affinity A_r in any region is defined as

$$A_r = -\sum_j \nu_{jr}\bar{\mu}_j \tag{13.7}$$

Combining Eqs. (13.6) and (13.7),

$$\Phi = -\sum_j \dot{n}_j^{\text{in}} \Delta \tilde{\mu}_j + \sum_r J_r A_r^{\text{ex}} \tag{13.8}$$

Equation (13.8) can be used to develop a phenomenological description of the system taking explicit account of all the transport processes that are occurring. A comprehensive treatment of this type has been given by Hill (1980) and will not be dealt with here.

If we consider the stationary state in which both $\Delta\tilde{\mu}_H$ and J_H ($-\dot{n}_H^{\text{in}}$) are nonzero, then $\dot{n}_j^{\text{in}} = 0$ for all $j \neq H$, and Eq. (13.8) gives

$$\Phi = J_P A_P^{\text{ex}} + J_H \Delta\tilde{\mu}_H + J_O A_O^{\text{ex}} \tag{13.9}$$

On the other hand, when the interior of the mitochondrion is in a stationary state for all components, $\dot{n}_j^{\text{in}} = 0$ for all j, and consequently for our example

$$\Phi = \sum_r J_r A_r^{\text{ex}} = J_P A_P^{\text{ex}} + J_O A_O^{\text{ex}} \tag{13.10}$$

Thus in the stationary state it suffices to measure changes in the external solution only, although in general J_P and J_O are given by the sum of the rates of change on both sides of the membrane. Equation (13.10) is an example of the general principle that if any forces are not controlled, the system will reach a stationary state in which their conjugate flows are zero. The flows J_P and J_O do not vanish, since A_P^{ex} and A_O^{ex} are assumed to be fixed experimentally.

In the approach to the above stationary states Eq. (13.8) applies; that is, each flow is a function of several forces. However, in the stationary state of Eq. (13.9), each flow is a function of only three forces. If $\Delta\tilde{\mu}_H$ is not directly controlled, J_H becomes zero, giving the stationary state of Eq. (13.10), in which each flow is a function of only two forces. Although neither flow is now an explicit function of $\Delta\tilde{\mu}_H$, the existence of the H^+ transport mechanism influences the values of the phenomenological coefficients (Katchalsky and Spangler, 1968).

As indicated earlier, application of the dissipation function of Eq. (13.3) requires deciding whether flows and affinities should be evaluated internally or externally. Equation (13.9) indicates that the reaction flows and affinities should be evaluated externally. Since the mitochondrion is here considered in a stationary state, this might seem self-evident. However, the situation may nevertheless appear complicated in that there is nonconservative flow of H^+, so it is not intuitively evident whether the flow should be evaluated from the change in internal or external hydrogen ion content. This question is resolved by Eq. (13.8): if the affinities are defined externally, H^+ flow must be defined in terms of internal changes. Although this quantity would be difficult to measure, for most purposes such measurements are unnecessary.

Symmetry considerations show that we could have expressed the dissipation function in terms of external flows and internal affinities. Although this formulation is equally valid, it would be impracticable for several reasons. There are obvious difficulties in evaluating internal affinities. The mitochondrion is highly compartmentalized, and knowledge of internal activities is imprecise.[3] Further, even when internal concentrations are constant, it would be necessary to consider many flows and forces, since the external flows, unlike the internal flows, need not be zero for nonconserved species when the mitochondrion is in a stationary state.

13.3.2 The phenomenological relations

Equations (13.9) and (13.10) specify three-term and two-term dissipation functions which give rise, respectively, to phenomenological descriptions of the system involving three equations with six independent coefficients, or two equations with three independent coefficients. These descriptions are applicable to different experimental situations, as will be discussed below. The most general description arises from Eq. (13.9). The appropriate phenomenological equations are:

$$J_P = L_P A_P^{ex} + L_{PH} \Delta \bar{\mu}_H + L_{PO} A_O^{ex} \tag{13.11}$$

$$J_H = L_{PH} A_P^{ex} + L_H \Delta \bar{\mu}_H + L_{OH} A_O^{ex} \tag{13.12}$$

$$J_O = L_{PO} A_P^{ex} + L_{OH} \Delta \bar{\mu}_H + L_O A_O^{ex} \tag{13.13}$$

This formalism should be applicable, at least near equilibrium, whichever hypothesis is correct. (We have made the usual assumptions of Onsager symmetry and linearity.) We now consider each hypothesis in the light of these equations.

Chemical hypothesis. In this view, since substrate oxidation produces a high-energy intermediate, which can drive either phosphorylation or outward H^+ transport, J_O is positively coupled to both J_P and J_H as shown in Fig. 13.3. However, since outward H^+ transport may also result from the hydrolysis of ATP, J_P and J_H are negatively coupled. It is clear that for this system each flow will be influenced by every force; in phenomenological terms, all coefficients of Eqs. (13.11)–(13.13) are nonzero.

Chemiosmotic hypothesis. Eqs. (13.11)–(13.13) are again applicable. However, in this case we can characterize the overall coefficients in terms of those of the sites shown in Fig. 13.4.

Site I:

$$J_H^I = L_H^I \Delta\bar{\mu}_H + L_{OH}^I A_O^{ex} \tag{13.14}$$

$$J_O = L_{OH}^I \Delta\bar{\mu}_H + L_O^I A_O^{ex} \tag{13.15}$$

Site II:

$$J_P = L_P^{II} A_P^{ex} + L_{PH}^{II} \Delta\bar{\mu}_H \tag{13.16}$$

$$J_H^{II} = L_{PH}^{II} A_P^{ex} + L_H^{II} \Delta\bar{\mu}_H \tag{13.17}$$

At site I oxidation results in the transport of H^+ against $\Delta\bar{\mu}_H$, while at site II the spontaneous flow of H^+ results in phosphorylation.

The net flow of H^+ is given by

$$J_H = J_H^I + J_H^{II} \tag{13.18}$$

so that adding Eqs. (13.14) and (13.17) gives

$$J_H = L_{PH}^{II} A_P^{ex} + (L_H^I + L_H^{II})\Delta\bar{\mu}_H + L_{OH}^I A_O^{ex} \tag{13.19}$$

Equations (13.16), (13.19), and (13.15) correspond respectively to Eqs. (13.11), (13.12), and (13.13). Thus the straight coefficients L_O and L_P and the cross coefficients L_{OH} and L_{PH} are simply those of the appropriate elemental sites, while L_H is the sum $L_H^I + L_H^{II}$. The important observation is that

$$L_{PO} = 0 \quad \text{(chemiosmotic hypothesis)} \quad (13.20)$$

that is, if $\Delta\bar{\mu}_H$ were maintained constant experimentally, phosphorylation would be independent of oxidation, and vice versa. As we would expect from the model, if $\Delta\bar{\mu}_H = 0$ uncoupling is complete, and $J_P = 0$ if $A_P^{ex} = 0$, irrespective of J_O, while $J_O = 0$ if $A_O^{ex} = 0$, irrespective of J_P.

Parallel coupling hypothesis. As seen in Fig. 13.2, this picture embraces the chemical hypothesis at site I, and the chemiosmotic hypothesis at site II. Formally one cannot distinguish between the chemical and the parallel coupling hypotheses: all the coefficients of Eqs. (13.11)–(13.13) are again nonzero. The coefficient L_{PO} in the case under consideration is specifically associated with site I. However, studies by Griffiths (1976) suggest that a high-energy intermediate produced at site I may possibly diffuse to site II prior to bringing about phosphorylation. If this is so, L_{PO} includes kinetic parameters of both sites as well as the diffusion path.

It should be realized that while L_{PO} is nonzero in both the chemical and parallel coupling hypothesis, in the latter case it may be very small. This means that if $\Delta\bar{\mu}_H$ were maintained at or near zero experimentally, the phosphate potential obtained in state 4 would be correspondingly small. The poor coupling between oxidation and phosphorylation under these circumstances would, in this view, reflect the dissipation of energy stored in the high-energy intermediate through the reaction driving protons at level flow (as discussed earlier). As long as oxidation continues, however, the affinity of the postulated high-energy intermediate reaction could not actually be reduced to zero by this process and should be increased by increasing the oxidoreduction potential.[4]

Obviously there are substantial experimental difficulties in applying the phenomenological equations. However, it is possible to vary and measure the affinities A_O^{ex} and A_P^{ex} over a considerable range (Rottenberg, 1973). Furthermore, both the pH difference and the electrical potential difference may be evaluated and regulated in functioning mitochondria (Rottenberg, 1970). Although it may be difficult to fix $\Delta\bar{\mu}_H$ exactly at zero, it seems possible to make it very small.

13.3.3 The degree of coupling and effectiveness of energy utilization

For oxidative phosphorylation we are concerned with three degrees of coupling, q_{PH}, q_{OH}, and q_{PO}. As has been discussed in Chapter 6, these are readily derived from the inverse phenomenological equations, since for an n-flow system $q_{ij} = -R_{ij}/\sqrt{R_{ii}R_{jj}}$:

$$A_P^{ex} = R_P J_P + R_{PH} J_H + R_{PO} J_O \tag{13.21}$$

$$\Delta\tilde{\mu}_H = R_{PH} J_P + R_H J_H + R_{OH} J_O \tag{13.22}$$

$$A_O^{ex} = R_{PO} J_P + R_{OH} J_H + R_O J_O \tag{13.23}$$

If the chemiosmotic hypothesis holds, a special relationship applies,[5] that is:

$$q_{PO} = -q_{PH} q_{OH} \tag{13.24}$$

(Since J_P and J_H are negatively coupled, q_{PH} is negative.) Any factors which decrease q_{OH} or $-q_{PH}$ will decrease q_{PO} and hence the effectiveness with which oxidative energy is used for phosphorylation.

13.4 Applications of the NET analysis

13.4.1 Steady states

The above treatment should be applicable to the steady state analysis of oxidative phosphorylation and, in essence, photophosphorylation, irrespective of the detailed mechanisms. The system can be analyzed in a variety of experimental circumstances, all describable in terms of the NET formulation. It is simplest to consider situations in which the dissipation function is reduced to two terms, so that each flow depends on only two forces. The degree of coupling q which characterizes the resulting two-flow system is determined by the specific conditions considered (Caplan, 1966). Since in general we are primarily interested in the conversion of free energy of oxidation into free energy of phosphorylation, we shall consider specifically only those cases in which the term $J_H\Delta\tilde{\mu}_H$ is zero, that is, static head and level flow of H^+.

If A_0^{ex} is maintained constant, either by experimental manipulation or by the *in vivo* characteristics of the system, and $\Delta\tilde{\mu}_H$ is not controlled, J_H will become zero in the stationary state. For this case $q = q_{PO}$ (case 1 below). On the other hand it would be very useful, if it were possible, to carry out experiments in which $\Delta\tilde{\mu}_H$ is maintained at zero.[6] In this case $q = (q_{PO} + q_{PH}q_{OH})/\sqrt{(1 - q_{PH}^2)(1 - q_{OH}^2)}$ (case 2 below). The two cases are as follows:

1. $J_H = 0$ (static head). Equations (13.21) and (13.23) then become

$$A_P^{ex} = R_P J_P + R_{PO} J_O \qquad (13.25)$$

$$A_0^{ex} = R_{PO} J_P + R_O J_O \qquad (13.26)$$

Consequently, the effective degree of coupling is given by

$$q = \frac{-R_{PO}}{\sqrt{R_P R_O}} = q_{PO} \qquad (J_H = 0) \qquad (13.27)$$

2. $\Delta\tilde{\mu}_H = 0$ (level flow). Equations (13.21), (13.22), and (13.23) then give

$$A_P^{ex} = R_P(1 - q_{PH}^2) J_P - \sqrt{R_P R_O}(q_{PO} + q_{PH}q_{OH}) J_O \qquad (13.28)$$

$$A_0^{ex} = -\sqrt{R_P R_O}(q_{PO} + q_{PH}q_{OH}) J_P + R_O(1 - q_{OH}^2) J_O \qquad (13.29)$$

Consequently,

$$q = \frac{q_{PO} + q_{PH}q_{OH}}{\sqrt{(1 - q_{PH}^2)(1 - q_{OH}^2)}} \qquad (\Delta\tilde{\mu}_H = 0) \qquad (13.30)$$

Figures 13.5 and 13.6, which correspond to Figs. 4.2 and 4.3, respectively, apply to both these states. Introducing Eq. (13.24) shows that for the chemiosmotic hypothesis, setting $\Delta\tilde{\mu}_H = 0$ effectively uncouples oxidation from phosphorylation, so that in the range where the phenomenological equations are linear, J_P/J_O is directly proportional to A_P^{ex}/A_0^{ex}, as indicated in Fig. 13.5. With nonzero but con-

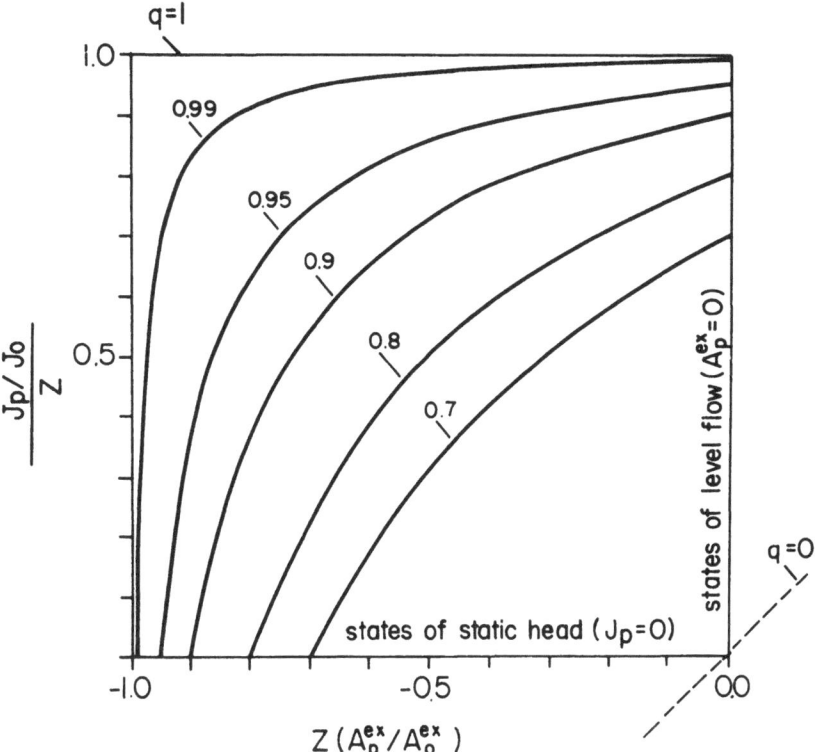

Fig. 13.5. Dependence of the P/O ratio on the ratio of reaction affinities; in mitochondria the pertinent reaction affinities are those in the external solutions. (For complete coupling, the P/O ratio is given by Z under all circumstances. Current studies suggest that when $q = 1$ the quantity $Z = 3$ for appropriate substrates. When $q \neq 1$ the value of Z may change.) Two cases are included in this diagram: (1) $J_H = 0$: $q = q_{PO}$; (2) $\Delta\bar{\mu}_H = 0$: $q = (q_{PO} + q_{PH}q_{OH})/\sqrt{(1 - q_{PH}^2)(1 - q_{OH}^2)}$.

stant $\Delta\bar{\mu}_H/A_O^{ex}$, proportionality would be lost, but linearity retained. To the extent that the relation between J_P/J_O and A_P^{ex}/A_O^{ex} becomes nonlinear when $\Delta\bar{\mu}_H = 0$ or $\Delta\bar{\mu}_H/A_O^{ex}$ is maintained constant (q differs from zero), the system deviates from the chemiosmotic theory, which is therefore seen to be a limiting case. If the system were not chemiosmotic, oxidation could drive phosphorylation even when $\Delta\bar{\mu}_H = 0$. In this case either the chemical mechanism or the parallel coupling mechanism might be operative.

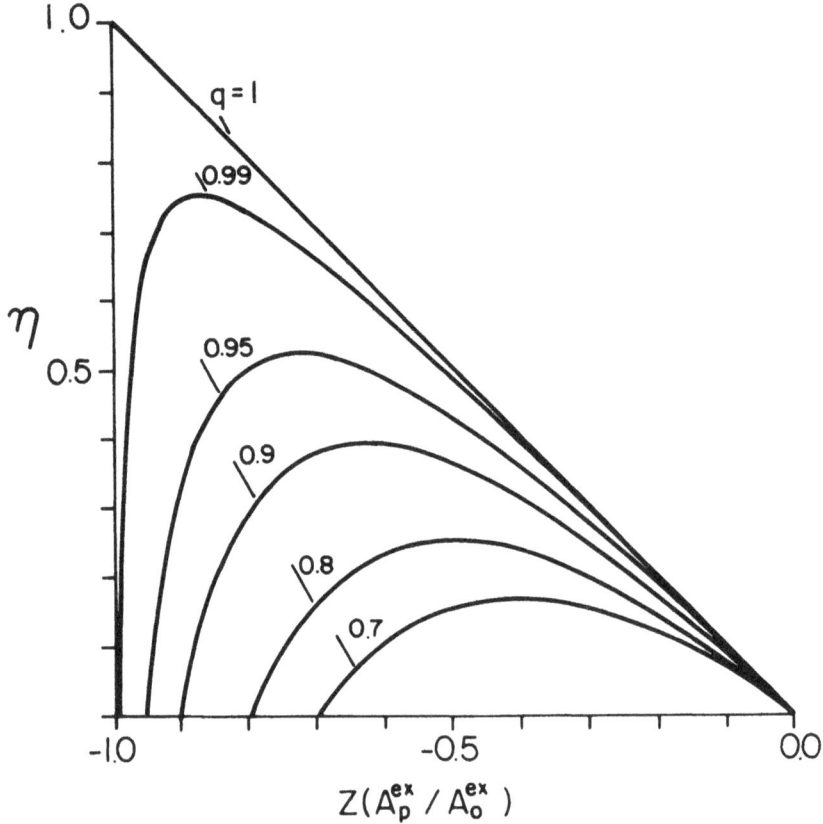

Fig. 13.6. Dependence of the efficiency on the ratio of reaction affinities; in mitochondria the pertinent reaction affinities are those in the external solutions. (Current studies suggest that when $q = 1$, the quantity $Z = 3$ for appropriate substrates. When $q \neq 1$ the value of Z may change.)

Rottenberg has tested linear relations such as (13.25) and (13.26) in mitochondria, both under normal conditions of oxidative phosphorylation (Rottenberg, 1973) and in the "reverse electron transport" mode, when the system is driven by hydrolysis of ATP (Rottenberg and Gutman, 1977). As shown in Fig. 13.7, which illustrates normal-mode operation, the system exhibited linearity over a considerable range of affinities. Moreover, within this range reciprocity held rather precisely. However, he pointed out that linearity failed to hold at higher or lower affinities. Consequently the linear regime depicted cannot be extrapolated to equilibrium, and indeed Rottenberg expressed these results in the form

$$J_P = L_P A_P + L_{PO} A_O + K_P \tag{13.31}$$

$$J_O = L_{PO} A_P + L_O A_O + K_O \tag{13.32}$$

where K_P and K_O are constants. The interpretation to be placed on these constants is not as yet clear. However, one possible approach to the problem has been discussed in the appendix to Chapter 6. Another, not necessarily inconsistent, is the following, in which it is taken for granted that the linearity is only apparent. In other words, what has been observed is (in all probability) a region in the vicinity of a multidimensional inflection point, which can be described as linear to a very good degree of approximation. We have referred to this situation as *kinetic* linearity in Chapter 6.

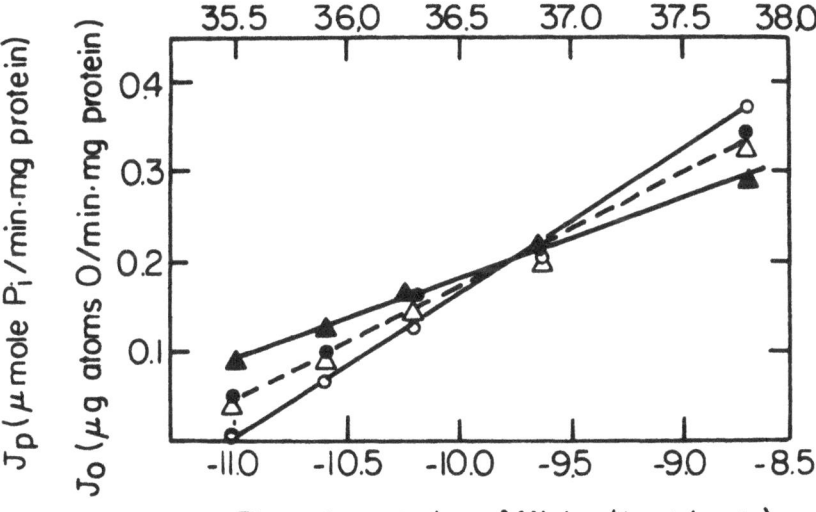

Fig. 13.7. Rates of oxidation and phosphorylation in rat liver mitochondria. J_P (—○—) and J_O (--△--) as functions of A_P^{ex}, with A_O^{ex} constant at 36.7 kcal/mole. J_P (--●--) and J_O (—▲—) as functions of A_O^{ex}, with A_P^{ex} constant at 9.7 kcal/mole. The coincidence of the dashed lines indicates symmetry of the cross effects. (From Rottenberg, 1973.)

368 Bioenergetics and linear nonequilibrium thermodynamics

The most general case of a nonlinear system far from equilibrium can be described in the following terms:

$$J_P = L_P A_P + L_{PO} A_O + F_P(A_P^2, A_P A_O, A_O^2, \cdots) \qquad (13.33)$$

$$J_O = L_{OP} A_P + L_O A_O + F_O(A_P^2, A_P A_O, A_O^2, \cdots) \qquad (13.34)$$

where F_P and F_O are functions consisting of a sum of second-order and higher-order terms in the affinities, each term including its appropriate phenomenological coefficient (compare Eqs. 5.25 and 5.26). It is quite conceivable that over a particular range of values of the affinities, the functions F_P and F_O become essentially, if not completely, constant. If this is true, as may be the case here, it is fortunate because the coefficients of the first-order terms are thereby rendered experimentally accessible and may be determined from the slopes of straight lines. It is also possible that suitable manipulation of the system could bring it to a hypothetical (or even real) state close to equilibrium such that the functions F_P and F_O vanish, while the coefficients of the first-order terms remain unchanged. This state can then be characterized by a degree of coupling. The reciprocity observed experimentally in this system suggests strongly, although of course it does not prove, that the coefficients of the first-order terms are constants within the limits of experimental error. From the slopes of the lines in Fig. 13.7, Rottenberg calculated q_{PO} to be 0.92. It should be stressed that in principle this value pertains only to the state we have just described, which may or may not be experimentally attainable. However, if the L's are indeed constants, Rottenberg's calculation gives a correct description of the system in this hypothetical corresponding equilibrium state, and the lines drawn in Fig. 13.7 are parallel to lines representing proper pathways.

13.4.2 Evaluation of the degree of coupling and apparent stoichiometry by using a reference state

In Chapter 7 it was shown that the degree of coupling for a system with thermodynamic linearity can be obtained from the ratio of the input flows at static head and level flow (Eq. 7.47), providing the input force is kept constant. Unfortunately, in mitochondria and chloroplasts, level flow cannot easily be maintained either for phosphorylation or for proton transport. In the former case the adenylate kinase reaction prevents the ATP/ADP ratio from falling sufficiently low, while in the latter the swift buildup of electrical potential cannot ordinarily be prevented even though the system may be highly buffered. However, static head may be reached within a few seconds. Rottenberg (1978) has suggested a method of determining the degree of coupling which overcomes the level-flow problem without drastic measures and which is based on the classical determination of the respiratory control ratio. It involves a comparison of static head (state 4 for oxidative phosphorylation) and a suitable reference state other than level flow (obviously a state 3 for oxidative phosphorylation). In principle the reference state should be allowed to change only to a small extent, if at all, while being characterized, but this may still be difficult to achieve in practice.

For convenience the following arguments are presented in terms of the resistance formulation for oxidative phosphorylation, assuming that $J_H = 0$ in both of the states considered. It is also assumed that the system can be studied sufficiently close to equilibrium so that the observed behavior is truly linear in the thermodynamic sense. However, if as shown in Fig. 13.7 linearity is merely kinetic, that is, the range of study requires introducing additive constants as in Eqs. (13.31) and (13.32), the treatment becomes more complex than Rottenberg has indicated. Thus an alternative, purely empirical expression of Eqs. (13.31) and (13.32) is in terms of constants additive to the affinities:

$$J_P = L_P(A_P + A_P') + L_{PO}(A_O + A_O') \tag{13.35}$$

$$J_O = L_{PO}(A_P + A_P') + L_O(A_O + A_O') \tag{13.36}$$

where

$$L_P A_P' + L_{PO} A_O' = K_P \tag{13.37}$$

$$L_{PO} A_P' + L_O A_O' = K_O \tag{13.38}$$

Both A_P' and A_O', as well as all the L's, may be determined, in principle, from an experiment such as that depicted in Fig. 13.7. For the following argument we will assume that if A_P' and A_O' are not zero,

370 Bioenergetics and linear nonequilibrium thermodynamics

they are included in the respective A's referring to any given state and that the phenomenological coefficients are constant. It should be noted that in this approach we are treating the quantities denoted by A' purely formally, as if they were additional terms to be added to the standard affinities. Thus up to this point the following relation has been implicit in our discussions of the affinities:

$$A = A^{\text{stan}} + A^{\text{conc}} \tag{13.39}$$

where A^{stan} denotes the standard value of the affinity appropriate for the experimental conditions chosen, and A^{conc} denotes the concentration-dependent part. In the remainder of this chapter we will consider the affinities to be given by

$$A = A' + A^{\text{stan}} + A^{\text{conc}} \tag{13.40}$$

It may be useful to refer to the affinity defined in this way as the "effective" affinity, remembering that it is only a device introduced for purposes of calculation and has no real thermodynamic significance; it does not appear in the dissipation function.

Bearing all these considerations in mind, the most general description of the system is given by Eqs. (13.25) and (13.26). Dropping the superscript ex for simplicity, and remembering that A_O is held constant, the two states of interest may be characterized as follows:

Static head (state 4):

$$A_P^S = R_{PO} J_O^S \tag{13.41}$$

$$A_O = R_O J_O^S \tag{13.42}$$

Reference state (state 3):

$$A_P^R = R_P J_P^R + R_{PO} J_O^R \tag{13.43}$$

$$A_O = R_{PO} J_P^R + R_O J_O^R \tag{13.44}$$

where the superscripts S and R represent static head and reference state, respectively. From Eqs. (13.41), (13.42), and (13.44),

$$\frac{A_P^S}{A_O} = \frac{R_{PO}}{R_O} = \frac{1 - (J_O^R/J_O^S)}{(J_P^R/J_P^S)(J_O^R/J_O^S)} \qquad (13.45)$$

It is useful to denote the respiratory control ratio J_O^R/J_O^S by R_c, and the flow ratio (or apparent stoichiometric ratio in the reference state) J_P^R/J_O^R by R_f. Then Eq. (13.45) becomes

$$\frac{A_P^S}{A_O} = \frac{1 - R_c}{R_f R_c} \qquad (13.46)$$

Thus the ratio A_P^S/A_O and hence R_{PO}/R_O may be determined from measurements of R_c and R_f. This ratio of course refers to effective affinities. Independent measurements of the true values of A_P^S and A_O will indicate whether or not the additive constants are present (except in the unlikely case that $A_P'/A_O' = R_{PO}/R_O$). If they are, the simplest approach would be to carry out a series of static head measurements, varying A_O and measuring A_P^S and J_O^S. A linear extrapolation of the data to $J_O^S = 0$ according to Eqs. (13.41) and (13.42) would then yield A_P' and A_O', respectively. With this information it becomes possible to determine q_{PO}. (Of course, q_{PO} can be obtained from an experiment of the type shown in Fig. 13.7, but since static head is so readily achievable, the type of experiment discussed here appears to be simpler.) Using the definition of R_c, it can readily be shown from Eqs. (13.41)–(13.44) that

$$q_{PO}^2 = \frac{R_{PO}^2}{R_P R_O} = \frac{1 - R_c}{(A_P^R/A_P^S) - R_c} \qquad (13.47)$$

Relations similar to (13.47) may be derived for q_{PH} and q_{OH}. Thus the degree of coupling may be obtained from measurements of the respiratory control ratio, providing the appropriate effective affinities have also been determined.

If the system were fully coupled, R_c would tend to infinity while R_f would give the true stoichiometry of the process. The question still arises as to how to evaluate the *maximal* apparent stoichiometry in the driving region for systems which are less than fully coupled. As was pointed out in Chapter 4, this quantity is simply the flow ratio at level flow (a state not readily obtainable in the systems under discussion) and is given by $q_{PO}Z$ (as can be seen in Fig. 13.5). In the present context, $A = \sqrt{R_O/R_P}$. Rottenberg (1978) has argued, on the grounds that Z gives the stoichiometry of the system if it is fully coupled, that even if it is not, Z still expresses its mechanistic stoichiometry. In this view Z represents the stoichiometry that would have been obtained if the system had been completely coupled, and the maximal apparent stoichiometry is just the mechanistic stoichiometry modified by the degree of coupling. This view is not unattractive, especially if one believes one has *a priori* knowledge of the mechanistic stoichiometry. As pointed out in Sec. 5.3, identity of the phenomenological stoichiometry Z and the mechanistic stoichiometry n when $q < 1$ is unlikely but not impossible; however, at high q the error in assuming $Z = n$ will not be great (for example, at $q = 0.9$, Z will be within 10% of n). In this case the degree of coupling could be estimated quite simply, as from a single static head measurement of the affinity ratio (or, if appropriate, the effective affinity ratio) A_P^S/A_O, which is readily seen to be nothing other than $-q_{PO}/Z$. The idea that Z may represent an intrinsic mechanistic stoichiometry even when $q_{PO} < 1$ was also implicit in a graphical representation of the flow ratio as a function of the affinity ratio in oxidative phosphorylation given by Rottenberg, Caplan, and Essig (1970, Fig. 6). It must, however, be stressed that there are no thermodynamic grounds whatsoever for assuming either (1) that factors which tend to uncouple a system would leave R_O, R_P, or their ratio unchanged, or (2) that a hypothetical, completely coupled state of a system must exist even if in practice it is never found, and that the ratio $\sqrt{R_O/R_P}$ relates to that state and is an integral number. Thus in Figs. 5.1 and 6.2, cycles b and c are just as intrinsic to the systems depicted as cycle a is.

It turns out that this problem is readily resolvable by means of the experimental technique just discussed, that is, the use of a reference state. Once q_{PO} has been determined by this method, one can legitimately make use of the affinity ratio at static head, this time writing

$$Z = -q_{PO}\left(\frac{A_O}{A_P^S}\right) \tag{13.48}$$

Alternatively, one has

$$Z = \frac{R_c R_f}{\sqrt{(1 - R_c)(A_P^R/A_P^S - R_c)}} \tag{13.49}$$

The maximal apparent stoichiometry is given by

$$q_{PO}Z = -q_{PO}^2 \left(\frac{A_O}{A_P^S}\right) = \frac{R_c R_f}{R_c - (A_P^R/A_P^S)} \quad (13.50)$$

Hence no assumptions as to the value of Z need be made, but once again the question of additive constants must be taken into account. It should be noted that at *high* values of q_{PO}, as can be appreciated from Fig. 13.5, R_f will approximate $q_{PO}Z$ rather well, even in reference states in which the affinity ratio A_P^R/A_P^S is well above its level flow value of zero. By the same token, at sufficiently high values of q_{PO}, R_f will be insensitive to changes in the reference state over a wide range of state 3 conditions. However, as state 3 tends to state 4, R_f falls precipitously (but without discontinuity, as is seen in Fig. 13.5). Rottenberg (1978) has pointed out that conventional experimental procedures which evaluate state 3 by averaging a continuum of states evolving toward state 4 are inadequate for this type of analysis. The average values of J_P^R and J_O^R obtained by such techniques may not be strictly comparable. This may be the reason for the discrepancies noted by Slater, Rosina and Mol (1973) between values of A_P^S measured directly and calculated by means of Eq. (13.36) in rat liver mitochondria. Another possible cause of discrepancy may be failure to take into account A_P' and A_O'.

Although the methods outlined above are potentially capable of yielding a great deal of information, it must be admitted that this potential is only lately being realized, as will be shown below. The main problem is that it is not always clear whether the experiments reported have been performed in a thermodynamically linear region or in a kinetically linear region such that due consideration must be given to the problem of additive constants. A case in point, which indicates in principle what might be accomplished, is a study described by Rottenberg (1978) involving the measurement, in rat liver mitochondria, of q_{PO}, q_{PH}, and q_{OH} under comparable conditions by methods based on Eq. (13.47). Unfortunately, the validity of the formulation considered here was assumed without taking any cognizance of the possible, or even probable, presence of additive constants. The results obtained were $q_{PO} = 0.96$, $q_{PH} = -0.92$, and $q_{OH} = 0.94$. It appears from this that Eq. (13.24) is not obeyed, since $-q_{PH}q_{OH} = 0.87$. Thus, if we take this result at face value, the chemiosmotic model per se appears to underestimate the overall degree of

374 Bioenergetics and linear nonequilibrium thermodynamics

coupling. This would have been a remarkable result had it been demonstrated unambiguously. For example, continuing this calculation on its own terms, and assuming the parallel coupling model to hold instead, Eq. (13.30) should give the degree of coupling of the "chemical" component at $\Delta\bar{\mu}_H = 0$, which turns out to be 0.71. Note that even if these results had been unambiguous, on the basis of information of this type alone it is impossible to distinguish between the chemical and parallel coupling models.

It should be emphasized once again that whenever the observed linearity is kinetic rather than thermodynamic, that is, additive constants are involved according to Eqs. (13.31) and (13.32), *all* of the relations (13.41)–(13.50) involve effective affinities as defined by Eq. (13.40).

13.4.3 Experimental examples

In Sec. 13.3.2 we commented in a footnote on the effect of adding uncoupler to the mitochondrial system. An experimental study based on this type of analysis, in which uncoupler was added at several different concentrations, was carried out by van Dam et al. (1978). The results are shown in Fig. 13.8. A striking linearity is observed in the dependence of J_P on J_O, with the lines increasing in slope as the level of uncoupler is raised. Van Dam et al. assumed that the sole source of uncoupling in the system was the leakage pathway induced by the uncoupler; that is, they considered that the chemiosmotic description held with complete coupling both at site I and site II. In this case it is readily shown from Eqs. (13.14)–(13.17), taking into account L_H^{leak} (and remembering that site II is negatively coupled), that

$$J_P = \sqrt{\frac{L_P^{II}}{L_O^{II}}}\left(\sqrt{\frac{L_H^I}{L_H^{II}}} + \frac{L_H^{\text{leak}}}{\sqrt{L_H^I L_H^{II}}}\right) J_O - \sqrt{\frac{L_P^{II} L_O^I}{L_H^I L_H^{II}}}\, L_H^{\text{leak}} A_O \quad (13.51)$$

Hence the lines would all be expected to intersect at a single point (independent of the magnitude of L_H^{leak}), under which conditions the flow ratio is given by

$$\frac{J_P}{J_O} = Z^I Z^{II} \quad (13.52)$$

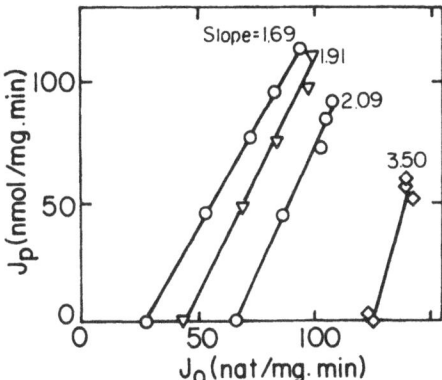

Fig. 13.8. Relation between rate of phosphorylation and rate of oxygen uptake in rat liver mitochondria at different degrees of uncoupling. Rat liver mitochondria (1 mg/ml) were incubated in a medium containing 10 mM succinate, 10 mM malate, 1 µg rotenone/ml, 15 mM KCl, 50 mM Tris-Cl, 20 mM phosphate buffer, 10 mM glucose, 5 mM $MgCl_2$, 2 mM EDTA; final pH = 7.4, temperature = 25° C. Oxygen uptake was recorded with a Clark electrode. Glucose-6-phosphate formation was measured enzymatically. The rate of oxygen uptake was varied with hexokinase; 2,4-dinitrophenol was present in a concentration of 0, 2, 5, and 10 µM from left to right. (After van Dam et al., 1978.)

and the corresponding force ratio is

$$\frac{A_P}{A_O} = \frac{(L_H^I/L_H^{II})}{Z^I Z^{II}} \qquad (13.53)$$

This point is somewhat analogous to the upper right-hand corner of Fig. 4.2. In this situation both phosphorylation and oxidation occur spontaneously, and $\Delta\bar{\mu}_H = 0$ (protons are pumped in at site II exactly as fast as they are pumped out at site I). However, if completeness of coupling is relaxed at sites I and II, a unique intersection point will *not* be found for plots such as Fig. 13.8 even though linearity is retained. It is seen that since complete coupling at individual sites is assumed in the work of van Dam et al., J_P/J_O at the intersection point should give a product of stoichiometries according to Eq. (13.52), and indeed a value of 1.45 is reported for this ratio. Unfortunately, at the time of writing no stringent statistical test of the uniqueness of the

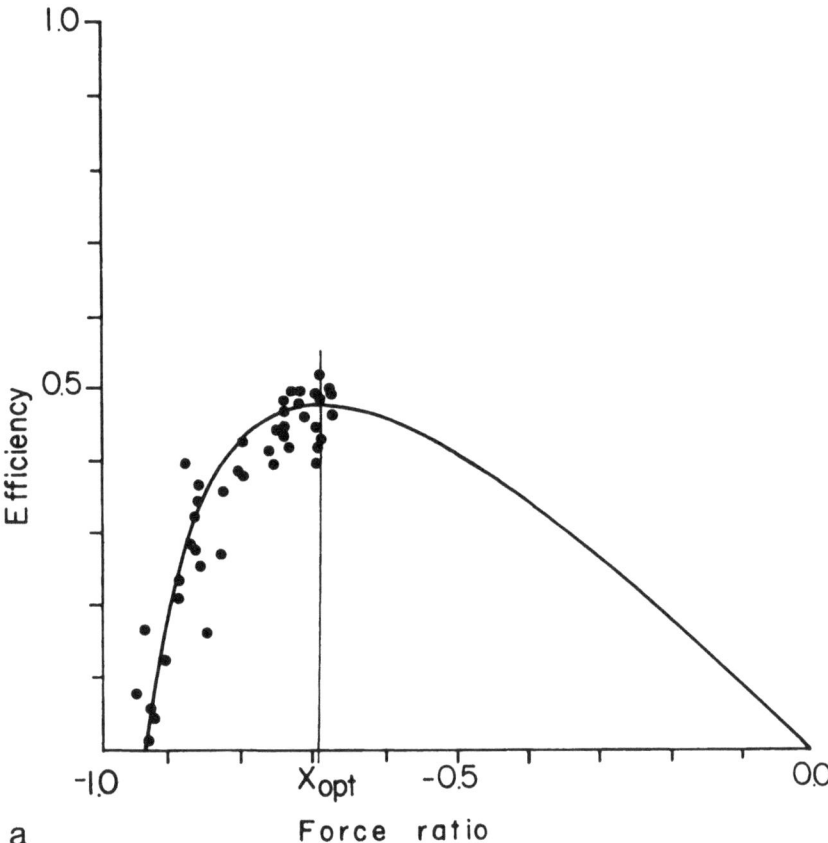

Fig. 13.9. Efficiency of oxidative phosphorylation with and without adenylate kinase. Rat liver mitochondria were incubated at 37° C. From the measured values of J_P, J_O, and A_P, and from the experimentally established value of A_O, the efficiency and the normalized force ratio were calculated. (The phenomenological stoichiometry Z was found to be 2.83.) (a) Pooled data from three experiments without diadenosine-pentaphosphate. (b) Pooled data from three incubations with diadenosine-pentaphosphate. The solid line in each panel represents the theoretical curve with $q = 0.94$. X_{opt} is the force ratio permitting optimal efficiency. (After Stucki, 1980.)

Energy coupling in mitochondria 377

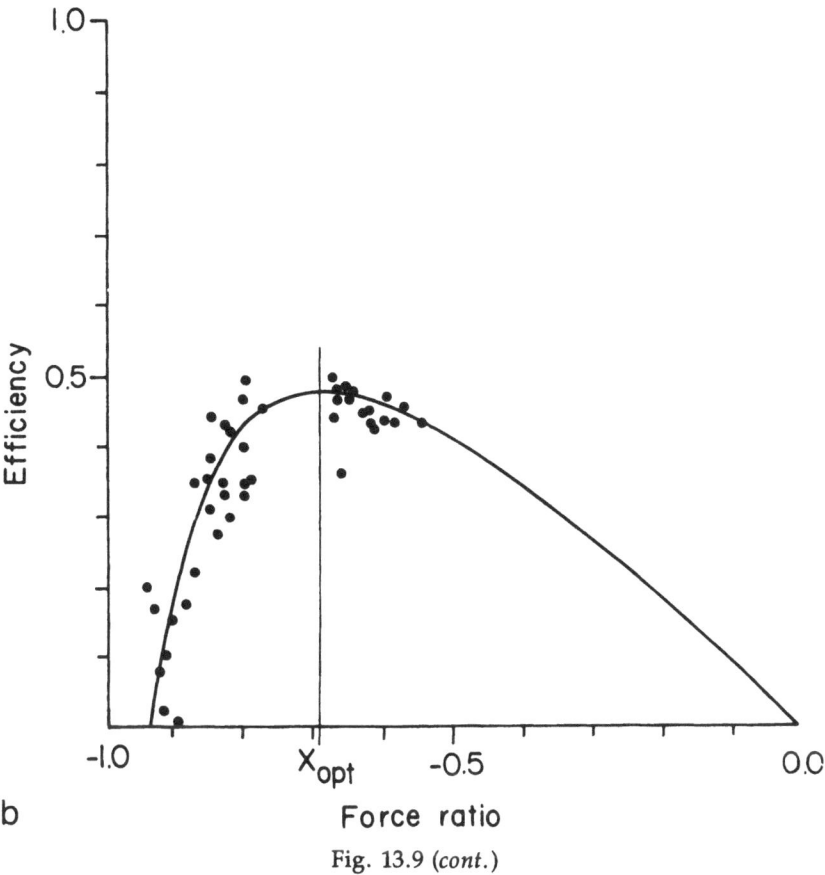

Fig. 13.9 (cont.)

crossing point has been presented, so it is not clear that the experimental findings establish the completeness of coupling.

The remarkable linearity between oxidative metabolism and sodium transport in epithelia, described in Chapter 8, suggests that linear behavior may well be a general characteristic of mitochondria. We have seen that experimental studies in recent years carried out in the laboratories of Rottenberg and van Dam give a great deal of support to this thesis, even though questions of interpretation remain to be resolved. It seems worthwhile, therefore, to comment on a series of studies which in a sense are more transparent than those discussed above. Figures 13.9 and 13.10 show the results of an inves-

378   Bioenergetics and linear nonequilibrium thermodynamics

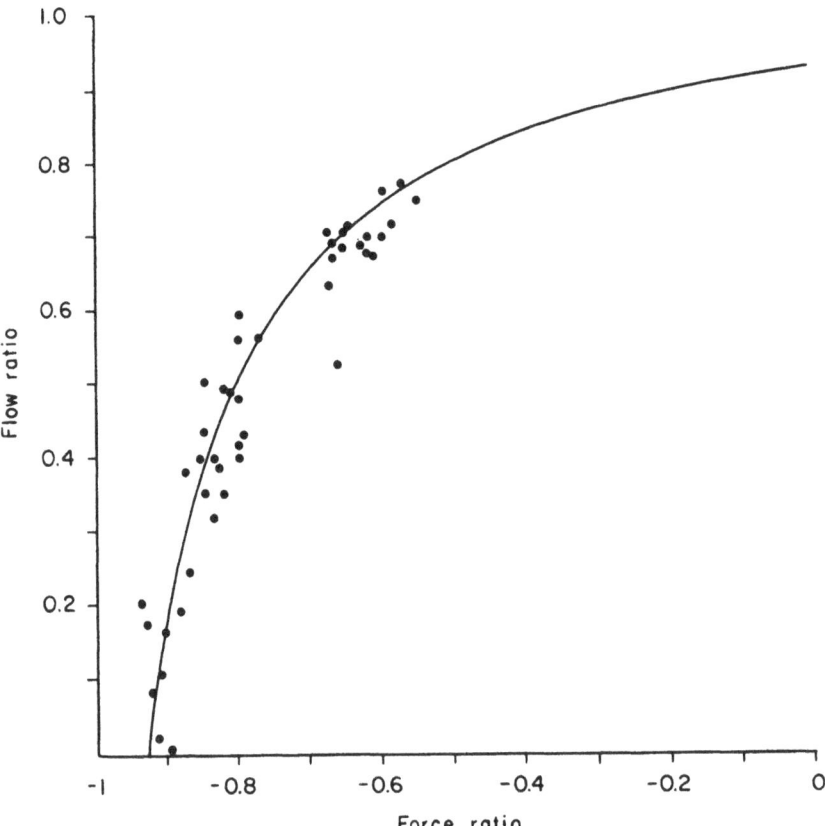

Fig. 13.10. Flow ratio versus force ratio for oxidative phosphorylation. Normalized plot of the data given in Fig. 13.9b. The solid line represents the theoretical curve with parameters as in Fig. 13.9. (J. W. Stucki, private communication.)

tigation of rat liver mitochondria by Stucki (1980) in the presence and absence of diadenosine-pentaphosphate, an adenylate kinase inhibitor. The phosphate potential (A_P) was altered by adding glucose and varying amounts of hexokinase to the incubation medium; it is seen that in the absence of adenylate kinase activity the force ratio could be reduced almost to half of its static head value. The obviously excellent fit of the data to theoretical curves of the type described in Figs. 13.5 and 13.6 suggests possible thermodynamic linearity of the flux–force dependence, from which the phenomenological coefficients can be obtained by a linear regression analysis. The relevant equations are (13.31) and (13.32); it was found that L_{PO} evaluated from the intercept at $A_P = 0$ according to Eq. (13.31), assuming $K_P = 0$, essentially agreed with L_{PO} evaluated from the slope according to Eq. (13.32). Since A_O was not varied in these experiments, one cannot eliminate the possibility of a nonzero value for K_O, although the high degree of coupling found by this procedure suggests that if present it must be small. On the other hand, the possibility of a nonzero value for K_P is clearly eliminated, assuming symmetry. (Under the conditions of Rottenberg's experiments, Fig. 13.7, this was not the case.) With this assumption a value for A_O of 45.27 kcal/mole with a standard deviation of ±3.09 kcal/mole is calculated from the data, which agrees very well indeed with the independent estimate of 45 kcal/mole based on the known composition of the incubation medium (J. W. Stucki, private communication). The reader will not fail to notice that this method of estimating the driving force is identical to that used in the case of epithelial tissues (compare Eq. 8.6).

Very similar results have been obtained in recent studies of the mitochondrial redox and ATP-driven proton pumps, each examined separately (G. F. Azzone, private communication). What is particularly significant in these studies is that incompleteness of coupling appears to be associated almost entirely with the redox proton pump, and that the value of $q(<1)$ is almost unaffected by increasing inhibition of the respiratory chain. This observation suggests that uncoupling due to proton leakage must be negligibly small.

13. 5 The light-driven proton pump of *halobacterium halobium*

Reference has already been made in Chapter 5 to the purple membrane of the halophile *H. halobium*, which enables this bacterium to survive under anaerobic conditions. The purple pigment is a single protein, bacteriorhodopsin, somewhat akin to the visual pigment found in the mammalian rod disc. This protein shows a broad absorption maximum at 570 nm (Oesterhelt and Stoeckenius, 1971; Blaurock and Stoeckenius, 1971). Absorption of light results in conversion of the 570-nm species, through a series of transient intermediates, to a second species which absorbs maximally at 412 nm and reconverts thermally to the 570-nm species within a few milliseconds. This is evidently accompanied by a conformational change in the molecule, which cycles between conformations at about 100 Hz, releasing protons on the exterior side of the membrane and taking them up on the interior. Thus in intact cells the bacteriorhodopsin acts as a light-driven proton pump, and as a consequence the bacterium is able to maintain the requisite ion gradients and phosphorylation rate (Stoeckenius, Lozier, and Bogomolni, 1979; Eisenbach and Caplan, 1979). There is good reason to believe that because of the relative simplicity of the system, this proton pump will eventually be the first example of an active transport mechanism to be elucidated at the molecular level.

As soon as one attempts to describe the bacteriorhodopsin system in terms of the formalism we have been using, one encounters substantial difficulties, both in the development of an appropriate dissipation function and in the expression of the corresponding phenomenological relations. As these are taken in turn, it will be appreciated that we do not have a straightforward means of characterizing the free energy carried and expended by the incident beam of light. Rottenberg (1978) has attempted such a characterization, assuming one can treat the light source as if it were a black body, that is, as if it had a continuous spectrum with a distribution corresponding to some given black-body temperature. In this case an entropy of the radiation can be estimated. But this approach is of questionable validity, and an efficiency calculation on this basis would not appear to be very meaningful, especially since (as pointed out by Rottenberg) the effective black-body temperature of the radiation source is very sensitive to scattering. Moreover, although it is well known (at least in photosynthesis) that at intensities well below saturation, processes such as electron transport, proton transport, phosphorylation, and CO_2 fixation are linear functions of light intensity (for example, see Kok, 1972), it must be remembered that photons are not conserved in such reactions and therefore Onsager symmetry cannot apply, even close to equilibrium. This question has been discussed in some detail in Sec. 5.3. As pointed out by Hill (1977), systems that absorb (or produce) radiant energy are exceptional in important respects and indeed do not fully fit into the formalism of the diagram method any more than they do into that of simple nonequilibrium thermodynamics. Nevertheless, the diagram method can cope very well with the essential *kinetic* behavior of such systems. We shall confine ourselves to this approach, but we draw the interested reader's attention to a comprehensive attempt to treat the special case of bacteriorhodopsin-containing liposomes using linear phenomenological equations (Hellingwerf et al., 1978b, 1979; Westerhoff, Scholte, and Hellingwerf, 1979). This series of studies covers both theory and experiment. Unfortunately, as in the case of the mitochondrial studies described in Sec. 13.4.3, incomplete coupling is attributed entirely to leaks, and the fundamental difficulties raised above are not satisfactorily resolved.

Figure 13.11 illustrates the current "classical" concept of the pho-

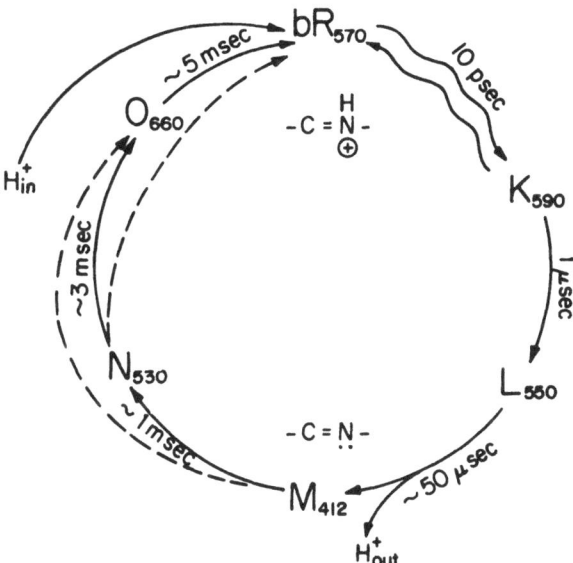

Fig. 13.11. The "classical" photocycle of bacteriorhodopsin, showing the stages at which proton binding (from the inside) or release (toward the outside) most probably occur. Conventional symbols are used; numerical subscripts denote wavelengths of maximum absorption (nm). The approximate reaction half-times for room temperature, neutral pH, and low ionic strength are indicated. The only photoreactions in the scheme are the forward and reverse transitions between bR_{570} and K_{590}; all the remaining reactions are thermal. (Adapted from Stoeckenius, Lozier, and Bogomolni, 1979.)

tocycle of light-adapted bacteriorhodopsin (Stoeckenius, 1979; Stoeckenius, Lozier, and Bogomolni, 1979). The diagram relates a series of spectroscopically identifiable states to one another; simple kinetic information is included in terms of the lifetimes of the states. The nature of the states in regard to molecular conformation is only poorly understood as yet, and the existence of some of the transitions (indicated by dashed lines) is not well established. Nevertheless, additional states and transitions, not indicated in Fig. 13.11, have been suggested (Eisenbach and Caplan, 1979), and the picture that is emerging is increasingly complex. It is also becoming increasingly probable that some of the presently recognized states may in fact represent more than one state, and there is little doubt that branched pathways are present. The additional complexity can readily be incorporated into a diagram such as Fig. 13.11, but it is advantageous, as will be seen, to slightly modify the presentation to conform with that introduced in Chapter 5.

A diagram which includes the most important transitions presently under discussion is given in Fig. 13.12 (compare Fig. 5.4). Shown separately are the major cycles to be considered. It will be seen that an excited state, bR^*, has been interposed between the lowest-energy state bR_{570} and the first phototransient K_{590}. This may possibly be a shared common excited state of both bR_{570} and K_{590}, participating in both the forward and reverse photoreactions indicated in Fig. 13.11, but at present there is disagreement on this (Goldschmidt, Ottolenghi, and Korenstein, 1976; Applebury, Peters, and Rentzepis, 1978). However, at low light intensities (such as are encountered under natural conditions) the photostationary state involves negligible depletion of bR_{570} (Eisenbach and Caplan, 1979), and the reverse photoreaction is therefore also negligible. Relaxation of bR^* apparently occurs thermally via two competing radiationless transitions: forward to K_{590} and back to bR_{570} (Goldschmidt, Ottolenghi, and Korenstein, 1976). We assume the system to be irradiated with monochromatic light far below saturation intensity, and we neglect cycles containing the state L'_{550} (and any other similar states not significant when irradiating with relatively low-intensity light of frequency ν, that is, at 570 nm). We also neglect the transition $M^l_{412} \to O_{660}$, which apparently does not add anything important to the present considerations. Then we have a diagram of thirteen lines and eight states, so that the number of independent, nonzero,

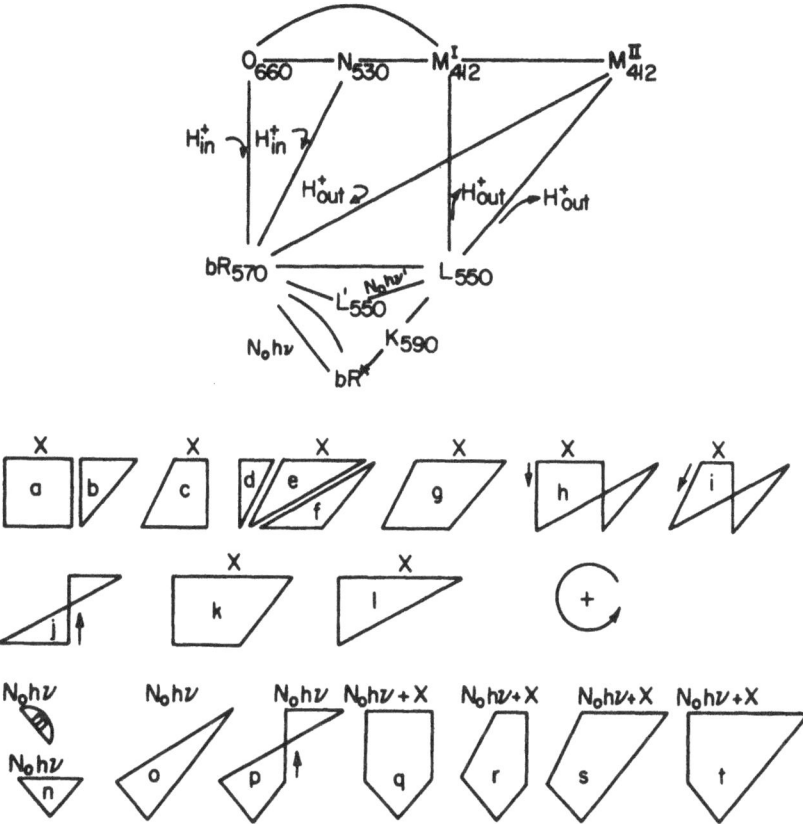

Fig. 13.12. Diagram corresponding to the photocycle of bacteriorhodopsin. The subscripts "in" and "out" refer to the side on which proton binding or release can occur in the given transition; the arrows signify (for convenience) the dominant proton movement during normal functioning. The twenty major cycles with their corresponding forces are represented separately (see text; the positive direction assigned to cycles h, i, j, and p is indicated). $X \equiv \Delta\bar{\mu}_{H^+} = \bar{\mu}_{H^+_{in}} - \bar{\mu}_{H^+_{out}}$. (After Caplan, 1980.)

steady-state transition fluxes is six. As we have seen in Chapter 5, the number of independent cycle fluxes may be larger than this, and indeed we can enumerate twenty cycles, as indicated in Fig. 13.12. Note that cycles f, h, i, j, o, and p require the release of a proton to the outside in one transition, followed by the binding of a proton *from the same side* in a subsequent transition. Of the twenty cycles, four are characterized by zero net force (b, d, f, and j). Thus we might expect to write a sixteen-term dissipation function of the type exemplified by Eq. (5.32). At this point, however, the formalism begins to break down for the reasons mentioned earlier. The proposed sixteen-term function is not a "thermodynamic" quantity, since as emphasized by Hill (1977), a photon beam is not a thermodynamic system. Nevertheless, this pseudo-dissipation function must be positive, since it represents the net rate of loss of photon energy plus free energy in the ensemble together with its baths. Accordingly, the efficiency of conversion of photon energy into electrochemical free energy is given by

$$\eta = \frac{(-X)(J_a + J_c + J_e + J_g + J_h + J_i + J_k + J_l + J_q + J_r + J_s + J_t)}{N_0 h\nu (J_m + J_n + J_o + J_p + J_q + J_r + J_s + J_t)}$$

$$= \frac{-X J_{H^+}}{N_0 h\nu J_r} \tag{13.54}$$

where $X \equiv \Delta\bar{\mu}_{H^+} = \bar{\mu}_{H^+_{in}} - \bar{\mu}_{H^+_{out}}$, J_{H^+} is the net flux of protons from inside to out, and J_r is the overall rate of the light-absorbing reaction $bR_{570} \rightarrow bR^*$. Evidently all cycles other than q, r, s, and t are sources of uncoupling.

A frequently measured property of such systems is the quantum yield, sometimes misnamed the "quantum efficiency." The importance of defining the conditions of measurement properly in order to obtain adequate reproducibility is brought out by the following consideration. At sufficiently low values of both X and the light intensity, it should be possible to find a region of linearity, for which one can write

$$J_{H^+} = L_+ X + L_{+r} N_0 h\nu, \quad J_r = L_{r+} X + L_r N_0 h\nu \tag{13.55}$$

where $L_{+r} \neq L_{r+}$ and indeed L_{r+} may possibly be zero. Nevertheless the maximal quantum yield is clearly given by $(J_{H^+}/J_r)_{X=0} = L_{+r}/L_r$. Hence quantum yields should be determined under level flow conditions. This requirement is satisfied in a suspension of purple membrane fragments, for which Goldschmidt et al. (1976, 1977) have estimated experimentally that the quantum yield of the forward photoreaction alone is only about 0.3 (molecules K_{590} formed per molecule bR_{570} excited). Thus the cycle m must account for a substantial proportion of the energy dissipated.

A schematic representation of the basic free energy levels corresponding to the kinetic diagram for bacteriorhodopsin (given in Fig. 13.12) is shown in Fig. 13.13. Apart from indicating the manifold sources of uncoupling already discussed, it illustrates rather well an important characteristic property of the system. The cycles m, n, o, and p (Fig. 13.12) are totally wasteful but may nevertheless play an important role. It may very well be the case that as purple membrane moves from level flow to static head, the decrease in $J_q + J_r + J_s + J_t$ is exactly or nearly balanced by an increase in $J_m + J_n + J_o + J_p$. This may account for the fact that the overall rate of light energy consump-

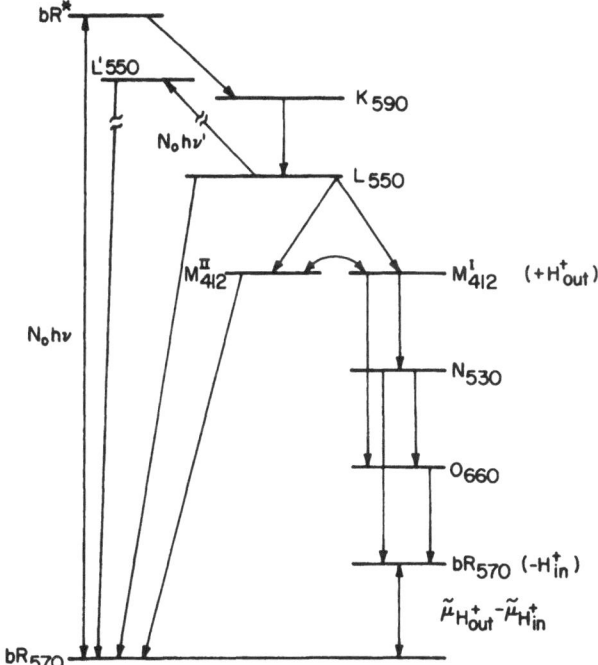

Fig. 13.13. Schematic representation of the basic free energy levels for bacteriorhodopsin corresponding to the kinetic diagram shown in Fig. 13.12. Proton transport is here taken to be at static head. Arrows indicate the predominant direction of cycling. The relative position of the L'_{550} level is not as yet known. (After Caplan, 1982.)

tion during this process appears to be virtually constant (Garty and Caplan, unpublished results). Even claims that a slowdown due to proton "backpressure" can be discerned (Hellingwerf et al., 1978a) are based on extremely small effects. Since a substantial decrease in the rate of the photoreaction $bR_{570} \rightarrow bR^*$ under monochromatic radiation at constant intensity may be out of the question, cycles m, n, o, and p may act as a "safety valve." In general, it is conceivable that multiple cycles may be associated with regulatory properties that are indispensable for effective pumping.

13.6 Conclusions

The thermodynamic considerations discussed in this chapter may be applied to the testing and construction of models. In practice, however, this may not be a simple matter. Even though the number of independent variables is reduced to a minimum, determining the affinities of reactions and electrochemical potential differences across the membrane involves difficulties. It is shown that in the steady state in mitochondria it is legitimate to measure forces (other than $\Delta\tilde{\mu}_H$) externally. Thus it seems that the determinations are possible in principle, at least under some experimental conditions. We stress the importance of measuring both the forces and the flows in the system simultaneously, as far as possible.

All the thermodynamic parameters relating to efficiency of energy conversion in mitochondria may be evaluated from experiments performed in two states: static head (state 4) and a reference state (state 3). It is emphasized that insofar as is practicable, the reference state should be a stationary or practically stationary state rather than a continuum of states through which the system may be evolving (toward state 4). All measurements in the reference state should therefore be made under comparable, if not identical, conditions. Measurements in more than one static head state may be necessary to establish whether the phenomenological equations are truly linear in the thermodynamic sense or involve additive constants which must be allowed for.

A distinction can be made on thermodynamic grounds between the chemiosmotic model on the one hand, and the chemical or parallel coupling models on the other, but not between the chemical and parallel coupling models.

13.7 Summary

1. Three models of energy coupling in oxidative phosphorylation and photophosphorylation are considered: the chemical hypothesis (sharing of a high-energy intermediate), the chemiosmotic hypothesis (coupling via a circulating proton current), and the parallel coupling hypothesis (a combination of the previous two). These three models are shown to be combinations of six possible coupling modes.

2. Focusing on mitochondrial oxidative phosphorylation, we show that the dissipation function for the system may be written $\Phi = J_P A_P^{ex} + J_H \Delta \tilde{\mu}_H + J_O A_O^{ex}$ where A_P^{ex} and A_O^{ex} are the affinities of the phosphorylation and oxidation reactions, respectively, measured externally to the mitochondrion; $\Delta \tilde{\mu}_H$ is the electrochemical potential difference for protons across the mitochondrial membrane, that is, $\tilde{\mu}_H^{in} - \tilde{\mu}_H^{ex}$; J_P and J_O are the rates of the phosphorylation and oxidation reactions, respectively; and J_H is the outward flux of protons, taken as the rate of loss of protons from the interior.

3. The phenomenological equations corresponding to the above dissipation function are interpreted in terms of the three models. While the chemical and parallel coupling hypotheses cannot be formally distinguished, since all the phenomenological coefficients are nonzero for both, the chemiosmotic hypothesis appears as a limiting case in that the cross coefficient L_{PO} is identically zero.

4. Evidence indicating that the mitochondrial system is both linear and symmetrical is presented. This behavior may be attributable to observation in the neighborhood of a multidimensional inflection point, as discussed in Chapter 6. The degree of coupling may be evaluated from studies at static head and in a suitably chosen reference state (since level flow cannot easily be maintained in mitochondria and chloroplasts). This method is based on the classical determination of the respiratory control ratio. A complication may arise here in that if the system is not sufficiently close to equilibrium it may nevertheless appear to behave linearly, but with additive con-

stants appearing in the phenomenological relations. If this is the case, appropriate precautions must be taken in treating the data.

5. The condition for the chemiosmotic hypothesis to hold requires a special relationship between the degrees of coupling: $q_{PO} = -q_{PH}q_{OH}$. This relationship has not as yet been demonstrated.

6. Several examples of linear behavior in mitochondria are presented and discussed, and it is shown that in appropriate circumstances the affinity of the oxidation reaction may be correctly determined by experimental procedures similar to those used in studying epithelial tissues.

7. An introductory account of bacteriorhodopsin is given. The problem of describing the light-driven proton pump in terms of conventional nonequilibrium thermodynamics is examined and shown to be intractable.

8. The photocycle of bacteriorhodopsin is represented as a kinetic diagram, and the component cycles are enumerated and related to the efficiency of energy transduction. The measurement of quantum yield is examined.

9. With the help of kinetic and free energy level diagrams, it is a simple matter to enumerate the possible sources of uncoupling of the photocycle. Moreover, one is led to suggest an important functional role to multiple cycling in the bacteriorhodopsin system.

Afterword

The formalisms presented here are evolving, and it must be anticipated that with additional experimental evidence, which at present is accumulating rapidly, emphases and interpretations are bound to change. NET arose from the attempt to eliminate error resulting from lack of explicit consideration of the influence of coupling between flows. For practical reasons the first treatments were necessarily linear (as in standard equivalent circuit analysis), although it was anticipated that true linearity could only be found near equilibrium. Experience has shown that in many systems linearity is found much further from equilibrium than could have been expected, permitting a systematic characterization over a wide range of conditions. The parameters evaluated by this means are of course phenomenological and must await detailed interpretation at the molecular level. Clearly, local nonlinearity, that is, nonlinearity of the elements making up a system, will have to be taken into account. The mechanisms whereby local nonlinearities combine to give rise to overall linearity are of fundamental interest (Nagel and Essig, 1982) but are largely beyond the scope of this book. This problem might well be approached with the powerful techniques of network thermodynamics. Whatever the nature of the regulatory mechanisms, it appears that linearity may confer important advantages relating to stability (Caplan, 1981), efficiency, and efficacy of force (Stucki, Compiani, and Caplan, 1982).

List of symbols
Notes
References
Index

List of symbols

A	affinity of chemical reaction
A^{conc}	concentration-dependent component of affinity
A'_i	constant additive to affinity
A_{reg}	affinity across nonlinear regulator
A^{stan}	standard affinity
A_{tot}	total fixed affinity of overall reaction sequence
a	mechanical constant of Hill equation
a_i	activity of species i
b	mechanical constant of Hill equation
C	concentration within membrane
$\bar{c}$	logarithmic mean bath concentration
c_i	mean concentration of species i
c_{iA} (c_{iB})	concentration of species i in bath A (B)
D	"bulk" diffusion coefficient
D^*	tracer diffusion coefficient
E	electrical potential difference between reversible electrodes

List of symbols

E, E_{Na}	electromotive force of Na^+ transport
E_H	electromotive force of H^+ transport
"E_{Na}"	apparent electromotive force of Na^+ transport
F	faraday
F	Helmholtz free energy
F_i	nonlinear component of flow equation
F_{ij}	force of frictional interaction between species i and j
f	flux ratio
f	regulator function relating flow and load
$\tilde{f}$	normalized regulator function f
G	Gibbs free energy
G_j	standard or basic free energy of state j of macromolecular system
$\Delta G'_{ij}$	basic free energy change for transition $i \to j$, whether isomeric or associated with ligand binding
g	partial conductance
g	gravitational constant
g	regulator function relating force and load
$\tilde{g}$	normalized regulator function g
H	enthalpy
h	Planck's constant
$h(R_L)$	programming function of muscle regulator
I	electrical current
J	influx
$\overset{\leftarrow}{J}$	efflux, outflux
J_D	diffusional flow
J_i	flow of species i
J'_i	J_i/n
J^∞_{ij}	net mean steady state transition flux between states i and j
J_κ	cycle flux
$J_\kappa^+(J_\kappa^-)$	unidirectional cycle flux in positive (negative) direction
j	ratio of output and input flows
j_i	reduced flows of carrier active transport model

List of symbols 395

K_i	constant component of flow equation
K_m	Michaelis constant
K_{ij}	equilibrium constant for transition between states i and j
K_{ij}^*	second-order equilibrium constant for transition between states i and j
k	rate constant for transport between sites
k_c, k	rate constant for unloaded and loaded carrier, respectively
k_1, k_{-1}	rate coefficients for ith step in reaction
L'	RTL
L	ligand
L_p	filtration coefficient
L_{ij}	phenomenological conductance coefficient relating ith flow to jth force
$\bar{L}_{ij}$	phenomenological coefficient along proper pathway
l	length
M_{in}	influx
M_{out}	efflux, outflux
m	mass
m_R	ratio between load resistance and internal resistance
N	number of units in ensemble
N	number of sites per unit area
N_0	Avogadro's number of photons
n	number of rows in lattice
n	mechanistic stoichiometry
n	number of electrons per mole of reaction
n	number of activated muscle bridges
n_i	mole number
P	tension
P_E	electroosmotic pressure
P_o	isometric tension
PC	phosphocreatine content
P/O	ratio of rates of phosphorylation and oxidation
PMF	protonmotive force

396 List of symbols

p	pressure
p_i^∞	steady state probability of ith state
Q	heat
Q	ratio of bulk and tracer diffusion coefficients
Q'	uncompensated heat
$\dot{Q}$	rate of heat production
Q_{O_2}	rate of oxygen consumption
q	degree of coupling
q_{ij}	degree of coupling between flows i and j
R	gas constant
R	resistance to net flow
R^*	exchange resistance
$\mathscr{R}$	electrical resistance
R_c	respiratory control ratio
R_f	flow ratio of oxidative phosphorylation in reference state
R_L	mechanical resistance; resistive load
R_L^*	distinguished load function
R_{ij}	phenomenological resistance coefficient
r_{ij}	local phenomenological resistance coefficient relating ith force to jth flow
S	entropy
d_eS	entropy exchanged with environment
d_iS	entropy produced within system
T	absolute temperature
t	time
U	internal energy
u	mobility
V	volume
V	velocity of contraction
V, V'	maximum forward and reverse velocities of reaction
$\bar{V}_i, \bar{v}_i$	partial molar volume of species i
V_m	velocity of contraction at zero load
v	velocity of reaction

W	work
W'	useful work
X	thermodynamic force
X	hypothetical metabolic intermediate
$\tilde{X}$	hypothetical metabolic intermediate; high energy state
$\overline{X}$	thermodynamic force along proper pathway
X_i'	nX_i
X_κ	sum of thermodynamic forces operative in cycle κ
x	position
x	fraction of occupied sites
x	ratio of output and input forces
Y	$\Delta G/\Delta H$
Z	phenomenological stoichiometry
Z_{+-}	$\sqrt{R_-^p/R_+^p}$
z_i	charge of species i
α	defined by Eq. (11.43)
α^0	second-order rate constant for adsorption
α_i	first-order rate constant for adsorption from bath i
α_i^*	nth-order rate constant for binding of ligand i
α_{ij}	first-order or pseudo-first-order rate constant for transition $i \to j$
α_{ij}^*	nth-order rate constant for transition $i \to j$
β	rate constant for desorption of ligand
β	partition coefficient
β	permeability
γ	activity coefficient
γ	ratio of external to internal binding constants for ligand
γ	defined by Eq. (10.32)
$\gamma_\pm$	mean ionic activity coefficient
ϵ_J	efficacy of flow
ϵ_X	efficacy of force
ζ	reduced phenomenological stoichiometry
η	efficiency

List of symbols

η_{max}	maximal efficiency
η_{mech}	mechanochemical efficiency
θ	defined by Eq. (10.42)
θ	parameter of Hill equation: a/P_o; b/V_m
θ_{ij}	index of membrane heterogeneity (Eq. 10.53)
κ, κ'	electric conductance
$\kappa \equiv J_{\kappa^-}$	unidirectional cycle flux in negative direction
κ^a	amiloride-sensitive "active" conductance
κ_H^a	H^+ conductance of active transport pathway
κ_{Na}^a	Na^+ conductance of active transport pathway
$\{\kappa\}_i$	set of cycles associated with flow J_i
$\{\kappa\}_{ij}$	subset of cycles in which both X_i and X_j act
μ_i	chemical potential of species i
$\bar{\mu}_i$	electrochemical potential of species i
μ_i^0	standard chemical potential of species i
μ_i^c	concentration-dependent part of μ_i
μ_j	chemical potential of component j in ensemble
$\Delta\mu'_{ij}$	gross free energy change for transition $i \to j$
ν	frequency of radiation
ν_i	stoichiometric coefficient of species i
ν_{jr}	stoichiometric coefficient of species j for reaction r
ξ	degree of advancement of chemical reaction
π	osmotic pressure
Π_{κ^+} (Π_{κ^-})	product of first-order or pseudo-first-order rate constants around cycle in positive (negative) direction
ρ_i	specific activity of species i
Σ	sum of directional diagrams of all states
Σ_k	sum of appendages feeding into cycle k
σ	reflection coefficient
σ_i	contribution to Σ of terms involving c_i
τ_i, τ'_i	transport number of species i
Φ	dissipation function
Φ_{chem}	chemical component of dissipation function

Φ_{mech}	mechanochemical component of dissipation function
ϕ	regulator function relating output power and load
ϕ_i	volume fraction of species i
χ	defined by Eqs. (11.39), (11.46)
ψ	electrical potential
ψ, ψ^{-1}	regulator function relating output force and flow
ω, ω'	permeability for net flow
ω^*	exchange permeability

Subscripts

c	control; contraction
e	experimental
i	species i; impermeant solute; pathway i; inside
m	mucosal; maximal (unloaded contraction)
o	outside; isometric
p	product
r	reaction
s	solute; permeant solute; substrate; serosal
t	time
v	volume
w	water
$+$	cation
$-$	anion

Superscripts

a	active
e	equilibrium; external
i	internal; individual muscle cross bridge
l	basolateral leak; level flow
o	optimal
p	passive
r	reference state
S	static head

List of symbols

ex	external
in	internal
∞	steady state
sb	suprabasal
0	static head; multidimensional inflection point
*	radioactive tracer
'	abundant isotope; distinguished species
obsd	observed

Notes

2. Fundamentals of equilibrium and nonequilibrium thermodynamics

1. It also enables the well-known relations of Maxwell to be written down at sight. For example,

$$\left(\frac{\partial p}{\partial S}\right)_V = -\left(\frac{\partial T}{\partial V}\right)_S$$

2. The electrical work produced by an electrochemical cell or battery, which is often considered in this context, seems to us to be more appropriately dealt with under open systems.

3. This quantity was termed by Clausius the "uncompensated heat."

4. It should be realized that stationary states will be reached by region (0) only if it possesses a very small capacity in comparison with the surrounding region (1). The intensive parameters of region (1) should be altered only minimally during the period of observation, although some slight changes are of course necessary. Such changes are precisely what one measures in characterizing the system.

5. An electrode is reversible with respect to a given ion if it establishes equilibrium with a solution of that ion extremely rapidly. Since we wish to

confine the irreversible processes to the membrane, we do not introduce salt bridges. For the same reason we place the electrodes as close to the membrane as possible.

6. In discontinuous treatments, people often adopt the sign convention given here in order to avoid the repetitive use of minus signs. However, continuous treatments lead naturally to the consideration of negative gradients ($-d\bar{\mu}_i/dx$) as local forces. Integration gives rise to overall forces with polarity the opposite of the sign convention used here. Rather than attempt to use a consistent convention throughout this book, we will follow in each chapter the convention used in related original papers so as to avoid confusion when comparisons are made with the literature. Similar considerations apply to parameters such as pressure, concentration, and electrical potential.

3. Relationships between flows and forces: the Kedem-Katchalsky equations

1. In matrix notation, assuming Onsager symmetry, $R_{ij} = |L_{ij}|/|L|,|L_{ij}|$ is the cofactor of the element L_{ij} in the matrix of L's and $|L|$ is the determinant of the matrix.

2. The quantity c_s can often be adequately approximated by the logarithmic mean, as can be seen from the following argument (Caplan and Mikulecky, 1966). We introduce the concept of a "corresponding" solution whose solute concentration c_s^x corresponds to the chemical potential of solute at point x within the membrane (Kedem and Katchalsky, 1963; Caplan and Mikulecky, 1966). Then, for dilute bathing solutions at constant temperature we have, using Eq. (2.23),

$$c_w^x d\mu_w^c + c_s^x d\mu_s^c = 0$$

Since by definition $d\mu_w^c = -\bar{V}_w d\pi$, and $c_w \bar{V}_w \simeq 1$,

$$d\pi = c_s^x d\mu_s^c \qquad (a)$$

For a salt that dissociates into ν_1 cations and ν_2 anions,

$$\mu_s^c = \nu_1 \mu_1 + \nu_2 \mu_2$$

and

$$\begin{aligned} d\mu_s^c &= \nu_1 RTd \ln a_1 + \nu_2 RTd \ln a_2 \\ &= \nu_1 RTd \ln (\nu_1 c_s \gamma_\pm) + \nu_2 RTd \ln (\nu_2 c_s \gamma_\pm) \\ &= \nu RTd \ln (c_s \gamma_\pm) \end{aligned} \qquad (b)$$

where we have introduced the mean ionic activity coefficients $\gamma_\pm$ and the relation $\nu = \nu_1 + \nu_2$. A corresponding expression holds for $\Delta\mu_s^c$. Thus, with Eq. (3.31)

$$c_s = \frac{\int_{(1)}^{(2)} c_s^x d\mu_s^c}{\Delta\mu_s^c} = \frac{\int_{(1)}^{(2)} c_s^x d\ln(c_s^x\gamma_\pm)}{\Delta\ln(c_s\gamma_\pm)}$$

$$= \frac{\int_{(1)}^{(2)} (dc_s^x + c_s^x d\ln\gamma_\pm)}{\Delta\ln c_s + \Delta\ln\gamma_\pm}$$

$$= \left(\frac{\Delta c_s}{\Delta\ln c_s}\right) \frac{\left[1 + \left(\int_{(1)}^{(2)} c_s^x d\ln\gamma_\pm\right)/\Delta c_s\right]}{[1 + (\Delta\ln\gamma_\pm)/\Delta\ln c_s]} \quad (c)$$

In most cases of practical interest, the right-hand terms in the brackets are small with respect to unity, and indeed they are frequently very close to one another in value, even though deviations from ideality may be appreciable (Kedem and Leaf, 1966). The definition given by Eq. (3.31) then becomes

$$c_s = \frac{\Delta c_s}{\Delta \ln c_s} \quad (d)$$

For example, for two solutions containing NaCl at 0.1 M and 0.01 M, the values of c_s given by Eqs. (3.31) and (d) differ by less than 2%. However, as the lower concentration approaches zero, both definitions become inappropriate. As $c_s^{(1)}/c_s^{(2)}$ approaches unity, the arithmetic mean becomes an increasingly good approximation to the logarithmic mean.

Equations (b) and (c) apply also to nonelectrolyte solutions if ν is set equal to 1, and $\gamma_\pm$ is replaced by γ. Since nonelectrolyte solutions are often near-ideal, Eq. (d) is often taken as a primary definition of c_s for nonelectrolytes.

3. In their original formulation Kedem and Katchalsky employed the coefficient P_E/κ, where the electroosmotic pressure $P_E = [(\Delta p - \Delta\pi)/E]_{J_s,\Delta\pi_s}$. We have used instead the equivalent quantity $(-\beta/L_p)$, since experimental measurements involve the electroosmotic permeability (or streaming potential) β more commonly than the electroosmotic pressure.

4. Effectiveness of energy conversion

1. The degree of coupling between a pair of flows may likewise be given for a system in which more than two flows interact (Caplan, 1966). In this

case the definition is analogous to Eq. (4.13); that is, $q_{ij} = -R_{ij}/\sqrt{R_{ii}R_{jj}}$. This measures the extent to which the ith flow is dragged by the jth flow when no other flows are present and the force conjugate to the ith flow is zero.

2. These considerations have been extended to multiple-flow systems (Caplan, 1966; Peusner, 1970) and to two-flow systems which are not necessarily time-invariant or symmetrical (Peusner, 1970). Stucki (1980) has drawn attention to the impossibility of obtaining a favorable output at maximal efficiency unless $q < 1$. He has derived four "distinguished" degrees of coupling (optimal or economical values of q) by optimizing four different physically meaningful output functions.

5. The diagram method

1. For convenience and consistency, the thermodynamic forces are defined here on a molar basis; that is, the gas constant R is used rather than the Boltzmann constant k (the latter appears in the original papers of Hill and coworkers).

2. Avogadro's number (N_0) of photons.

6. Possible conditions for linearity and symmetry of coupled processes far from equilibrium

1. In many systems the overall steady-state fluxes, for example, transport and oxygen consumption, are readily observable, whereas observation and measurement of cycle fluxes may be experimentally difficult. In some cases, however, it is possible by optical or other appropriate techniques to distinguish between the various states of the system and to measure their kinetics.

2. In Eq. (6.14), for simplicity the flows are normalized so that in the case of cycle a (complete coupling) their ratio is 1. The means of treating flows with a stoichiometric ratio other than unity is dealt with in Sec. 5.3.

3. It is important to note that the meaning of our term "reference steady state" differs from that of the term "reference state" as used by Sauer (1973). In his usage, "reference state" denotes the specification of the values of a minimum set of parameters which, when taken in conjunction with the values of the forces, suffices to define the values of all the flows. The choice of reference state is arbitrary, an infinite number being permissible. In these terms a proper pathway results from variation of the chosen reference state in such a way as to permit a linear description of the system. In our usage, "reference steady state" refers to a specific steady state of the system, around which perturbations are carried out.

4. The expansion of Rothschild et al. includes only a single second-order term for the case under consideration, namely, $\alpha \ln (c_i/c_i^0) \ln (c_j/c_j^0)$, where α is a constant (for a given flow) and c_i^0 and c_j^0 are the reference state concentrations. In our terms this corresponds to a term $\beta \delta X_i \delta X_j$, which can be neglected for perturbations in which at least one force remains very close to its reference value.

7. Energetics of active transport: theory

1. For completely coupled systems all R's become infinite, leading to indeterminacy. These systems may be treated by using the L formulation.

2. Recent evidence indicates that in the frog skin both passive transport at the apical cell surface and active transport at the basal lateral surface are nonlinear processes (Nagel and Essig, 1982), whereas transepithelial active transport and oxidative metabolism are linear. For phenomenological analysis the mechanisms by which nonlinear elements interact so as to give overall linearity are irrelevant. As indicated above, given linearity, the phenomenological equations of the composite transport system can be written directly without looking inside the black box. The important questions of the nature of the phenomenological coefficients and A remain to be investigated. This will necessarily involve the use of models and may be facilitated by the techniques of network thermodynamics.

3. Although the phenomenological equations derived in this way indicate linearity, remember that the resistance coefficients are functions of state. The effect of X_+ on intracellular concentrations is unknown. The concentration profile of the leak pathway should in general be affected more by variations of the concentration difference across the membrane than by thermodynamically equivalent variations of the electrical potential difference, particularly with coupling of ion and water flows. Therefore we might expect Eq. (7.14) to be more useful in describing the effect of $\Delta\psi$ than that of the concentration difference Δc_+. In the absence of an electrical field and coupled flows, $J_+^p = X_+/R_+^p = -RT\Delta \ln c_+/R_+^p$, and is given also by $J_+^p \simeq -\omega_+^p RT\Delta c_+$, where for ideal dilute solutions ω_+^p is approximately constant. Thus, $R_+^p \simeq \Delta \ln c_+/\omega_+^p \Delta c_+$.

4. The electrical potential difference between two solutions, $\Delta\psi$, is usually measured with calomel electrodes and salt bridges. The measurement of the electrochemical potential difference requires the use of reversible electrodes. If the two solutions are identical, $-X_+/F = \Delta\psi$; if they differ, this is not the case. In these circumstances $\Delta\psi$ is imprecisely defined owing to the uncertain contribution of diffusion potentials at the bridge tips. However, X_+ is always precisely defined, even in the presence of highly charged colloidal particles or macromolecules as in intracellular media, where estimates of $\Delta\psi$ may be seriously in error. Reversible electrodes are available for sev-

eral ions; electrodes of high specificity for Na^+, K^+, H^+, Ca^{++}, and Cl^- are commercially available, and further developments are to be anticipated.

5. The affinity has been designated by the symbols A, "A", and A_{app} (apparent affinity) in different publications.

6. See note 2, above.

7. The affinity of a reaction is defined as $A = -\Sigma_i \nu_i \mu_i$ (De Donder; see Chapter 2) where the stoichiometric coefficients ν_i are considered positive for products and negative for reactants. Here the rate of the reaction is defined as $J_r = (1/\nu_i)(dn_i/dt)$, where dn_i/dt is the rate of appearance of the ith component (it is obviously immaterial which reactant or product is considered). The rate of expenditure of free energy by the reaction is $-\Sigma_i \mu_i(dn_i/dt) = \Sigma_i \nu_i\mu_i(1/\nu_i)(dn_i/dt) = AJ_r$. However, experimentalists customarily choose to determine the rate of consumption or production of some easily measured species, for example, dn_j/dt. If we call this measured rate of reaction J_r^j, we have $J_r^j = |\nu_j|J_r$. Since the rate of expenditure of free energy is independent of the convention, the affinity must now be expressed as kcals/mole of species j. Thus $A^j = A/|\nu_j|$ and $A^j J_r^j = AJ_r$, irrespective of which species j is chosen. Furthermore, the validity of the Onsager relation is independent of the choice of species, since Eqs. (7.3) and (7.4) become

$$X_{Na} = R^a_{Na}J^a_{Na} + (R^a_{Na,r}/|\nu_j|)J^j_r = R^a_{Na}J^a_{Na} + R^{aj}_{Na,r}J^j_r,$$

$$A^j = A/|\nu_j| = (R^a_{Na,r}/|\nu_j|)J^a_{Na} + (R^a_r/\nu_j^2)J^j_r = R^{aj}_{Na,r}J^a_{Na} + R^{aj}_r J^j_r$$

Note that the degree of coupling is also unaffected, since

$$\frac{-R^{aj}_{Na,r}}{\sqrt{R^a_{Na}R^{aj}_r}} = q^a$$

Obviously, if the metabolism of oxygen is completely coupled to that of glucose ($J_r = J_r^{O_2}/|\nu_{O_2}| = J_r^G/|\nu_G|$), it is immaterial which is represented by j. Furthermore, since the set of stoichiometric coefficients for a reaction may be multiplied by an arbitrary factor without changing the value of $J_r A$, we may set $|\nu_j| = 1$ for any desired representative species j, for example, oxygen. Then $J_r^j = J_r^{O_2} = J_r$, and $A^j = A^{O_2} = A$.

8. Energetics of active transport: experimental results

1. Our stress on symmetry and 6-min duration of perturbations in the above studies was based on the observation of Vieira et al., discussed above, that prolonged clamping of $\Delta\psi$ at any given value influenced the rate of metabolism on subsequent return to the short-circuit state. Other studies have shown that even with these precautions, exposure to negative values of

$\Delta\psi$ may introduce ambiguities into the I^a–$\Delta\psi$ relationships, owing to effects on both cellular and paracellular pathways (Voûte and Ussing, 1968; Wolff and Essig, 1980; Bobrycki et al., 1981). Accordingly, it may be desirable in future studies to limit $\Delta\psi$ to positive values, symmetric with respect to the equilibration potential.

2. As is indicated, the application of Eq. (8.6) requires that A remain constant on perturbation of $\Delta\psi$. Nevertheless, in this chapter, which deals with experimental manipulations, we shall employ total rather than partial derivatives, since the value of A is not under direct experimental control.

3. Labarca, Canessa, and Leaf (1977) disagree with the above analysis, having demonstrated that $J^a_{Na}/J^{sb}_{CO_2}$ was statistically the same at $+50$ mV and -50 mV. This difference of view appears to be based on our different perceptions of the meaning of the term "steady state." In their study, since it was considered that changes in $\Delta\psi$ caused time-dependent changes in both J^a_{Na} and $J^{sb}_{CO_2}$ lasting 20–40 min, comparisons were based on measurements made at least 20 min following perturbation of $\Delta\psi$, at which time the ratio had become constant. We suggest that a "quasi-steady state," in which J^a_{Na} is conservative and appropriately related to J^{sb}_r, may be achieved much sooner. Initially our reason for this belief was the finding discussed above, that with symmetric perturbations of $\Delta\psi$ for 6 min, both I^a and J_{O_2} are linear in $\Delta\psi$ over an extended range. More recently, toad urinary bladders were studied in a continuously perfused chamber in which well-defined mixing characteristics allowed bathing solution CO_2 concentration to be related continuously to the rate of tissue CO_2 efflux, J_{CO_2} (Rosenthal, King, and Essig, 1979, 1981). It was found that although a step change in $\Delta\psi$ resulted in long-term transients in bath CO_2 concentration, calculations correcting for system capacity demonstrated that J_{CO_2} became constant (within the noise of the system) within about 3 min and remained so until about 7 min after perturbing $\Delta\psi$. This permitted a test for stoichiometry by perturbation of $\Delta\psi$ over an appropriate range (see Fig. 8.16). (Assuming, for example, the mean parameters found by Lang et al. in the 11 hemibladders discussed above, viz. $q = 0.86$; $Z = \sqrt{L^a_{Na}/L_r} = 16.8$; and $A = 57.1$ kcal · mole^{-1} O_2, setting $\Delta\psi = -50$ mV and $+50$ mV would give values of $J^a_{Na}/J^{sb}_{O_2}$ of 15.6 and 12.4 respectively, a decrease of 21% over a range of 100 mV. On the other hand, setting $\Delta\psi = 0$ mV and $+100$ mV would give values of $J^a_{Na}/J^{sb}_{O_2}$ of 12.3 and 7.2 respectively, a decrease of 41%.) In studies of 11 hemibladders during the early plateau some 4–7 min after setting $\Delta\psi$ at $+100$ mV, the mean decrement of $I^a/FJ^{sb}_{CO_2}$ was 46.9 ± 6.1(S.D.)%. Presuming that during the early plateau $I^a/F = J^a_{Na}$, these findings are consistent with incomplete coupling, demonstrable prior to changes of tissue parameters with long-term perturbation of $\Delta\psi$.

4. Clearly, the evaluation of q and J_{Na0}/J^{sb}_{r0} requires that under the conditions of our studies basal and suprabasal metabolism must be discrete and that the rate of basal metabolism must not be affected significantly by alteration of the rate of active Na transport. Al-Awqati, Beauwens, and Leaf

(1975); Coplon, Steele, and Maffly (1977); and Lang, Caplan, and Essig (1977a, b) have presented evidence that these conditions are satisfied in the toad urinary bladder. Several other observations suggest that in the presence of adequate concentrations of amiloride the rate of metabolism in both the frog skin and the toad urinary bladder is constant and not attributable to epithelial-cell active Na transport: J_{Nao}/J_{r0}^{sb} calculated on the basis of this assumption remains constant as J_{Nao} changes under diverse conditions; administration of amiloride lowers J_{r0} to the same level observed on substitution of Mg for external Na; in the absence of external Na, amiloride has no effect on J_{r0} (Lau, Lang, and Essig, 1979) and, in the toad urinary bladder, recirculation of Na is not demonstrable (Canessa, Labarca, and Leaf, 1976).

5. In the CO_2-free system, although J_H^a is highly sensitive to change of mucosal pH, it is not noticeably affected by large changes of serosal pH.

6. Studies of toad urinary bladder metabolism in a chamber with well-defined flow and mixing characteristics demonstrate that J_{CO_2} is near-constant about 3–7 min following a perturbation of $\Delta\psi$, whereas the bathing solution CO_2 concentration becomes constant only after ≥15 min (Rosenthal, King, and Essig, 1981). The published data of Al-Awqati et al. do not permit the calculation of J_{CO_2} during the period when bath CO_2 concentration is changing.

7. Although the above findings speak for a high degree of coupling of H^+ transport to oxidative metabolism, further study may be necessary to determine whether coupling is quite complete. One factor of possible significance is the 10–15 min required for bathing solution $^{14}CO_2$ and CO_2 concentrations to become constant following perturbation of pH, presumably partially as a consequence of chamber capacity factors, which can give an erroneous impression of the time required for the tissue CO_2 efflux rate to become constant (Rosenthal, King, and Essig, 1979, 1981). As for the Na transport system, it would be pertinent to examine coupling on the basis of measurements over as short an interval as possible, thereby minimizing the likelihood of change in tissue parameters. Similarly, it will be of interest to reexamine this issue when further information is available concerning the mechanisms whereby CO_2 and inhibitors affect pump function.

9. Kinetics of isotope flows: background and theory

1. One early example of the useful application of radioisotope techniques was the investigation of membrane permeability to sodium. For years it had been felt that the marked difference in ionic composition inside and outside typical animal cells—in which the concentration of sodium is substantially lower than in the extracellular fluid—might be attributable to the cell membrane's impermeability to this ion. The demonstration that cell membranes are highly permeable to ^{24}Na, however, necessitated a major revision in concepts concerning regulation of cellular ionic composition.

2. Schwartz (1971) has pointed out that homogeneity in the plane of the membrane is unnecessary in the absence of carriers and coupling of flows.

3. For the three-flow case the quantity $1/R_{22}$ appearing in Eq. (3.46) is identifiable with $c_s\omega$ in Eq. (3.50). Although Eq. (9.18) represents a formally analogous relationship for the n-flow case, it should be appreciated that the nature of the conjugate forces and flows in the two cases is different. In the latter case we are dealing entirely with material flows.

4. Although, as mentioned above, it was natural to use a resistance formulation in developing our general relationships, in applying them it is frequently more convenient to consider permeability coefficients. The equivalence of the two formulations is apparent. In the following we shall deal usually with permeabilities except where it is also useful, or perhaps preferable, to consider resistances.

5. Dawson (1977) has presented an ingenious combined thermodynamic and kinetic derivation which leads simply to a relationship between the flux ratio and the forces promoting transport analogous to our Eq. (9.38). Since the coupling of flows considered includes that between abundant and tracer species, the formulation incorporates the influence of isotope interaction without, however, explicitly identifying the ratio R^*/R. It is instructive to note that the equivalence of coupling (isotope interaction) and discrepancy between tracer and net permeability coefficients can be demonstrated simply for systems near equilibrium, where linearity in X (and Δc) and Onsager reciprocity can be assumed (Essig, Kedem, and Hill, 1966). For tracer flow we write

$$J^* = A\delta c^* + B\delta c \quad (i)$$

where $B \neq 0$ introduces isotope interaction. For net flow, $J = C\delta c$. In order to consider appropriate conjugate flows and forces, we substitute $J^* + J'$ for J, and $c^* + c'$ for c, where the superscript ' refers to the abundant isotope. Then

$$J^* = (A + B)\delta c^* + B\delta c' \quad (ii)$$

$$J' = (C - A - B)\delta c^* + (C - B)\delta c' \quad (iii)$$

For systems near equilibrium in the absence of a hydrostatic pressure gradient, $X^* \simeq RT\,\delta\ln c^* \simeq RT\delta c^*/c^*$ and $X' \simeq RT\delta c'/c'$, so that

$$J^* = \frac{(A + B)}{RT} c^* X^* + \frac{B}{RT} c' X'$$

and

$$J' = \frac{(C - A - B)}{RT} c^* X^* + \frac{(C - B)}{RT} c' X'$$

Onsager reciprocity now gives $Bc' = (C - A - B)c^*$. Replacing c' by $c - c^*$ and rearranging,

$$\frac{A}{C} = 1 - \frac{c}{c^*}\frac{B}{C}$$

But $A/C = \omega^*/\omega$, $c/c^* = 1/\rho$, and $B/C = (J^*/J)_{\delta c^* = 0}$, so

$$\frac{\omega^*}{\omega} = 1 - \left(\frac{J^*}{J}\right)_{\delta c^* = 0} \tag{iv}$$

Thus we see that here ω^* differs from ω if and only if there is coupling between net flow and tracer flow, and that the magnitude of the discrepancy is determined by the extent of such isotope interaction. (We note that although the above proof required that both the total and tracer species be near equilibrium, it is in fact not important that $\delta c^* \ll c^*$, since the tracer permeability $J^*/RT\delta c^*$ is highly insensitive to c^*.)

10. Kinetics of isotope flows: mechanisms of isotope interaction

1. Strictly, the normal flux ratio is

$$f = \exp(X/RT) = \exp[\ln(c^I/c^{II}) - \overline{V}_s \Delta p/RT]$$
$$= (c^I/c^{II}) \exp(-\overline{V}_s \Delta p/RT)$$

However, introducing Eq. (10.30) shows that

$$\exp(-\overline{V}_s \Delta p/RT) \simeq 1 - \overline{V}_s \Delta c_s (\Sigma\, \sigma_i L_{pi}/L_p)$$

Since for $0 \leq \sigma_i \leq 1$, $\Sigma\, \sigma_i L_{pi} \leq L_p$ and $\overline{V}_s \Delta c_s \ll 1$, $\exp(X/RT)$ differs insignificantly from (c^I/c^{II}).

11. Kinetics of isotope flows: tests and applications of thermodynamic formulation

1. Equations (11.19) and (11.20) are equivalent to relationships first derived by Chen and Walser (1974), their Eqs. (12)–(15). Equation (11.16) is their Eq. (7), Eq. (11.24) is their Eq. (22), Eq. (11.27) is their Eq. (30), and Eqs. (11.53) and (11.54) are their Eqs. (19) and (18), respectively. It should be

noted that their polarity conventions are such that the current I and net flows J are of opposite sign to ours, whereas the flux ratio is the same as ours.

12. Muscular contraction

1. The essential energy-yielding reaction in striated muscle *in vivo* is the splitting of phosphocreatine; the concentration of ATP remains constant as a consequence of rapid reconstitution (Carlson and Siger, 1960; Cain, Infante, and Davies, 1962). However, ATP is the only energy source in muscles pretreated with 2,4-dinitrofluorobenzene to inhibit creatine phosphoryltransferase. The efficiency based on the usage of ATP in such muscles has been measured directly by Kushmerick and Davies (1969).

2. Note that the use of J_{r_0} to denote the rate of metabolic reaction at static head, while consistent with accepted usage in the muscle literature, should not be confused with the convention used in our transport studies, where J_{r_0} denotes a level flow quantity.

13. Energy coupling in mitochondria, chloroplasts, and certain halophilic bacteria

1. Under many conditions respiration can take place at reasonable rates without net phosphorylation. The uncoupling may be the consequence of side reactions or leaks, as will be discussed later.

2. It is clear that electroneutrality necessitates that a net flow of H^+ be accompanied by an equivalent flow of anions and/or cations. If these flows occur through leak pathways, they are not coupled to the metabolic processes and so need not be included in this dissipation function; that is, each flow of Eq. (13.3) is a function of only three forces. If some compensating flow happens to be coupled to metabolism, its contribution may be completely eliminated from the dissipation function by making its conjugate force zero. It seems that the important compensatory ion in mitochondria is potassium, the flow of which appears not to be coupled to metabolism (Rottenberg, Caplan, and Essig, 1970).

3. Except at equilibrium, internal and external reaction affinities will be unequal. Consider an enclosed region into which a reactant α diffuses and forms a product β, which diffuses outward. Then $\mu_\alpha^{ex} > \mu_\alpha^{in}$ and $\mu_\beta^{in} > \mu_\beta^{ex}$. Adding these, $\mu_\alpha^{ex} - \mu_\beta^{ex} > \mu_\alpha^{in} - \mu_\beta^{in}$; that is, $A^{ex} > A^{in}$. A similar consideration applies to oxidative phosphorylation.

4. Note that the addition of an uncoupler which provides a proton leak through the coupling membrane can be considered analogous to the use of a voltage clamp as described in Chapter 8. Then the net flow J_H in the formalism

does *not* include the flow through the uncoupler site, but $\Delta\bar{\mu}_H$ is decreased. We can continue to use the original phenomenological coefficients, remembering that they characterize the original membrane (to which a parallel pathway has been added). Alternatively, if the total proton flux J_H^{tot}, which includes leakage flux, is to be measured, it may be preferable to characterize the membrane as a whole. In this case the only effect of adding the parallel leakage pathway is that Eq. (13.12) becomes

$$J_H^{tot} = J_H + J_H^{leak} = L_{PH}A_P^{ex} + (L_H + L_H^{leak})\Delta\bar{\mu}_H + L_{OH}A_O^{ex}$$

and Eq. (13.19) takes an analogous form. Here L_H^{leak}, the conductance coefficient of the leak, may be very large (leading to correspondingly small static head values of $\Delta\bar{\mu}_H$ as compared with identical conditions in the absence of an uncoupler). Other uncoupling techniques, such as the use of valinomycin in the presence of high levels of potassium to reduce $\Delta\psi$ essentially to zero, may in principle result in changes in all six conductance coefficients, since the state of the system under stationary conditions may now be very different (the phenomenological coefficients corresponding to Eq. 13.9 are functions of all those corresponding to Eq. 13.8, and changes in the latter can affect the former, even though the flows most directly involved are at static head or equilibrium).

5. $L_{ij} = R'_{ij}/D$, where R'_{ij} is the cofactor of R_{ij}, and D the determinant of the resistance matrix. Since $L_{PO} = 0$ for the chemiosmotic hypothesis, $R'_{PO} = R_{PH}R_{OH} - R_{PO}R_H = 0$. Dividing through by $R_H\sqrt{R_PR_O}$ leads to Eq. (13.24).

6. Here we refer to techniques which may be considered to affect the force but not the elemental phenomenological coefficients.

References

2. Fundamentals of equilibrium and nonequilibrium thermodynamics

Caplan, S. R., and Mikulecky, D. C. 1966. In *Ion exchange*, vol. 1, ed. J. A. Marinsky. New York: Dekker.
Denbigh, K. 1966. *The principles of chemical equilibrium.* London: Cambridge University Press.
Katchalsky, A., and Curran, P. F. 1965. *Nonequilibrium thermodynamics in biophysics.* Cambridge, Mass.: Harvard University Press.
Lakshminarayanaiah, N. 1969. *Transport phenomena in membranes.* New York: Academic Press.
Mikulecky, D. C. 1969. In *Transport phenomena in fluids*, ed. H. J. M. Hanley. New York: Dekker.
Prigogine, I., and Defay, R. 1954. *Chemical thermodynamics*, trans. D. H. Everett. London: Longmans.
Tisza, L. 1966. *Generalized thermodynamics.* Cambridge, Mass.: M.I.T. Press.

3. Relationships between flows and forces: the Kedem-Katchalsky equations

Blumenthal, R., Caplan, S. R., and Kedem, O. 1967. *Biophys. J.* 7: 735.
Caplan, S. R. 1973. *Quaderni dell' Istituto di Ricerca Sulla Acque* 10: 7.

Caplan, S. R., and Mikulecky, D. C. 1966. In *Ion exchange*, vol. 1, ed. J. A. Marinsky. New York: Dekker.
Curie, P. 1894. *Journal de Physique* ser. 3, 3: 393.
DeGroot, S. R., and Mazur, P. 1963. *Non-equilibrium thermodynamics*. Amsterdam: North-Holland.
DeSimone, J., and Caplan, S. R. 1973. *J. Theoret. Biol.* 39: 523.
Finlayson, B. A., and Scriven, L. E. 1969. *Proc. Roy. Soc. London*, ser. A, 310: 183
Hearon, J. Z. 1950. *Bull. Math. Biophys.* 12: 57, 135.
Katchalsky, A., and Curran, P. F. 1965. *Nonequilibrium thermodynamics in biophysics*. Cambridge, Mass.: Harvard University Press.
Katchalsky, A., and Kedem, O. 1962. *Biophys. J.* 2: 53.
Kedem, O. 1961. In *Membrane transport and metabolism*, ed. A. Kleinzeller and A. Kotyk. New York: Academic Press.
——— 1973. *J. Phys. Chem.* 77: 2711.
Kedem, O., and Katchalsky, A. 1958. *Biochim. Biophys. Acta* 27: 229.
——— 1961. *J. Gen. Physiol.* 45: 143.
——— 1963a. *Trans. Faraday Soc.* 59: 1918.
——— 1963b. *Trans. Faraday Soc.* 59: 1931.
——— 1963c. *Trans. Faraday Soc.* 59: 1941.
Kedem, O., and Leaf, A. 1966. *J. Gen. Physiol.* 49: 655.
Mazur, P., and Overbeek, J. T. 1951. *Rec. trav. chim.* 70: 83.
Michaeli, I., and Kedem, O. 1961. *Trans. Faraday Soc.* 57: 1185.
Miller, D. 1960. *Chem. Rev.* 60: 15.
Prigogine, I. 1947. *Etude thermodynamique des phénomènes irreversibles*. Liège, Belgium: Desoer.
——— 1961. *Thermodynamics of irreversible processes*. New York: Wiley.
Staverman, A. J. 1951. *Recueil trav. chim.* 70: 344.
Weinstein, J. N., and Caplan, S. R. 1968. *Science* 161: 70.
——— 1973. *J. Phys. Chem.* 77: 2710.
Weinstein, J. N., Bunow, B. J., and Caplan, S. R. 1972. *Desalination* 11: 341.

4. Effectiveness of energy conversion

Awan, M. Z., and Goldspink, G. 1972. *J. Mechanochem. Cell Motility* 1: 97.
Blumenthal, R., Caplan, S. R., and Kedem, O. 1967. *Biophys. J.* 7: 735.
Caplan, S. R. 1966. *J. Theoret. Biol.* 10: 209; 11: 346.
Essig, A., and Caplan, S. R. 1968. *Biophys. J.* 8: 1434.
Hodgkin, A. L., and Keynes, R. D. 1955. *J. Physiol.* 128: 61.
Hoshiko, T., and Lindley, B. D. 1967. *J. Gen. Physiol.* 50: 729.
Jardetzky, O., and Snell, F. M. 1960. *Proc. Natl. Acad. Sci. USA* 46: 616.
Kedem, O. 1961. In *Proc. Symp. Transport and Metabolism*. New York: Academic Press.

Kedem, O., and Caplan, S. R. 1965. *Trans. Faraday Soc.* 61: 1897.
Kedem, O., and Essig, A. 1965. *J. Gen. Physiol.* 48: 1047.
Kushmerick, M. J., Larsen, R. E., and Davies, R. E. 1969. *Proc. Roy. Soc. B.* 174: 293.
Miller, D. G. 1960. *Chem. Rev.* 60: 15.
Peusner, L. 1970. Ph.D. thesis, Harvard University.
Rosenberg, T. 1954. In *Active transport and secretion*, ed. R. Brown and J. F. Danielli. New York: Academic Press.
Rottenberg, H. 1973. *Biophys. J.* 13: 503.
Rottenberg, H., Caplan, S. R., and Essig, A. 1967. *Nature* 216: 610.
Stucki, J. W. 1980. *Eur. J. Biochem.* 109: 269.
Ussing, H. H. 1960. In *The alkali metal ions in biology*, ed. H. H. Ussing et al. Berlin: Springer-Verlag.
Ussing, H. H., and Zerahn, K. 1951. *Acta Physiol. Scand.* 23: 110.
Zerahn, K. 1958. Ph.D. thesis, Copenhagen University, Aarhus, Denmark.

5. The diagram method

Hill, T. L. 1977. *Free energy transduction in biology—the steady-state kinetic and thermodynamic formalism.* New York: Academic Press.
Hill, T. L., and Eisenberg, E. 1981. *Quart. Rev. Biophys.* 14: 463.
King, E. L., and Altman, C. 1956. *J. Phys. Chem.* 60: 1375.
Mikulecky, D. C. 1977. *J. Theor. Biol.* 69: 511.
Mikulecky, D. C., and Thomas, S. R. 1978. *J. Theor. Biol.* 73: 697.
Oster, G. F., Perelson, A. S., and Katchalsky, A. 1973. *Quart. Rev. Biophys.* 6: 1.

6. Possible conditions for linearity and symmetry of coupled processes far from equilibrium

Bunow, B. 1978. *J. Theor. Biol.* 75: 51.
Canessa, M., Labarca, P., and Leaf, A. 1976. *J. Membrane Biol.* 30: 65.
Caplan, S. R. 1981. *Proc. Natl. Acad. Sci. USA* 78: 4314.
Caplan, S. R., and Essig, A. 1977. In *Current topics in membranes and transport*, vol. 9, ed. F. Bronner and A. Kleinzeller. New York: Academic Press.
Chen, J. S., and Walser, M. 1975. *J. Membrane Biol.* 21: 87.
Dawson, D. C., and Al-Awqati, Q. 1978. *Biochim. Biophys. Acta* 508: 413.
DeGroot, S. R., and Mazur, P. 1962. *Non-equilibrium thermodynamics.* Amsterdam: North-Holland.
Essig, A., and Caplan, S. R. 1981. *Proc. Natl. Acad. Sci. USA* 78: 1647.
Garay, R. P., and Garrahan, P. J. 1973. *J. Physiol.* 231: 297.

Glansdorff, P., and Prigogine, I. 1971. *Thermodynamic theory of structure, stability, and fluctuations*. London: Wiley-Interscience.
Hill, T. L. 1977. *Free energy transduction in biology*. New York: Academic Press.
―――― 1982a. *J. Chem. Phys.* 76: 1122.
―――― 1982b. Unpublished studies.
Marmor, M. F. 1971. *J. Physiol.* 218: 599.
Mikulecky, D. C. 1977. *J. Theor. Biol.* 69: 511.
Nicolis, G., and Prigogine, I. 1977. *Self-organization in nonequilibrium systems*. New York: John Wiley and Sons.
Oster, G., and Perelson, A. 1974. *IEEE Trans. Circuits and Systems, CAS* 21: 709.
Oster, G. F., Perelson, A. S., and Katchalsky, A. 1973. *Quarterly Rev. Biophys.* 6: 1.
Prigogine, I. 1955. *Introduction to thermodynamics of irreversible processes*. New York: Interscience, John Wiley and Sons.
Rothschild, K. J., Ellias, S. A., Essig, A., and Stanley, H. E. 1980. *Biophys. J.* 30: 209.
Rottenberg, H. 1973. *Biophys. J.* 13: 503.
Rottenberg, H., and Gutman, M. 1977. *Biochemistry* 16: 3220.
Sauer, F. 1973. In American Physiological Society (Washington), *Handbook of physiology: renal physiology*, ed. J. Orloff and R. W. Berliner. Baltimore: Williams and Wilkins.
Stucki, J. W. 1978. *Prog. Biophys. Molec. Biol.* 33: 99.
―――― 1980a. *Eur. J. Biochem.* 109: 257.
―――― 1980b. *Eur. J. Biochem.* 109: 269.
Stucki, J. W., Compiani, M., and Caplan, S. R. 1982. Unpublished studies.
Westerhoff, H. V., and van Dam, K. 1979. In *Current topics in bioenergetics*, vol. 9, ed. D. R. Sanadi. New York: Academic Press.
Wolff, D., and Essig, A. 1977. *Biochim. Biophys. Acta* 468: 271.

7. Energetics of active transport: theory

Blumenthal, R., Caplan, S. R., and Kedem, O. 1967. *Biophys. J.* 7: 735.
Boulpaep, E. 1972. *Am . J. Physiol.* 222: 517.
Civan, M. M., Kedem, O., and Leaf, A. 1966. *Am. J. Physiol.* 211: 569.
Essig, A., and Caplan, S. R. 1968. *Biophys. J.* 8: 1434.
Handler, J. S., Preston, A. S., and Orloff, J. 1969. *J. Biol. Chem.* 244: 3194.
Heinz, E. 1975. In *Current topics in membranes and transport*, vol. 5, ed. F. Bronner and A. Kleinzeller. New York: Academic Press.
―――― 1979. *Mechanics and energetics of biological transport*. Berlin: Springer-Verlag.
Hess, B., and Brand, K. 1965. In *Control of energy metabolism*, ed. B. Chance, R. W. Esterbrook, and J. R. Williamson. New York: Academic Press.

Hoshiko, T., and Lindley, B. D. 1967. *J. Gen. Physiol.* 50: 729.
Jardetzky, O., and Snell, F. M. 1960. *Proc. Natl. Acad. Sci. U.S.A.* 46: 616.
Katchalsky, A., and Spangler, R. 1968. *Quart. Rev. Biophys.* 1: 127.
Kedem, O. 1961. In *Membrane transport and metabolism*, ed. A. Kleinzeller and A. Kotyk. New York: Academic Press.
Kedem, O., and Caplan, S. R. 1965. *Trans. Faraday Soc.* 61: 1897.
Kedem, O., and Katchalsky, A. 1958. *Biochim. Biophys. Acta* 27: 229.
Mikulecky, D. C., Huf, E. G., and Thomas, S. R. 1979. *Biophys. J.* 25: 87.
Miller, D. G. 1960. *Chem. Rev.* 60: 15.
Nagel, W., and Essig, A. 1982. *J. Membrane Biol.*, 69: 125.
Prigogine, I. 1961. *Thermodynamics of irreversible processes.* New York: John Wiley and Sons.
Rapoport, S. I. 1970. *Biophys. J.* 10: 246.
Stucki, J. W. 1980. *Eur. J. Biochem.* 109: 269.
Ussing, H. H., and Windhager, E. E. 1964. *Acta Physiol. Scand.* 61: 484.
Ussing, H. H., and Zerahn, K. 1951. *Acta Physiol. Scand.* 23: 110.
Veech, R. L., Raijam, L., and Krebs, H. A. 1970. *Biochem. J.* 117: 499.

8. Energetics of active transport: experimental results

Al-Awqati, Q. 1977. *J. Clin. Invest.* 60: 1240.
Al-Awqati, Q., Beauwens, R., and Leaf, A. 1975. *J. Membrane Biol.* 22: 91.
Al-Awqati, Q., Mueller, A., and Steinmetz, P. R. 1977. *Am. J. Physiol.* 233: F502.
Al-Awqati, Q., Norby, L. H., Mueller, A., and Steinmetz, P. R. 1976. *J. Clin. Invest.* 58: 351.
Beauwens, R., and Al-Awqati, Q. 1976. *J. Gen. Physiol.* 68: 421.
Biber, T. U. L., and Mullen, T. L. 1977. *Am. J. Physiol.* 232: C67.
Bobrycki, V. A., Mills, J. W., Macknight, A. D. C., and DiBona, D. R. 1981. *J. Membrane Biol.* 60: 21.
Canessa, M., Labarca, P., DiBona, D. R., and Leaf, A. 1978. *Proc. Natl. Acad. Sci. USA* 75: 4591.
Canessa, M., Labarca, P., and Leaf, A. 1976. *J. Membrane Biol.* 30: 65.
Civan, M. M. 1970. *Am. J. Physiol.* 219: 234.
Coplon, N. S., Steele, R. E., and Maffly, R. H. 1977. *J. Membrane Biol.* 34: 289.
Corcia, A., Lahav, J., and Caplan, S. R. 1980. *Biochim. Biophys. Acta* 596: 264.
Danisi, G., and Vieira, F. L. 1974. *J. Gen. Physiol.* 64: 372.
Dixon, T. E., and Al-Awqati, Q. 1979a. *Kidney Int.* 16: 41A.
——— 1979b. *Proc. Natl. Acad. Sci. USA* 76: 3135.
Fanestil, D. D., Herman, T. S., Fimognari, G. M., and Edelman, I. S. 1968. In *Regulatory functions in biological membranes*, ed. J. Järnefelt. Amsterdam: Elsevier.
Handler, J. S., Preston, A. S., and Orloff, J. 1969. *J. Biol. Chem.* 244: 3194.
Handler, J. S., Preston, A. S., and Rogulski, J. 1968. *J. Biol. Chem.* 243: 1376.

Helman, S. I., and Miller, D. A. 1971. *Science* 173: 146.
Hong, C. D., and Essig, A. 1976. *J. Membrane Biol.* 28: 121.
Labarca, P., Canessa, M., and Leaf, A. 1977. *J. Membrane Biol.* 32: 383.
Lahav, J., Essig, A., and Caplan, S. R. 1976. *Biochim. Biophys. Acta* 448: 389.
Lang, M. A., Caplan, S. R., and Essig, A. 1977a. *Biochim. Biophys. Acta* 464: 571.
—— 1977b. *J. Membrane Biol.* 31: 19.
Lau, Y. T., Lang, M. A., and Essig, A. 1979. *Biochim. Biophys. Acta* 545: 215.
—— 1981. *Biochim. Biophys. Acta* 647: 177.
Leaf, A., Anderson, J., and Page, L. B. 1958. *J. Gen. Physiol.* 41: 657.
Nishiki, K., Erecińska, M., and Wilson, D. F. 1979. *Am. J. Physiol.* 237: C221.
Owen, A., Caplan, S. R., and Essig, A. 1975a. *Biochim. Biophys. Acta* 389: 407.
—— 1975b. *Biochim Biophys. Acta* 394: 438.
Rosenthal, S. J., King, J. G., and Essig, A. 1979. *Am. J. Physiol.* 236: F413.
—— 1981. *J. Membrane Biol.* 63: 157.
Saito, T., and Essig, A. 1973. *J. Membrane Biol.* 13: 1.
Saito, T., Essig, A., and Caplan, S. R. 1973. *Biochim. Biophys. Acta* 318: 371.
Saito, T., Lief, P. D., and Essig, A. 1974. *Am. J. Physiol.* 226: 1265.
Schultz, S. G., Frizzell, R. A., and Nellans, H. N. 1977. *J. Theor. Biol.* 65: 215.
Steinmetz, P. R. 1974. *Physiol. Rev.* 54: 890.
Stucki, J. W. 1980. *Eur. J. Biochem.* 109: 269.
Ussing, H. H., and Zerahn, K. 1951. *Acta Physiol. Scand.* 23: 110.
Varanda, W. A., and Vieira, F. L. 1978. *J. Membrane Biol.* 39: 369.
Vieira, F. L., Caplan, S. R., and Essig, A. 1972a. *J. Gen. Physiol.* 59: 60.
—— 1972b. *J. Gen. Physiol.* 59: 77.
Voûte, C. L., and Ussing, H. H. 1968. *J. Cell. Biol.* 36: 625.
Wilson, D. F., Stubbs, M., Veech, P. L., Ericińska, M., and Krebs, H. 1974. *Biochem. J.* 140: 57.
Wolff, D., and Essig, A. 1980. *J. Membrane Biol.* 55: 53.
Zerahn, K. 1956. *Acta Physiol. Scand.* 36: 300.

9. Kinetics of isotope flows: background and theory

Chen, T. S., and Walser, M. 1975. *J. Membrane Biol.* 21: 87.
Dawson, D. C. 1977. *J. Membrane Biol.* 31: 351.
Dawson, D. C., and Al-Awqati, Q. 1978. *Biochim. Biophys. Acta* 508: 413.
Essig, A. 1968. *Biophys. J.* 8: 53.
Essig, A., Kedem, O., and Hill, T. L. 1966. *J. Theor. Biol.* 13: 72.
Hodgkin, A. L., and Keynes, R. D. 1955. *J. Physiol.* 128: 61
Kedem, O., and Essig, A. 1965. *J. Gen. Physiol.* 48: 1047.
Koefoed-Johnsen, V., Levi, H., and Ussing, H. H. 1952. *Acta Physiol. Scand.* 25: 150.

Koefoed-Johnsen, V., and Ussing, H. H. 1953. *Acta Physiol. Scand.* 28: 60.
Levi, H., and Ussing, H. H. 1948. *Acta Physiol. Scand.* 16: 232.
Ling, G. N. 1962. *A physical theory of the living state: The association-induction hypothesis.* New York: Blaisdell.
Minkoff, L., and Damadian, R. 1973. *Biophys. J.* 13: 167.
Sauer, F. A. 1978. In *Membrane transport in biology*, ed. G. Giebisch, D. C. Tosteson, and H. H. Ussing. Berlin: Springer-Verlag.
Schultz, S. G., and Frizzell, R. A. 1976. *Biochim. Biophys. Acta* 443: 181.
Schwartz, T. L. 1971. *Biophys. J.* 11: 596.
Teorell, T. 1949. *Arch. Sci. Physiol.* 3: 205.
Ussing, H. H. 1947. *Nature* 160: 262.
——— 1949. *Physiol. Rev.* 29: 127.
——— 1952. *Adv. in Enzymology* 13: 21.
——— 1960. *The alkali metal ions in biology.* Berlin: Springer-Verlag.
Ussing, H. H., and Zerahn, K. 1951. *Acta Physiol. Scand.* 23: 110.
Wolff, D., and Essig, A. 1977. *Biochim. Biophys. Acta* 468: 271.

10. Kinetics of isotope flows: mechanisms of isotope interaction

Biber, T. U. L., and Curran, P. F. 1968. *J. Gen. Physiol.* 51: 606.
Blumenthal, R., and Kedem, O. 1969. *Biophys. J.* 9: 432.
Diamond, J. M. 1979. *J. Membrane Biol.* 51: 195.
Essig, A. 1966. *J. Theor. Biol.* 13: 63.
Essig, A., Kedem, O., and Hill, T. L. 1966. *J. Theor. Biol.* 13: 72.
Franz, T. J., Galey, W. R., and Van Bruggen, J. T. 1968. *J. Gen. Physiol.* 51: 1.
Hodgkin, A. L., and Keynes, R. D. 1955. *J. Physiol.* 128: 61.
Hoshiko, T., and Lindley, B. D. 1964. *Biochim. Biophys. Acta* 79: 301.
Kedem, O., and Katchalsky, A. 1961. *J. Gen. Physiol.* 45: 143.
Lee, M. H., Berker, A. N., Stanley, H. E., and Essig, A. 1979. *J. Membrane Biol.* 50: 205.
Levi, H., and Ussing, H. H. 1948. *Acta Physiol. Scand.* 16: 232.
Li, J. H., and Essig, A. 1976. *J. Membrane Biol.* 29: 255.
Lief, P. D., and Essig, A. 1973. *J. Membrane Biol.* 12: 159.
Mackay, D., and Meares, P. 1959. *Trans. Faraday Soc.* 55: 1221.
Patlak, C. S., and Rapoport, S. I. 1971. *J. Gen. Physiol.* 57: 113.
Pietras, R. J., and Wright, E. M. 1974. *Nature* 247: 222.
Smulders, H. P., Tormey, J. McD., and Wright, E. M. 1972. *J. Membrane Biol.* 7: 164.
Spiegler, K. S. 1958. *Trans. Faraday Soc.* 54: 1408.
Ussing, H. H. 1952. *Adv. in Enzymology* 13: 21.

Ussing, H. H., and Johansen, B. 1969. *Nephron* 6: 317.
Wright, E. M., Smulders, A. P., and Tormey, J. McD. 1972. *J. Membrane Biol.* 7: 198.

11. Kinetics of isotope flows: tests and applications of thermodynamic formulation

Chen, J. S., and Walser, M. 1974. *J. Membrane Biol.* 18: 365.
────── 1975. *J. Membrane Biol.* 21: 87.
DeSousa, R. C., Li, J. H., and Essig, A. 1971. *Nature* 231: 44.
Essig, A., and Lang, M. 1975. *J. Membrane Biol.* 24: 401.
Essig, A., and Li, J. H. 1975. *J. Membrane Biol.* 20: 341.
Gottlieb, M. H., and Sollner, K. 1968. *Biophys. J.* 8: 515.
Helman, S. I., O'Neil, R. G., and Fisher, R. S. 1975. *J. Physiol.* 229: 947.
Krämer, H., and Meares, P. 1969. *Biophys. J.* 9: 1006.
Labarca, P., Canessa, M., and Leaf, A. 1977. *J. Membrane Biol.* 32: 383.
Lang, M. A., Caplan, S. R., and Essig, A. 1977. *J. Membrane Biol.* 31: 19.
Li, J. H., DeSousa, R. C., and Essig, A. 1974. *J. Membrane Biol.* 19: 93.
Li, J. H., and Essig, A. 1977. *Biochim. Biophys. Acta* 465: 421.
Meares, P., Dawson, D. G., Sutton, A. H., and Thain, J. F. 1967. *Berichte der Bunsengesellschaft für Physikalische Chemie* 71: 765.
Meares, P., and Sutton, A. H. 1968. *J. Colloid-Interface Sci.* 28: 118.
Meares, P., and Ussing, H. H. 1959a. *Trans. Faraday Soc.* 55: 142.
────── 1959b. *Trans. Faraday Soc.* 55: 244.
Saito, T., Lief, P. D., and Essig, A. 1974. *Am. J. Physiol.* 226: 1265.
Ussing, H. H. 1960. *The alkali metal ions in biology.* Berlin: Springer-Verlag.
Wolff, D., and Essig, A. 1977. *Biochim. Biophys. Acta* 468: 271.

12. Muscular contraction

Abbott, B. C. 1951. *J. Physiol.* (London) 112: 438.
Abbott, B. C., Aubert, X. M., and Hill, A. V. 1950. *J. Physiol.* (London) 111: 41P.
Alberty, R. A. 1968. *J. Biol. Chem.* 243: 1337.
Ashby, W. R. 1963. *An introduction to cybernetics.* New York: Science Editions.
Atkinson, D. E. 1965. *Science* 150: 851.
Aubert, X. 1956. *Le couplage energètique de la contraction musculaire.* Brussels: Éditions Arscia.
Bárány, M. 1967. *J. Gen. Physiol.* 50: 197.
Bornhorst, W. J., and Minardi, J. E. 1969. *Biophys. J.* 9: 654.

――― 1970a. *Biophys. J.* 10: 137.
――― 1970b. *Biophys. J.* 10: 155.
Cain, D. F., Infante, A. A., and Davies, R. E. 1962. *Nature* 196: 214.
Caplan, S. R. 1966. *J. Theoret. Biol.* 11: 63.
――― 1968a. *Biophys. J.* 8: 1146.
――― 1968b. *Biophys. J.* 8: 1167.
Carlson, F. D., and Siger, A. 1960. *J. Gen. Physiol.* 44: 33.
Carlson, F. D., and Wilkie, D. R. 1974. *Muscle physiology.* Englewood Cliffs, N. J.: Prentice-Hall.
Chaplain, R. A., and Frommelt, B. 1971. *J. Mechanochem. Cell Motility* 1: 41.
Cohen, C., and Longley, W. 1966. *Science* 152: 794.
Curtin, N. A., and Davies, R. E. 1973. *Cold Spring Harbor Symposia* 37: 619.
――― 1975. *J. Mechanochem. Cell Motility* 3: 147.
Curtin, N. A., Gilbert, C., Kretzschmar, K. M., and Wilkie, D. R. 1974. *J. Physiol.* 238: 455.
Daemers-Lambert, C. 1964. *Angiol.* 1: 249.
Daemers-Lambert, C., and Roland, J. 1967. *Angiol.* 4: 69.
Eisenberg, E., and Hill, T. L. 1978. *Progr. Biophys. Mol. Biol.* 33: 55.
Eisenberg, E., Hill, T. L., and Chen, Y. 1980. *Biophys. J.* 29: 195.
Elliott, G. F., Lowy, J., and Millman, B. M. 1965. *Nature* 206: 1357.
Ernst, E. 1963. *Biophysics of the striated muscle.* Budapest: Akadémiai Kiadó.
Essig, A., and Caplan, S. R. 1970. *Federation Proc.* 29: 655 (abstract).
Fenn, W. O. 1923. *J. Physiol.* (London) 58: 175.
Fenn, W. O., and Marsh, B. S. 1935. *J. Physiol.* (London) 85: 277.
Gasser, H. S., and Hill, A. V. 1924. *Proc. Roy. Soc. B.* 96: 398.
Gilbert, C., Kretzschmar, K. M., and Wilkie, D. R. 1973. *Cold Spring Harbor Symp.* 37: 613.
Hess, B. 1975. *Ciba Foundation Symposium* 31: 369.
Hill, A. V. 1938. *Proc. Roy. Soc. B.* 126: 136.
――― 1939. *Proc. Roy. Soc. B.* 127: 434.
――― 1949. *Proc. Roy. Soc. B.* 136: 399.
――― 1964a. *Proc. Roy. Soc. B.* 159: 297.
――― 1964b. *Proc. Roy. Soc. B.* 159: 319.
――― 1965. *Trails and trials in physiology.* London: Edward Arnold.
Hill, A. V., and Woledge, R. C. 1962. *J. Physiol.* (London) 162: 311.
Hill, T. L. 1974. *Progr. Biophys. Mol. Biol.* 28: 267.
――― 1975. *Progr. Biophys. Mol. Biol.* 29: 105.
――― 1977. *Free energy transduction in biology.* New York: Academic Press.
Huxley, A. F. 1957. *Progr. Biophys. Chem.* 7: 255.
Huxley, H. E., Brown, W., and Holmes, K. C. 1965. *Nature* 206: 1358.
Infante, A. A., Klaupiks, D., and Davies, R. E. 1964a. *Nature* 201: 620.
――― 1964b. *Biochim. Biophys. Acta* 88: 215.
――― 1964c. *Science* 144: 1577; (see also Davies, R. E. 1964. *Proc. Roy. Soc. B.* 160: 480).

Kedem, O., and Caplan, S. R. 1965. *Trans. Faraday Soc.* 61: 1897.
Kushmerick, M. J., and Davies, R. E. 1969. *Proc. Roy. Soc. B.* 174: 315.
Lundholm, L., and Mohme-Lundholm, E. 1962. *Acta Physiol. Scand.* 55: 45.
Macpherson, L. 1953. *J. Physiol.* (London) 122: 172.
Maitra, P. K., and Chance, B. 1965. In *Control of energy metabolism*, ed. B. Chance et al. New York: Academic Press.
Maréchal, G. 1964. *Le métabolisme de la phosphorylcréatine et de l'adénosine triphosphate durant la contraction musculaire*. Brussels: Éditions Arscia.
Mommaerts, W. F. H. M., Brady, A. J., and Abbott, B. C. 1961. *Ann. Rev. Physiol.* 23: 529.
Oplatka, A. 1972. *J. Theoret. Biol.* 34: 379.
Paul, R. J., Peterson, J. W., and Caplan, S. R. 1973. *Biochim. Biophys. Acta* 305: 474.
——— 1974. *J. Mechanochem. Cell Motility* 3: 19.
Polissar, M. J. 1952. *Am. J. Physiol.* 168: 766.
Prigogine, I. 1961. *Thermodynamics of irreversible processes*. New York: John Wiley and Sons.
Pringle, J. W. S. 1961. *Symp. Soc. Exptl. Biol.* 14: 41.
——— 1967. *Progr. Biophys. Mol. Biol.* 17: 1.
Rüegg, J. C. 1971. *Physiol. Rev.* 51: 201.
Sandow, A. 1961. In *Biophysics of physiological and pharmacological actions*, ed. A. M. Shanes. Washington: American Association for the Advancement of Science.
Seraydarian, K., Mommaerts, W. F. H. M., and Wallner, A. 1962. *Biochim. Biophys. Acta* 65: 443.
Stucki, J. W. 1978. *Progr. Biophys. Mol. Biol.* 33: 99.
Vieira, F. L., Caplan, S. R., and Essig, A. 1972. *J. Gen. Physiol.* 59: 60.
Walter, C. 1972. In *Biochemical regulatory mechanisms in eukaryotic cells*, ed. F. Kun and S. Grisolia. New York: Wiley.
Wilkie, D. R. 1950. *J. Physiol.* (London) 110: 249.
——— 1960. *Progress in biophysics* 10: 259.
——— 1967. *Symp. Biol. Hung.* 8: 207.
Wilkie, D. R., and Woledge, R. C. 1967. *Proc. Roy. Soc. B.* 169: 17.
Wilson, D. F., Stubbs, M., Veech, R. L., Erecińska, M., and Krebs, H. 1974. *Biochem. J.* 140: 57.
Yagi, K., and Mase, R. 1962. *J. Biol. Chem.* 237: 397.
——— 1965. In *Molecular biology of muscular contraction*, ed. S. Ebashi et al. New York: Elsevier.

13. Energy coupling in mitochondria, chloroplasts, and certain halophilic bacteria

Applebury, M. L., Peters, K. S., and Rentzepis, P. M. 1978. *Biophys. J.* 23: 375.

Arnon, D. I., Allen, M. B., and Whatley, F. R. 1954. *Nature* 174: 394.
Avron, M. 1971. In *Structure and function of chloroplasts*, ed. M. Gibbs. Berlin: Springer-Verlag.
Bakker, E. P., van den Heuvel, E. J., and van Dam, K. 1974. *Biochim. Biophys. Acta* 333: 12.
Blaurock, A. E., and Stoeckenius, W. 1971. *Nature New Biol.* 233: 152.
Caplan, S. R. 1966. *J. Theoret. Biol.* 10: 209; errata, 11: 346.
―――― 1982. In *Dynamic aspects of biopolyelectrolytes and biomembranes*, ed. F. Oosawa, N. Imai, I. R. Miller, S. Sugai, and Y. Kobatake. Amsterdam: Elsevier.
Caplan, S. R., and Essig, A. 1969. *Proc. Natl. Acad. Sci. USA* 64: 211.
Chance, B., and Montal, M. 1971. In *Current topics in membranes and transport*, vol. 2, ed. F. Bronner and A. Kleinzeller. New York: Academic Press.
Chance, B., and Williams, G. R. 1956. In *Advances in enzymology*, vol. 17, ed. F. F. Nord. New York: Interscience.
Eisenbach, M., and Caplan, S. R. 1979. In *Current topics in membranes and transport*, vol. 12, ed. F. Bronner and A. Kleinzeller. New York: Academic Press.
Goldschmidt, C. R., Ottolenghi, M., and Korenstein, R. 1976. *Biophys. J.* 16: 839.
Goldschmidt, C. R., Kalinsky, O., Rosenfeld, T., and Ottolenghi, M. 1977. *Biophys. J.* 17: 179.
Griffiths, D. E. 1976. *Biochem. J.* 160: 809.
Hellingwerf, K. J., Arents, J. C., Scholte, B. J., and Westerhoff, H. V. 1979. *Biochim. Biophys. Acta* 547: 561.
Hellingwerf, K. J., Schuurmans, J. J., and Westerhoff, H. V. 1978. *FEBS Lett.* 92: 181.
Hellingwerf, K. J., Tegelaers, F. P. W., Westerhoff, H. V., Arents, J. C., and van Dam, K. 1978. In *Energetics and structure of halophilic microorganisms*, ed. S. R. Caplan and M. Ginzburg. Amsterdam: Elsevier/North-Holland.
Hill, T. L. 1977. *Free energy transduction in biology*. New York: Academic Press.
Hill, T. L. 1980. *Proc. Natl. Acad. Sci. USA* 77: 2681.
Katchalsky, A., and Spangler, R. 1968. *Quarterly Rev. Biophys.* 1: 127.
Kok, B. 1972. In *Horizons of bioenergetics*, ed. A. San Pietro and H. Gest. New York: Academic Press.
Lehninger, A. L. 1964. *The mitochondrion: molecular basis of structure and function*. New York: W. A. Benjamin.
Lehninger, A. L., Carafoli, E., and Rossi, C. S. 1967. In *Advances in enzymology*, vol. 29, ed. F. F. Nord. New York: Interscience.
Mitchell, P. 1961. *Nature* 191: 144.
―――― 1966. *Biol. Rev.* 41: 445.
Oesterhelt, D., and Stoeckenius, W. 1971. *Nature New Biol.* 233: 149.
Rottenberg, H. 1970. *Europ. J. Biochem.* 15: 22.

——— 1973. *Biophys. J.* 13: 503.
——— 1978. In *Progress in surface and membrane science*, vol. 12, ed. J. F. Danielli et al. New York: Academic Press, pp. 245–325.
——— 1979. *Biochim. Biophys. Acta* 549: 225–253.
Rottenberg, H., Caplan, S. R., and Essig, A. 1967. *Nature* 216: 610.
——— 1970. In *Membranes and ion transport*, vol. 1, ed. E. E. Bittar. New York: Wiley-Interscience.
Rottenberg, H., and Gutman, M. 1977. *Biochemistry* 16: 3220.
Slater, E. C. 1966. In *Comprehensive Biochemistry*, vol. 14, ed. M. Florkin and E. H. Stotz. New York: Elsevier.
——— 1971. *Quart. Rev. Biophys.* 4: 35.
Slater, E. C., Rosing, J., and Mol, A. 1973. *Biochim. Biophys. Acta* 292: 534.
Stoeckenius, W. 1979. *Soc. Gen. Physiol. Ser.* 33: 39.
Stoeckenius, W., Lozier, R. H., and Bogomolni, R. A. 1979. *Biochim. Biophys. Acta* 505: 215.
Stucki, J. W. 1980. *Eur. J. Biochem.* 109: 257.
van Dam, K., Casey, R. P., van der Meer, R., Groen, A. K., and Westerhoff, H. V. 1978. In *Frontiers of biological energetics*, vol. 1, ed. P. L. Dutton, J. Leigh, and A. Scarpa. New York: Academic Press.
Walz, D. 1979. *Biochim. Biophys. Acta* 505: 279.
Westerhoff, H. V., Scholte, B. J., and Hellingwerf, K. J. 1979. *Biochim. Biophys. Acta* 547: 544.

Afterword

Caplan, S. R. 1981. *Proc. Natl. Acad. Sci. USA* 78: 4314.
Nagel, W., and Essig, A. 1982. *J. Membrane Biol.* 69: 125.
Stucki, J. W. Compiani, M., and Caplan, S. R. 1982. Unpublished studies.

Index

Acetazolamide, turtle bladder, 207
Active stretch, 328
Active transport, 28, 34, 132–213
  background of, 55
  carrier model, 255–259
  chemical potential dependence of, 179–181, 205
  effect of model compounds on, 183–189, 201, 207, 210
  electrical potential dependence of, 173–179, 205
  experimental characterization, 143–149
  flux ratio and, 234–236
  kinetics of isotope flows, 294–299
  linearity of, 173, 175, 205
  mechanisms of stimulation of, 189
  and metabolism, 162–167
  model, 49–52, 117, 133–143
  proton, 203–213, 351, 354, 379–386
  sodium, 104–105, 141, 153–155, 164, 168–181, 193, 196, 237, 377
  unidirectional fluxes, potential dependence of, 296
  See also Degree of coupling; Electromotive force of sodium transport; Equivalent circuit model; Oxygen consumption
Adenylate kinase, 369, 376–377, 378
Advancement, degree of, 18
Affine transformation, 336
Affinity, 17, 136, 144, 358, 406
  of ATP hydrolysis, 147, 182, 188, 191, 212
  effective, 370
  effect of model compounds, 183–189, 201
  evaluation of, 146–148, 181–183, 190, 379
  flows and, 156
  global, 165
  in muscle, 317–318
  near-constancy of, 170, 171, 179
  potential dependence of, 175–176
  of proton transport system, 207–208
  regulation of, 167
  for series reactions, 156
  of sodium transport system, 182–191
  standard, 370
Affinity regulation, 318, 331

Index

Aldosterone, 176, 177, 178, 189, 190, 199, 208, 210
Amiloride, in epithelia, 175, 183, 186, 187, 199
Amino acid transport, sodium-linked, 133
Anisotropy, 29, 30
Antidiuretic hormone (ADH), 189, 191, 199, 264
Apical surface, 134
Arrays:
  parallel, 236–243
  series, 243–244
Association-induction hypothesis, 241
ATPase system, proton-translocating, 354
ATP formation, oxidative and glycolytic, 188
ATP hydrolysis, affinity of, 147, 182, 188, 191, 212
Autocatalytic systems, 113
Axons, giant, 221

Bacteria, halophilic, 351
Bacteriorhodopsin, 379–386
  photocycle of, 382–386
Basal metabolism, 178, 407
Basolateral surface, 134
Binding sites, 75
Black-body radiation, 380
Black box, 60, 303–304
Bladder, see Toad urinary bladder; Turtle urinary bladder
Bomb calorimetry, 58
Bovine mesenteric vein, 312, 313
Brusselator, 113
Bulk flow, 219. See also Volume flow

Calcium ions, 320, 329
Caloric catastrophe, 242
Canonical regulator equation, 316
Carbon dioxide production, 203–205
Carnot cycle, 71–72
Carrier mechanism, 299
Carrier models:
  active transport, 255–259
  isotope interaction, 248–259, 282
  passive transport, 248–255

Catch mechanism, 328–329
Cells:
  blood, 140
  muscle, 140
  symmetrical, 140, 151
Charge-mosaic membranes, 48
Chemical hypothesis of oxidative phosphorylation, 351–352, 354, 356, 361
Chemical reactions, 10, 112
Chemiosmotic hypothesis, 351–352, 354–356, 361–362, 364
Chen and Walser treatment of isotope flows, 286, 287, 292, 296–298
Chloroplasts, 350, 368
Chromatophores, 350
Classical thermodynamics, see Equilibrium thermodynamics
Closed systems, 8–9, 13–14
Compartmentation, local, 320
Concentration:
  local, 319
  logarithmic mean, 36–37, 99–102, 402
Concentration polarization, 277
Conductance:
  active, 154, 199, 202, 208
  electric, 42
  internal, 338
  partial, 285, 292
  passive, 154, 175, 199, 208, 287
Conformational change, 75–76, 380
Constant-current source, 144, 333, 337
Constant-field equation, 101
Constants, additive, 369, 373
Constant-voltage source, 333, 337
Contraction, see Muscular contraction
Cooperativity, isotope interaction, 255
Cotransport, Na–Cl, 134
Countertransport, 282
Coupled water flow, 218. See also Volume flow
Coupling, 28–29, 42, 46, 92–94, 133, 140, 287, 348–387
  complete, 207
  cyclic, 357
  extent of, 58, 64, 105, 133, 149, 150, 205, 255
  incomplete, 195–198, 307–313, 316
  of isotope flows, 281–283

Coupling (*Continued*)
  isotope interaction and, 257, 259
  of metabolism and transport, 162–167
  models of, 352–356
  parallel, 352, 356, 362
  positive, 137
  scalar-vector, 30
  of sodium and potassium fluxes, 160–161
  of solute and volume flow, 158–160, 261–269, 272
  stationary-state, 30, 31–34, 50, 52, 320
  of water flows, 218
  See also Active transport; Degree of coupling
Coupling sites, 355
Cross coefficient, 26
Curie-Prigogine principle, 29–31
Current-voltage relationship, 173, 175
Cybernetic interpretation of regulator program, 347
Cycle flux, 78–84, 383
  unidirectional, 80, 83, 92, 128–131
Cycles, 75–84, 113, 165, 255, 257, 258
  dominant, 129
  leakage, 129
  photo, 382–386
Cyclic coupling process, 357

Dashpots, 330
De Donder's notation, 18
Degree of advancement, 18
Degree of coupling, 60–67, 70–72, 120, 145, 149–150, 162, 258, 328, 331, 341, 368, 371, 403–404, 406
  apparent, 130
  of converter, 316, 318
  evaluation of, 148–149, 193–198, 368–374
  global, 166
  local, 166
  in muscle, 307–314, 316, 318–319, 322, 324, 326
  in nonisothermal systems, 70–72
  optimal, 141
  overall, 141, 322
  for oxidative phosphorylation, 363
  of proton transport system, 206–207

  of sodium transport system, 193
  See also Active transport; Coupling
Degrees of freedom, 24, 30, 327, 334
Deoxy-D-glucose, in epithelia, 186–188, 199, 210
Deoxygenation, in turtle bladder, 210
Diagram method, 74–94, 258
Diagrams:
  cyclic, 78
  directional, 76–77
  flux, 78–80
  Hill, 112, 113
  indicator, 335
  input-output, 335–338
  multicycle, 78
  partial, 76
Diffusion:
  exchange, 221, 242, 254, 257, 260, 272, 356
  single-file, 221, 252, 260, 267, 268, 297, 299
Diffusion coefficients:
  integral bulk, 286
  tracer, 286
Dihydroergotamine, in muscle, 313
Dinitrophenol, 210, 221
Discontinuous systems, 303
Dissipation function, 15, 24, 83, 88, 137, 164, 356–360
  input and output terms for, 60
  for muscle, 303–307
  for oxidative phosphorylation, 357, 359
  transformations of, 34–40
Dissymmetry, 30
Driving reactions:
  local, 162
  multiple, 166
Driving region, 66, 67, 336

Efficacy, 69–70, 151–153
Efficiency, 66–68, 69, 141, 151, 325, 349, 366
  entropic, 70
  maximum, 67, 321
  mechanical, 316
  mechanochemical, 316
  of oxidative phosphorylation, 376–377

Efficiency (*Continued*)
  thermal, 70
  velocity and, in muscle, 327–328
Efflux, 219, 230
Electrodes:
  Ag/AgCl, 35
  current-passing, 36
  oxygen, 176
  reversible, 19, 35, 401–402
  sensing, 35
Electrokinetic phenomena, 38–39
Electromotive force of sodium transport, 154–155, 198–202, 235, 236, 295–298
  flux ratio and, 56, 219–220, 235–236, 294–298
Electroneutrality, 34
Electron transport, 348
Energetics of active transport:
  classical analysis of, 56–59
  experimental results, 168–213
  flux ratio and, 234–236
  NET formulation of, 59–60
  theory of, 132–167
Energy:
  conservation of, 5–6
  dissipation of, 15–22
  free, 7, 9–11, 15–18, 84–92, 133, 136, 147, 302, 349, 385
  internal, 6
  kinetic, 5–6
  potential, 6
Energy conversion:
  effectiveness of, 54–72, 151–153
  mechanochemical, 301
  regulation of, 314–326
  See also Metabolism; *entries for particular systems*
Energy converters:
  linear, 334
  self-regulated, 315, 316, 333–347
Energy coupling in organelles and halophilic bacteria, 348–387
  models of, in oxidative phosphorylation, 352–356
Energy expenditure without work, 68–70
Energy utilization, effectiveness of, 151–153, 162

Enthalpy, 7, 8–9
Entropy, 13–15
Entropy production:
  excess, 112
  internal, 14, 304
  minimum, 307
Epinephrine, in muscle, 313
Epithelial tissue, 124, 140, 144, 153, 169, 202–203, 237, 260, 264, 320, 377. *See also* Frog skin; Toad skin; Turtle urinary bladder
Equilibrium, reversible, 67
Equilibrium constants, 85
Equilibrium thermodynamics:
  vs. nonequilibrium thermodynamics, 1–2
  principles and functions of, 5–12
Equivalent circuit model, 56–59, 153–155, 198–199, 200
  proton transport, 203–212
  sodium transport, 153–155, 198–199, 200
  See also Electromotive force of sodium transport; Protonmotive force
Ergometer, Levin-Wyman, 318, 330
Exchange diffusion, 221, 242, 254, 257, 260, 272, 356
Extent of reaction, 18

Facilitated diffusion, 356
Fibrils, 303, 306, 330
Filament mechanism, sliding, 307
Filaments, 303–304
  myosin, 308
Filtration coefficient, 42
Flow:
  actual, 36
  conservative, 210, 224
  dependence on affinity, 156
  diffusional, 40
  efficacy of, 70
  heat, 71
  ion, in oxidative phosphorylation, 356–363
  level, 57, 67, 68–69, 140, 142, 150, 153, 310, 325, 339, 364, 368
  nonconservative, 175, 357, 360
  proton, 203–212

Index 429

Flow (*Continued*)
  salt, 35–36
  sodium, *see* Active transport
  solute, coupling with volume flow, 158–160, 261–269, 272
  thermodynamic, 24–25
  virtual, 36, 37, 47
  volume, 37, 39, 46, 47–48
  volume, coupling with solute flow, 158–160, 261–269, 272
  See also Active transport; Isotope flow kinetics
Flow ratio, 62, 149, 258, 371, 378
  reduced, 64, 65
  variability of, 198
Fluxes:
  cycle, 78–84, 383
  cycle, unidirectional, 80, 83, 92, 128–131
  operational, 83
  steady-state, 81–82
  tracer, 265–266, 287
  transition, 76, 77
  unidirectional, 218, 229, 230, 236, 241, 242, 265–268, 272, 284–285, 288–293, 294, 296
  See also Flow
Flux ratio, 56, 218–222, 230, 234–236, 238, 239, 240, 259, 265–269, 272, 275–276, 280, 283, 294–298
  electromotive force of sodium transport and, 56, 219–220, 235–236, 294–298
  and leak, 56, 239, 240
  metabolism and, 219–220, 234–236, 239–240, 294–298
Force:
  conjugate, 26
  driving, 68
  efficacy of, 70
  electromotive, 35, 56, 65
  electromotive, of sodium transport, 154–155, 198–202, 235, 236, 297–298
  evaluation by isotope techniques, 218, 230, 298
  protonmotive, 209–210
  thermodynamic, 24–25, 87
  See also Static head

Force ratio, 62, 378
  reduced, 64, 65, 67
Force–velocity relation, 302, 308, 314, 315, 321, 331
Freedom, degrees of, 24, 30, 327, 334
Free energy:
  of ATP hydrolysis, 147, 182, 188, 191, 212
  basic, 84, 89–90, 385
  Gibbs, 7, 9–11, 15–18, 85, 86, 133, 136
  gross, 84
  Helmholtz, 7, 9
  standard, 84
Free energy change:
  basic, 86–87
  gross, 86–87
  standard, 86
Free energy levels, 84–92, 302, 385
Frictional models, of isotope interaction, 246–248
Frog rectus abdominis muscle, 307, 312
Frog sartorius muscle, 306
Frog skin, 57, 144, 169, 170, 175, 178, 185, 187, 188, 190, 191, 192, 194, 197, 260
Functions of state, 7, 156, 179, 227

Gibbs-Duhem equation, 11
Gibbs equation, 10, 11, 16, 304
Gibbs free energy, 7, 9–11, 15–18, 85, 86, 133, 136
Glucose, in turtle bladder, 210
Glycolysis, 188, 190
Gravity, acceleration of, 304

Halobacterium halobium, 351
  light-driven proton pump of, 379–386
Heat, 6, 11, 13
  maintenance, 306
  shortening, 306, 314
  stable, 306
Heat content, 9. *See also* Enthalpy
Heat engines, 70–72
Helmholtz free energy, 7, 9
Heterogeneity, 48, 139–143, 236–243, 259–273, 283
High-energy intermediate, 351, 354–355, 357

Hill diagram, 112, 113
Hill equation, 314–318
  derivation of, 333–347
  inverse, 319
Histamine, 313
Hormones, 189–192. *See also entries for specific hormones*

Influx, 219, 230
Input, 60, 133, 136, 165
Input power, peak, 338
Input space, 335–337, 339
Intestinal mucosa, 159
Ion flows and oxidative phosphorylation, NET analysis of 356–363
Ionophores, 356
Ion pump model, electrogenic, 117–122
Ion transport and oxidative phosphorylation, 352, 356–363
Irreversible processes, 1, 9
  scalar, 15–18, 21–22
  vectorial, 18–22
Isotope effects, 217
Isotope exchange, 227, 228, 232, 233
Isotope flow kinetics:
  background of, 215–222
  coupling, 281–283
  mechanisms of, 247–273
  NET analysis of, 222–236
  in synthetic membranes, 275–283
  tests and applications of thermodynamic formulation, 275–299
  theory of, 222–244
  tracer fluxes, 265–266, 287
  tracer permeability, 217, 252–253, 261, 269
Isotope interaction, 57, 222, 236, 238, 243, 287
  absence of, 225–230
  in absence of concentration gradient, 276–277, 281
  in active transport pathway, 298–299
  attributable to allosteric cooperativity, 255
  attributable to heterogeneity, 259–273
  carrier models of, 248–259, 282
  coupling and, 257, 259
  direct demonstration of, 282
  frictional models of, 246–248
  in heterogeneous membrane, 283
  lattice models of, 248–255
  in passive pathway, 290–291
  presence of, 230–234
Isotropy, 29–30

Jacobian matrix, 112

Kedem-Katchalsky equations, 24–52
Kidney, proximal tubule, 140, 150
Kinetic formalism, 248, 255
Kinetic linearity, 97, 137, 182, 367

Lattice models, 248–255
Leak, 56, 58, 139–143, 150, 236–242
  effects of, 58, 142, 143, 144, 150, 151, 152, 153, 174
  evaluation of, 175, 203, 288–293
  and flux ratio, 56, 239, 240
  proton, in oxidative phosphorylation, 354, 374, 379
  as source of incomplete coupling, 58, 133, 141
Level flow, 57, 67, 68–69, 140, 142, 150, 153, 310, 325, 339, 364, 368
Levin-Wyman ergometer, 318, 330
Linearity, 96–131, 139–140, 150, 156, 160, 177, 361, 405
  active transport, 173, 175, 205
  intrinsic, 126
  kinetic, 97, 137, 182, 367
  local, 222
  muscle and, 307–314
  oxidative phosphorylation and, 366
  thermodynamic, 97, 137, 182, 368
Linear phenomenological equations, *see* Phenomenological equations
Linear relationships, 25
Liposomes, bacteriorhodopsin-containing, 381
Load, optimal, 333–335
Load lines, 335, 337, 344
Locus, 336, 345
  self-regulated, 339, 344

Logarithmic mean concentration, 36–37, 99–102, 402

Matching, 333, 337
Matrix transformation, 148
Maxwell relations, 401
Membrane heterogeneity, 48, 139–143, 236–243, 259–273, 283
Membrane potential, 42
Membrane processes, 19
Membranes:
  charge-mosaic, 48
  composite, 37–38. *See also* Membranes, heterogeneous; Membranes, series
  heterogeneous, 48, 139–143, 236–243, 259–273, 283
  mosaic, 48
  purple, 89, 379, 384
  selectivity of, 46, 142, 143, 153, 278
  series, 138–143, 158–160, 243–244
  synthetic, 275–283
Memory effect, 172
Metabolic reaction:
  global, 144
  local, 144
Metabolism:
  basal, 178, 407
  chemical potential dependence of, 179–181
  coupling with transport, 162–167
  effects of 2-deoxy-D-glucose, 186
  electrical potential dependence of, 176–179
  flux ratio and, 219–220, 234–236, 239–240, 294–298
  intermediary, 313
  ouabain-resistant, 178
  oxidative, 176, 377, 408
  specificity, 178
  suprabasal, 171
Michaelis-Menten kinetics, 156–157
Mitochondria, 123, 358–360, 367, 368, 374–378. *See also* oxidative phosphorylation
Mode of operation, 336
Molluscan muscles, 328–329
Mucosa, intestinal, 159

Mucosal surface, 196
Multidimensional inflection point (MIP), 111–116, 123–124, 367
  and Onsager's reciprocal relations, 111, 112, 123
Muscle:
  affinity in, 317–318
  cytoplasm, 308
  dissipation function for, 303–307
  energy conversion in, 314–326
  frog rectus abdominis, 307, 312
  frog sartorius, 306
  linearity in, 307–314
  molecular constants, 325
  molecular parameters, 325–326
  molluscan, 328–329
  NET analysis of, 303–314
  NET, usefulness of, 326–329
  phenomenological equations for, 307–314
  self-regulation in, 315, 329–330
  skeletal, 312, 325
  smooth, 331
  striated, 303
  vascular smooth, 311–312
Muscle twitches, 304
Muscular contraction, 301–347
  control of, 302
  Hill equation, deviation, 333–347
  isometric, 301, 306, 310, 328
  isotonic, 306
  regulation of, 302, 315–316, 347
  tetanic, 304, 306
  unloaded, 301, 310
  velocity of, 306
Myofibril, 303
Myosin filaments, 308

Nernst-Planck equation, 98–99
NET, *see* Nonequilibrium thermodynamics
Network thermodynamics, 94, 155
Nonequilibrium thermodynamics:
  basic concepts of, 12–15
  general considerations, 1–4
  See also entries for particular systems
Nonisothermal systems, 70–72

Nonlinearity, 175, 181, 206, 322, 368, 389, 405
Nonstoichiometry, 149, 198
Norepinephrine, in muscle, 313

Onsager's reciprocal relations:
application of, 3, 26, 38, 61, 133, 136, 137, 171, 206, 231, 258, 310, 361
evidence for validity of, 39, 50, 62, 82, 123, 366, 368, 379
inapplicability of, 380
kinetic basis of, 82, 96–131
and multidimensional inflection point, 111, 112, 123
and proper pathway, 136, 171, 206, 258
Onsager symmetry, see Onsager's reciprocal relations
Open-circuit potential, 42, 144–145, 154
Open systems, 9–11, 14–15, 327, 401
Ouabain, in epithelia, 183, 184–185, 199, 284, 294
Ouabain-resistant metabolism, 178
Outflux, 230
Output, 60, 133, 136, 325
Output space, 335–337, 339
Oxidation, see Active transport; Metabolism; Oxidative phosphorylation; Oxygen consumption
Oxidation-reduction reactions, 350
Oxidative phosphorylation, 4, 348, 349–352
degrees of coupling for, 363
dissipation function, 357, 359
efficiency of, 376–377
evaluation of degree of coupling in, 368–374
flow ratio vs. force ratio for, 378
linearity in, 366
models of energy coupling in, 352–356
NET analysis of, 356–379
proton transport in, 353–363
rate of, 367
succinate in, 355
See also Chemical hypothesis of oxidative phosphorylation; Chemiosmotic hypothesis; Coupling, parallel
Oxidoreduction chain, 354
Oxygen consumption, 136, 171, 172, 177, 178, 184–185
Oxygen electrodes, 176

Parallel arrays, 236–243. See also Heterogeneity
Parallel coupling hypothesis, 356, 362
Parameters:
molecular, of muscle, 325–326
NET, of active transport system, 171–173, 181–202
Parameters of state, 156, 227
Passive transport:
carrier models, 248–255
evaluation in epithelia, 175
isotope flows in biological membranes, 284–293
Kedem-Katchalsky equations, 24–52
lattice models, 248–255
See also Leak
Pathways:
active, 196, 237, 240
branched, 382
passive, evaluation of, 175, 203, 288–293
See also Leak; Proper pathways
Performance, index of, 315
Permeability:
carrier model, 253
electroosmotic, 42
evaluation by isotope techniques, 216
exchange, 233
lattice model, 250, 252
solute, 42
tracer, 217, 252–253, 261, 269
water, anomalous nature of, 217
Permeability barrier, passive, 196
Permeability coefficient, 45, 216, 217, 222, 225, 244, 248, 253, 258, 260–262, 269, 278
See also Permeability
Permselectivity, 46, 142, 143, 153, 278
Phenomenological coefficients, 26, 29, 133, 136–137, 138, 359

Phenomenological coefficients (*Continued*)
  composite membrane, 139–140
  conductance, 136, 148, 192–193
  of energy converter, 341
  evaluation of, in epithelia, 148
  as functions of state, 156, 179, 227
  linear, 128–131
  molecular parameters in muscle and, 325–326
  near-constancy, 171, 179
  potential dependence of, 175–176
  practical, 40–47
  resistance, 136, 137–138, 145, 148, 223, 224, 405
  scalar, 294
  vectorial, 294
Phenomenological equations, 25–28
  for epithelia, 138, 140
  for muscle, 307–314
  for oxidative phosphorylation, 360–362
  transformations of, 40–47
  validation of, 169–181
Phenomenological relations, *see* Phenomenological equations
Phenomenological stoichiometry, 64, 92, 94, 120–121, 376–377
Phosphocreatine breakdown, 311, 312, 321
Phosphorylation, *see* Oxidative phosphorylation
Photon absorption, 91, 380
Photophosphorylation, 350–351, 379–386
Polarity, 136, 177, 402
P/O ratio, 148, 182, 191, 208, 213, 349, 365
Potential:
  chemical, 10
  electrical, 20, 405
  electrochemical, 20, 248
  membrane, 42
  open-circuit, 42, 144–145, 154
  oxidoreduction, 350
  phase-boundary, 117–118
  phosphate, 350
  reversal, 130–131
  streaming, 42

Potential dependence:
  chemical, of active transport, 179–181, 205
  electrical, of active transport, 173–179, 205
Potential difference, electrochemical, 136, 205, 405
Power output, 333
Pressure:
  electroosmotic, 42
  osmotic, 36, 158–159
Program, regulator, 347
Programming function, 341
Proper pathways, 132, 136, 137, 147, 148, 171, 176, 179, 182, 258, 331, 368
  coupled processes, 103–111
  at multidimensional inflection point, 123–124
  proton transport, 206
  sodium transport, 171, 176, 179, 182
  uncoupled processes, 98–102
Proton flow:
  conservative, 210
  nonconservative, 360
  See also Proton transport
Proton leakage, 354, 374, 379
Protonmotive force, 209–210, 356
Proton pump, *see* Proton transport
Proton translocation, *see* Proton transport
Proton transport, 203–213, 348, 351, 354, 385
  affinity of, 207–208
  biochemical correlates of, 212–213
  degree of coupling in, 206–207
  effect of model compounds, 207, 210
  equivalent circuit analysis of, 208–212
  light-driven, 379–386
  NET analysis of, 205–206
  in oxidative phosphorylation, 353–363
  protonmotive force, 209–210, 356
  in turtle urinary bladder, 203–205
Pump:
  complex, 160–161
  ion, 88, 117–122, 134, 135–139
  proton, 207, 208, 212, 351, 379–386
  sodium, *see* Active transport, sodium

Pump-leak systems, 160, 161, 236, 237, 241, 242
Purple membrane, 89, 379, 384

Quantum yield, 384
Quasi-chemical notation, 133
Quasi-stationary state, 173, 176, 306
Quasi-steady state, 173, 176, 306

Rabbit colon, 203
Radiationless transitions, 382
Rate constants:
 first-order, 76, 86, 90
 second-order, 80
Rat liver mitochondria, 367, 375, 376–377, 378
Reaction coordinate, 18
Reciprocity, see Onsager's reciprocal relations
Recirculation, 195
Reference bath, 82
Reference state, 98, 114, 370–371, 372
Reflection coefficient, 42, 45–46, 159, 271
Regulator, 144. See also Muscular contraction, regulation of
Regulator function, 341
Regulator program, 347
Resistance:
 electrical, 145–146
 exchange, 234
 internal, 334, 338
 load, 334
 mechanical, 330
 phenomenological, 136, 137–138, 145, 148, 223, 224
Resistance coefficients, 223, 224, 405
Resistance formulation, 136
Respiratory chain, 350
Respiratory control ratio, 350, 369, 371
Respiratory quotient, 198, 205, 208
Rest length, 306
Reverse electron transport, 366
Reversible electrodes, 19, 35, 401–402
Reversible processes, 1, 6, 9
Rotenone, in frog skin, 187

Salt flow, 35–36
Sarcomere, 307
Saxén relations, 39
Scalar processes, 15–18, 21–22
Self interaction, 287. See also Isotope interaction
Self-regulation, 315, 316
 of energy converters, 333–347
 implications of, 329–330
Series membrane, 138–143, 158–160, 243–244
Series reactions, 156–158
Serosal surface, 196
Servomechanisms, 333–334
Short-circuit current, 144–145, 172, 186
Single-valued function loci, 340–342
Skin:
 frog, 57, 144, 169, 170, 175, 178, 185, 187, 188, 190, 191, 192, 194, 197, 260
 toad, 179, 180, 181, 260
 See also Epithelial tissue
Sodium flux, nonconservative, 175
Sodium-linked amino acid transport, 133
Sodium pump, 134, 135–139, 320
Sodium transport, see Active transport
Solute flow, coupling with volume flow, 158–160, 261–269, 272
Solute permeability, 42. See also Permeability
Solvent drag, 219. See also Solute flow
Specificity of metabolism, 178
Spontaneous process, 66
Stability, local asymptotic, 125
State probabilities, 76
States, 350, 369, 373
 excited, 382
 reduction of, 117
 See also Stationary state
Static head, 57, 68, 69, 91, 111, 121, 124, 128–131, 140, 141, 143, 145, 150, 152, 310, 325, 328, 339, 350, 364, 368, 370, 385
Stationary state, 16, 334
 coupling, 30, 31–34, 50–52
 in epithelia, 173, 205
 of minimal entropy production, 30–31

Stationary State (*Continued*)
  in mitochondria, 359, 363–368
  in muscle, 306, 321, 330
Steady state, *see* Stationary state
Steady-state shortening, 321
Stoichiometric coefficient, *see* Stoichiometry
Stoichiometric ratio, 60, 62, 149, 205, 349. *See also* Stoichiometry
Stoichiometry, 4, 17, 57–58, 92–94, 149–150, 193–198, 407
  apparent, 371–373
  in epithelia, 133, 136, 137, 141, 149–150, 193–198, 208, 407
  mechanistic, 93–94, 112, 114
  in mitochondria, 371–373, 376–377
  nonintegral, 78
  phenomenological, 64, 92, 94, 120–121, 376–377
  regulation of, 323–325, 331
Straight coefficient, 26
Streaming current, 39, 42
Streaming potential, 42
Stretch, active, 328
Substrate depletion, turtle bladder, 210
Succinate, and oxidative phosphorylation, 355
Suprabasal oxygen consumption, 171. *See also* Metabolism
Symmetry, 29. *See also* Onsager's reciprocal relations
Symport, 74
Systems, 12
  closed, 8–9, 13–14
  open, 9–11, 14–15, 327, 401

Tension, isometric, 309
Thermocouple, 70–71
Thermodynamic flows, 24–25
Thermodynamic forces, 24–25, 87
Thermodynamic linearity, 97, 137, 182, 368
Thermodynamics, laws of, 5–12
  second law of, and oxidative phosphorylation, 349
  See also Equilibrium thermodynamics; Network thermodynamics; Nonequilibrium thermodynamics

Thylakoid space, 348
Toad skin, 179, 180, 181, 260
Toad urinary bladder, 104, 141, 143, 144, 169, 170, 174–175, 192, 193, 195, 200, 264, 284, 287, 289, 290–291, 294, 296, 297, 298–299, 408
Tracer fluxes, 265–266, 287
Tracer permeability, 217, 252–253, 261, 269
Transition probabilities, 90
Transitions, 76
  spontaneous, 88
Transport, *see* Active transport; Passive transport; Proton transport
Transport coefficients, practical, 41, 42, 403
Transport numbers, 42, 142, 285
Tropomyosin, 320, 329
Turtle urinary bladder, 203–212
Twitches, muscle, 304

Uncouplers, 351, 355–356, 374
Uncoupling, sources of, 105

Vectorial coupling coefficient, 29
Vectorial processes, 18–22
Velocity and efficiency, in muscle, 327–328. *See also* Force–velocity relation
Visual pigment, 379
Volume flow, 37, 46, 261–269
  coupling with solute flow, 158–160, 261–269, 272
  definition of, 47–48
  electroosmotic, 39
  observed, 37, 47
Volume fraction, 37

Wastage, 72
Water flow, 25. *See also* Volume flow
Water permeability, anomalous nature of, 217. *See also* Permeability
Work, 6, 13, 321
  electroosmotic, 151, 236–237
  internal, 57, 328
  useful, 8–9
Working element, 60, 303

Bei Fragen zur Produktsicherheit wenden Sie sich bitte an:
If you have any questions regarding product safety,
please contact:

Walter de Gruyter GmbH
Genthiner Straße 13
10785 Berlin
productsafety@degruyterbrill.com